Advances in Earthquake Research and Engineering

Edited by Ali Ismet Kanlı

Published in London, United Kingdom

Advances in Earthquake Research and Engineering
http://dx.doi.org/10.5772/intechopen.1004510
Edited by Ali Ismet Kanlı

Contributors
Abdallah Azizi, Alonso Gómez-Bernal, Darioush Motazedian, David Emanuel Andersson, Evgenia Kozhoukharova, Oliver F. Shyr, Parisa Asgharzadeh, Shutian Ma, Takumi Onuma, Wei Li, Yasuto Itoh, Yiyang Luo, Şakir Şahin

First published in London, United Kingdom, 2025 by IntechOpen
IntechOpen is the global imprint of INTECHOPEN LIMITED, registered in England and Wales, registration number: 11086078, 167-169 Great Portland Street, London, W1W 5PF, United Kingdom

For EU product safety concerns: IN TECH d.o.o., Prolaz Marije Krucifikse Kozulić 3, 51000 Rijeka, Croatia, info@intechopen.com or visit our website at intechopen.com.

British Library Cataloguing-in-Publication Data
A catalogue record for this book is available from the British Library

Advances in Earthquake Research and Engineering
Edited by Ali Ismet Kanlı
p. cm.

This title is part of the Earth Sciences Book Series, Volume 5
Topic: Geology
Series Editor: Maurizio Lazzari

Print ISBN 978-1-83634-281-6
Online ISBN 978-1-83634-280-9
eBook (PDF) ISBN 978-1-83634-282-3
ISSN 3049-8848

If disposing of this product, please recycle the paper responsibly.

IntechOpen Book Series

Earth Sciences

Volume 5

Aims and Scope of the Series

The world of Earth Sciences, considering the interactions within the geosphere and between the geosphere–biosphere, is a place in which a large number of scientists find and have found over time their own relationship or sector of application precisely because it is absolutely transversal to many disciplines and subdisciplines, which do not necessarily fall within the Geosciences. The objective of this book series is to welcome original scientific contributions both in consolidated contexts and in new frontiers of research, as well as review papers included in the various disciplines of Earth Sciences, but above all, those that show a modern and transversal vision of applications and impacts on the community in a particular historical context, which, following the COVID-19 pandemic, has shifted global attention to sectors that were previously more neglected. In particular, those of mining research and fossil and renewable energy sources, environmental geology and the sustainable use of natural resources and impacts on the built environment, land consumption, geoarchaeology, forensic geology, geotourism/geoheritage, georisks and climate and environmental changes, considered at different scales, up to new applications of geostatistical and geospatial analysis, GIS and artificial intelligence for the definition of forecasting models and scenarios in various sectors of basic and applied research.

Meet the Series Editor

Dr. Maurizio Lazzari has a Ph.D. in Earth Science and is a researcher at the Italian National Research Council, Institute of Cultural Heritage Sciences. Since 2001, Dr. Lazzari has been a Professor of Pedology at the University of Basilicata (Italy) and a geoarchaeologist at the University of Salento (Italy). His research activities are focused on natural and anthropic hazards and risk factors, aimed at safeguarding and conserving settlements and the historical–monumental heritage of the Mediterranean, with particular attention to landslide processes, susceptibility maps, monitoring, and modelling. Since 2004, he has been working as a scientific coordinator for several national research projects studying landslides and triggering factors, natural and anthropogenic risks, geological and geomorphological mapping, soil erosion, preservation of historical and archaeological sites, enhancement of degraded areas, geo-touristic use, and the protection of the landscapes. He is the author of about 150 scientific publications in national and international journals, monographs, book chapters, and conference proceedings concerning applied geology, geomorphology, dynamics of artificial reservoirs, soil erosion, landslides, geoarchaeology, hydrogeological instability, natural hazards, monitoring, cultural landscape, UNESCO Heritage, geoarchaeology, and geo-tourism.

Meet the Volume Editor

Prof. Dr. Ali Ismet Kanlı received his undergraduate degree from the Faculty of Engineering, Department of Geophysical Engineering, Istanbul University in Turkey, in 1989. He obtained an MSc and Ph.D. from Istanbul University, Institute of Science, in 1994 and 1998, respectively. He began his academic career as a research assistant at the Faculty of Engineering, Department of Geophysical Engineering, Division of Applied Geophysics, Istanbul University in 1992. He became an assistant professor in 2001, an associate professor in 2010, and a professor in 2016. Dr. Kanlı is the head of the Applied Geophysics Division at the Geophysical Engineering Department at Istanbul University. He has carried out and directed many international and national projects and has several national and international scientific publications to his credit. He is an editorial board member and reviewer for numerous journals. He has also served as a referee for many international and national projects. He is a member of three national and eight international scientific associations. Dr. Kanlı's scientific interests include applied and near-surface geophysics, engineering and environmental geophysics, engineering seismology, exploration seismology, structural geophysics, earthquake engineering, geotechnical geophysics, borehole geophysics and well logging, alternative energy and geothermal exploration.

Contents

Preface

Earthquakes have long been a subject of significant interest, especially in exploring their causes, characteristics, and how seismic waves are used to study them. Additionally, seismic data are combined with other information to analyze the plate tectonic processes responsible for earthquakes.

Earthquake risk assessment is inherently multidisciplinary, involving studying earthquake sources and mechanisms, various methods for characterizing earthquake input motion, and soil-structure interaction effects.

This book covers a wide range of topics, including earthquake disasters by computational gravimetry, seismic resistant buildings, statistical seismology, geomagnetic and ionospheric anomalies as earthquake precursors, crustal velocity and source rupture models, ground motion amplification, socio-economic impacts of earthquakes, SAR differential interferometry, and paleoseismic zones of the Earth's crust.

This book will interest a wide range of readers, including civil engineers, geophysical engineers, geologists, and other professionals involved in earthquake-related subjects.

Ali Ismet Kanlı
Department of Geophysical Engineering,
Istanbul University-Cerrahpasa,
Istanbul, Türkiye

Chapter 1

Analysis of Rayleigh and Love Waves in the Valley of Mexico and their Effect on Ground Motion Amplification

Alonso Gómez-Bernal

Abstract

The propagation of Rayleigh and Love waves in the basin of Mexico City during the two catastrophic earthquakes of September 19, the events of 1985 and 2017 are investigated. The analysis of the strong ground motion characteristics observed during the Michoacán Earthquake of September 19, 1985, M8.1, included accelerations recorded in stations near to the source, in the path, and in Mexico City. The motions consist of a group of superficial waves of different periods: the fundamental mode of Rayleigh wave is identified with a period about 12 sec, with retrograde elliptical motion; later waves of periods with 5–2 sec with prograde motions are identified. Besides, a Love wave with period of 10 sec is identified. As a complement to the observations, we use a sedimentary layer model of the Valley of Mexico. In the analysis of the Puebla-Morelos earthquake of September 19, 2017, M7.1, with its epicenter roughly 120 km to the SE of Mexico City, shear and surface waves were compared in the acceleration waveforms. Additionally, analyses of microearthquakes occurred on December 12, 2023, with epicenters inside Mexico City were included in this work, showing the generation of surface waves inside the basin during local events.

Keywords: Rayleigh and love waves, Mexico earthquakes, surface wave amplification, September 19, microearthquakes in Mexico City

1. Introduction

Although there will always be uncertainty in the spatial distribution of ground motion, this variability in response can be attributed to geological features such as underground topographic irregularities. Shallow sedimentary strata can cause resonance and amplification of seismic waves when they propagate vertically towards the surface. To quantify the amplification of ground motion at a site, it has become common practice to use one-dimensional (1D) shear wave velocity profiles. Although the 1D method of shear wave propagation is a practical procedure that can provide a simple framework to engineers, it is accepted that it does not capture the full effect of the variability of 3D wave propagation effects, because it is necessary to include the other amplification component, that is, that due to surface waves.

Although the existence of surface waves during earthquakes has been always accepted, their importance in local effects has generally not been recognized and their relevance in seismic amplification has often been minimized, partly due to the assumption that this amplification is challenging to model. It has been argued on other occasions that surface waves are most significant at periods not important for ordinary buildings, for instance, Swanger and Boore [1] and Joyner [2].

The content of significant surface waves within a broadband of periods (3–20 sec) of some earthquakes that occurred in the subduction zone of Mexico is detected when arrive at the basin of Mexico City, as has been shown in some works (Gomez-Bernal and Saragoni [3]; Gómez-Bernal [4]). Therefore, it is necessary to study the strong influence that these waves have on local effects. Ideally, it is required of earthquakes of magnitudes greater to 7, which produce Rayleigh waves on an ample range of periods (Ewing et al. [5]). Great earthquakes produce important surface waves, specially, in sedimentary zones (Bard and Bouchon [6]). Tanaka et al. [7] observed that during intense earthquakes, surface waves with period around 8 sec dominated the strong ground motion records in the sedimentary soils of Tokyo. Pei and Papageorgiou [8], when analyzing the Loma Prieta Earthquake of 1989, identified the fundamental and superior mode of Rayleigh waves in the valley of Santa Clara. Recently, Meza-Fajardo et al. [9] have presented results on the analysis of Rayleigh and Love waves propagating in the Nobi and Kanto basins during the 2011 Tohoku earthquake, and other events. It is clearer that stratigraphy is responsible for practically all the effects observed during the earthquakes.

Figure 1.
Epicenters of the studied earthquakes, of September 19 (1985 and 2017), and microearthquake of 2023.

The Michoacán Earthquake of September 19, 1985 (Ms = 8 1), was an event that left great lessons to specialists of different areas of knowledge in Mexico and in the world. Regarding to the strong ground motion, and to the site effects, the knowledge has been considerably advanced. Since the 1985 earthquake, there have been a large number of records of strong ground motions that have allowed us to understand in more detail the characteristics of the seismic waves that arrive to the Mexico City Valley, as is the case of the Puebla-Morelos Earthquake (**Figure 1**) of September 19, 2017 (M = 7.1), which occurred as a result of the tectonism process of the Cocos plate, which is subducted by the North American Plate in the north-east direction by continuing the process of subduction of the Cocos plate under the continental plate [10]. In this region of Mexico, Intraslab or Normal fault earthquakes has been occurred in this same area, such as the events of 1928, 1945, 1973, 1980, and 1999. It is possible that all these earthquakes have been caused by the sinking of the subducted plate, which is estimated with little angle of inclination, almost horizontal.

2. Observations on the strong ground motion characteristics detected during the Michoacan earthquake of September 19, 1985, M8.1

The network of stations that recorded the earthquake of September 19, 1985, throughout the country, is indicated in **Figure 2**, where graphs of the East-West acceleration time histories of 23 sites at the same scale are illustrated. The stations close to the epicenter, Caleta de Campos (CALE), La Union (UNIO), La Villita (VILE), Airport Zihuatanejo (AZIH), and Zacatula (ZACA), registered relatively moderate values of the absolute peak ground acceleration (PGA), considering the magnitude from the event. The maximum value of PGA in the horizontal component was of 300 cm/seg^2, and it was registered in ZACA, whereas the maximum vertical PGA was over 250 cm/s^2, also in ZACA. Some studies based on the records in these stations indicated the possible existence of two different sub-events, corresponding to two different asperities, in which was released most of the energy (Anderson et al., [11]);

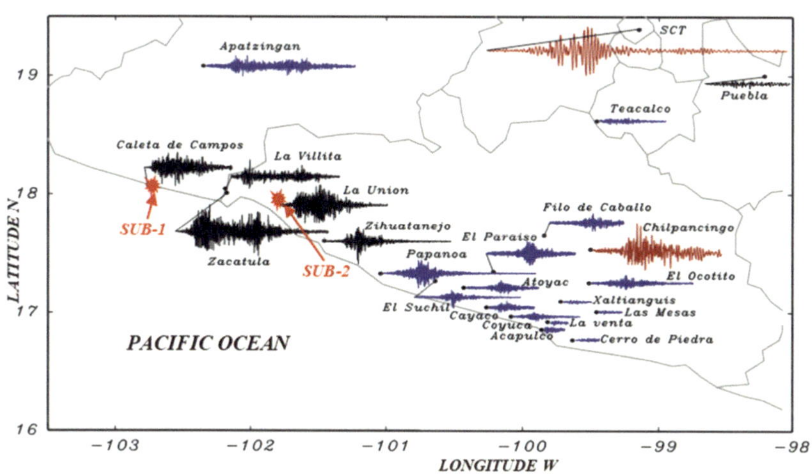

Figure 2.
East–West (EW) accelerations at stations that recorded the earthquake of September 19, 1985. They are to the same scale horizontally and vertically.

the first of these sub-events, SUB-1, it was originated near CALE close to the epicenter, and the second one, SUB-2, close to UNIO site, with 26 sec later and located at 70 km from SUB-2 (**Figure 2**).

The stations located over the rupture zone registered an important content of high periods (or low frequencies), as they show the displacement spectrum of **Figure 2** of stations CALE, ZACA, and UNIO, where the ordinates of the displacements in the range periods between 5 and 13 sec (0.07 to 0.2 Hz) stand out, in the 3 components of motion, mainly in the vertical direction.

2.1 Analysis of displacements and particle motion

With the purpose of investigating with detail the strong ground motion features during the 1985 earthquake, the integration of all acelerogramas was carried out to obtain the ground displacements. Radial component is defined as the direction of the epicenter, and tangential as the orthogonal to the epicenter; horizontal components were rotated. **Figure 3a** presents the vertical displacements in all stations in Mexico City, which are aligned according to the most remarkable wave. The displacements graph from Teacalco station (TEAC) is included; this station was the nearest to Mexico City, in which the same surface wave is also detected. The similar form of the curves indicates that the magnitude of the vertical displacements is independent of the type of soil on which is located the station. In some studies, related to the 1985 earthquake, some authors indicated this characteristic of the vertical motion of the soil (p. e. Bard et al., [12]). The period of this arrival was about 12 sec; this phase is described in some works (Campillo et al., [13]) like a surface wave with 12 sec and with ripples of 3 sec.

Each component of displacement (Radial, tangential, and vertical) is compared simultaneously in four stations (**Figure 3b**), two located on firm soil (TACY, and CU01), one on transition soil (SXVI), and one more in TEAC. In this four graphics is observed a similar evolution of the particle motion. The phase of the vertical wave goes in front of the radial one, and when both components are combined simultaneously, they define a trajectory of the particle with a typical retrograde evolution of a Rayleigh wave, in this case with an approximated period of 12 sec. Odograms of these four sites (**Figure 3b**) describes the trajectory of the classic particle motion of Rayleigh wave, in the time range between the points A and B, indicated in the superior part of **Figure 3b**.

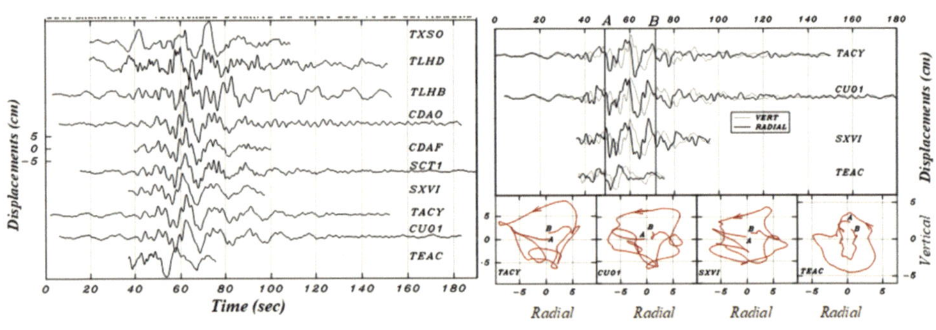

Figure 3.
Vertical displacements at stations in Mexico City during the September 19, 1985, earthquake (left). Vertical and Radial displacements only on firm soil sites, their odograms are included between the A and B time range (Right).

2.2 Retrograde particle motion in the frequency band between 0.065 and 0.15 Hz

When passband filters are used in the estimation of displacements, the evolution of the particle trajectory between certain frequency bands is appraised much better. The superior part of **Figure 4** shows the same Rayleigh wave of **Figure 3b**, but after applying a passband filter between 0.065 and 0.13 Hz. The similarity of the curves in the 4 stations in the 3 directions of motion is clear; this condition allows align exactly the curves in the time axis. The first important arrival is in the tangential component; it begins according to the line marked in the figure with number 1, and in the radial-tangential (RT) odogram attached in the right part, it begins in that same time 1 and finishes in point 2; the period of this surface wave is about 8 sec; in addition, in TEAC and SXVI stations, this wave was the one that began the record. In the radial-vertical (RV) odogram, attached in **Figure 4** show two complete cycles of the Rayleigh wave with retrograde motion. The range time is indicated between the lines A and B. The period of this wave oscillates between 11.0 and 12.8 sec.

Following the same criterion used in the construction of **Figure 4**, in **Figure 5** also the displacements of the stations near the epicenter were drawn: CALE, VILE, ZACA, UNIO, and AZIH; besides, two stations located in the path PAPN and OCTT were included, and also the one of SCT1 in Mexico City. The displacements were calculated using passband filters between 0.065 and 0.15 Hz. Also were drawn, in the right side, the respective odograms in the planes radial-vertical (RV) and radial-tangential (RT) of subevents SUB-1 and SUB-2, in which it is possible to detect the retrograde particle motion that is characteristic of Rayleigh waves (like in **Figure 4**). In stations VILE and ZACA, arrivals of both subevents are identified clearly; thus, SUB1 is observed between the range time of 5.5 and 33.6 sec, while in SUB2 from the time of 34 sec.

2.3 Prograde and retrograde particle motion in the frequency bands between 0.25–0.50 Hz and between 0.15–0.25 Hz

The same procedure in the analysis of displacements followed in the section previous was made here, but now with passband filters between 0.25 and 0.50 Hertz. In this case, something interesting is observed: when drawing odograms, it is possible to identify another surface wave with period of 2.7 sec, but now, this surface wave has a

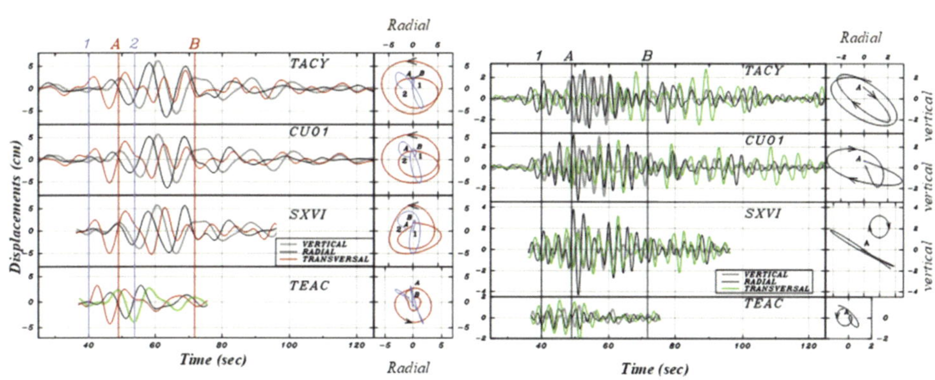

Figure 4.
Radial (R), tangential (T), and vertical (V) displacements in the same stations of **Figure 2**, *with their respective odograms R-T and R-V, but using band-pass filters between: 0.065–0.15 Hz (left), and 0.2–0.5 Hz (right). In the first case, the particle motion is retrograde but in the second case is prograde.*

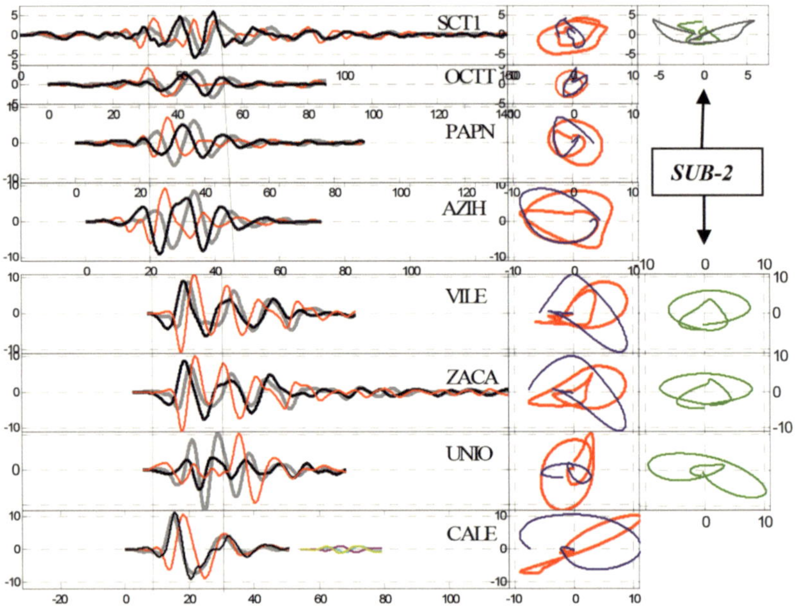

Figure 5.
Radial (R), tangential (T), and vertical (V) displacements for SUB-1, using band-pass filters between 0.065 and 0.15 Hz, in stations close to the source and at Mexico City, with their respective R-T and R-V odograms. In the last column (at right) is shown for SUB-2. The particle motion is retrograde in all cases.

prograde evolution in stations of Mexico City, CU01, TACY, and SXVI (**Figure 4**). In addition, this wave initiates their motion at the same time that the wave of long period (line A), according to the synchronization of **Figure 5**. In stations AZIH, PAPN, and OCCT, the motion also has a prograde way, but in ZACA and UNIO, it has retrograde sense (**Figure 6**). In the case of station TEAC, the motion also is retrograde. In the

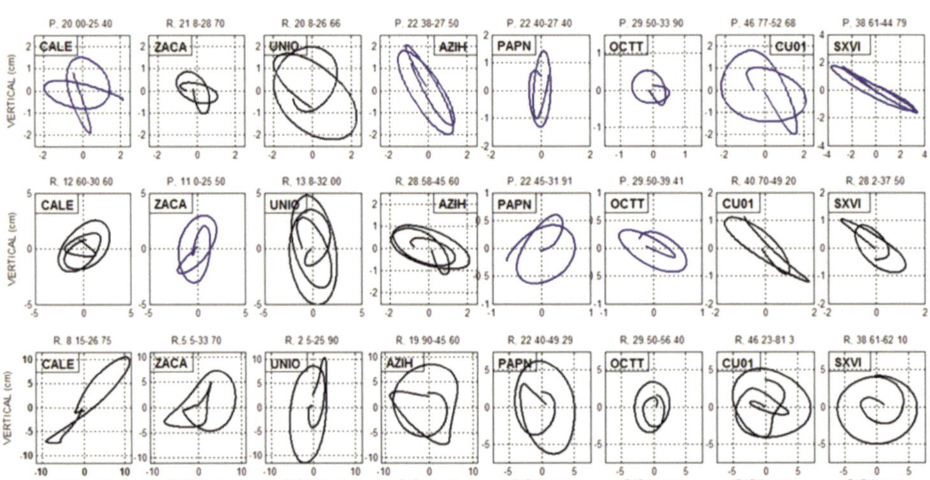

Figure 6.
Odograms RV for three different passband filters: 0.25–0.50 Hz (top), 0.15–0.25 (center), and 0.06–0.15 (bottom). In each block, R indicates Retrograde motion and P Prograde motion.

case of the frequency range of 0.15–0.25 Hz, in some sites like ZACA, PAPN, and OCTT, the motion is prograde, as shown in the middle part of **Figure 6**, while in the other sites, it is retrograde.

3. Theoretical modes of the fundamental mode of Rayleigh waves

In a previous work, Gomez Bernal [4] studied some models of Rayleigh wave propagation, which were calculated using a 1D model of finite element according with

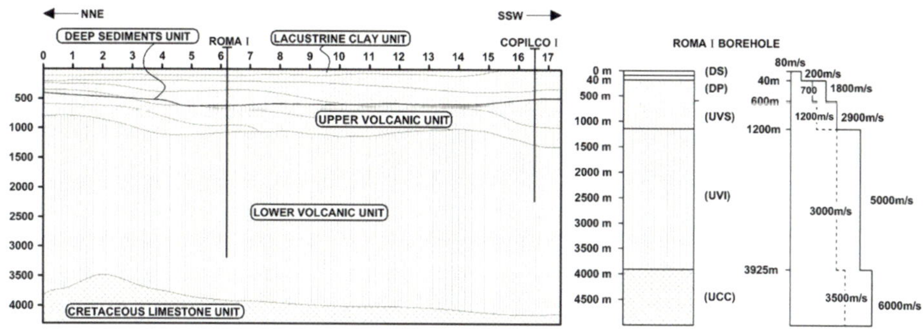

Figure 7.
Schematic geological sections of the subsoil of Mexico City modified after Perez Cruz [14], showing the Roma I borehole, and the P and S wave velocities of the adopted model.

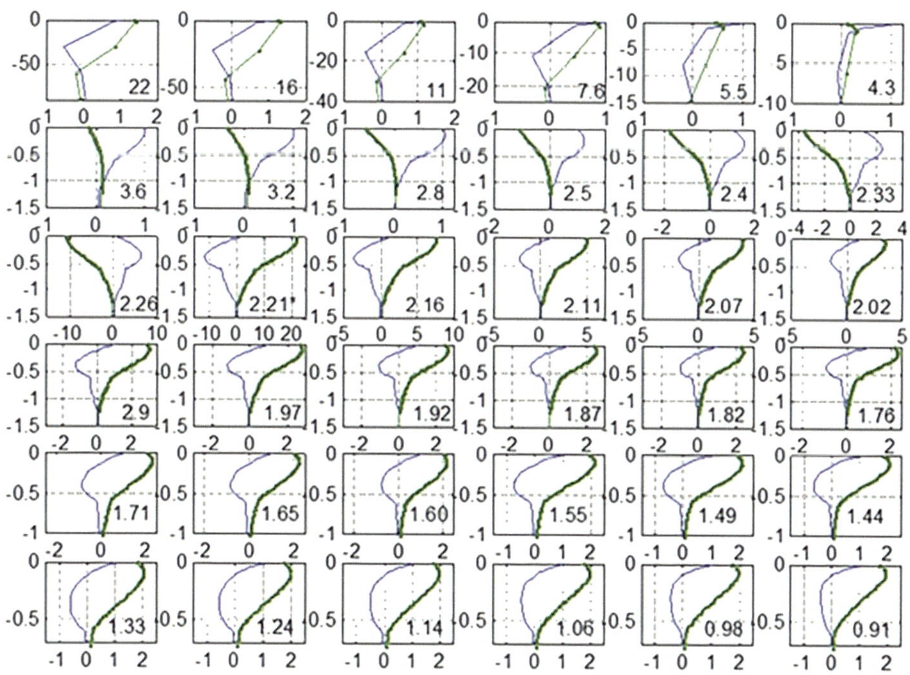

Figure 8.
Thirty six configurations of the fundamental mode of Rayleigh wave, H (thin line) y V (thick line) calculated with ROMA I model. The number in each box is the period in sec. The particle trajectory is prograde in the range periods between 3.6 and 2.26 sec, while in the rest of periods, it is retrograde.

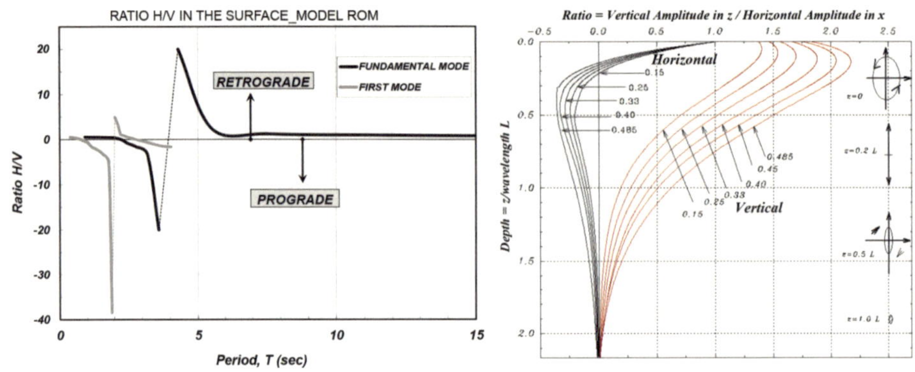

Figure 9.
Left: *Ratios of amplitude H/V in the surface, of ROM model, the fundamental mode and the first upper mode.*
Right: *An illustration of a retrograde motion in surface for different values of Poisson ratio.*

the stratigraphic geological illustrated in **Figure 7**. This model considers the different strata, not including the thin lacustrine deposits.

Figure 8 shows 36 configurations corresponding to 36 periods of the fundamental mode of Rayleigh waves from model ROMA I. The graphics were ordered from major to minor; in each box, the number refers to the period; Modal shapes of the range of periods between 3.6 and 2.26 sec, show a pattern different from the rest, as they are amplified more than the rest, and in this range of the vertical and horizontal mode periods have contrary signs at the free surface, indicating that the evolution of the particle has a prograde motion contrary to the rest of the periods in which the particle motions are retrograde. And when the period is about 2.21 sec, the motion reaches the maximum vertical value.

Figure 9a proposes an interpretation of the result of the ROMA I model, which represents the H/V ratio of each period (z = 0). In this graph, the difference is made between the prograde and retrograde elliptical motion. **Figure 9b** shows the normalized amplitudes of Rayleigh waves as a function of dimensionless depth (z/L) for six different values of Poisson's ratio, v (from 0.15 to 0.485). L indicates the wavelength. The wavelength for different depths, z, is also indicated. The curves of the horizontal components change sign with depth, but not the vertical ones.

4. Earthquake ground motion amplification for surface waves detected during the Puebla-Morelos earthquake of September 19, 2017, M7.1

Strong ground motion records of the Puebla-Morelos (P-M) Earthquake were obtained at a significant number of stations in Central México, as well as in the México City Network. RABO and HMTT, two acceleration stations closest to the epicenter recording a horizontal PGA of 15.9% g and 17.9% g, respectively, and a vertical PGA of 12.3% g and 9.2% g. TACY station of Mexico City recorded a horizontal PGA of 6.5% g and a vertical PGA of 3.6% g; this station registered a value of 8.0% g in the September 19, 1985, earthquake (M = 8.1) (**Figure 10**).

An analysis of recorded ground motions of the P-M Earthquake in 2017, which occurred SE of Mexico City Basin, exhibits differences in relative amplification between shear waves and surface waves. Acceleration waveforms comparing shear and surface waves are show in **Figure 11**, for the unfiltered acceleration records in

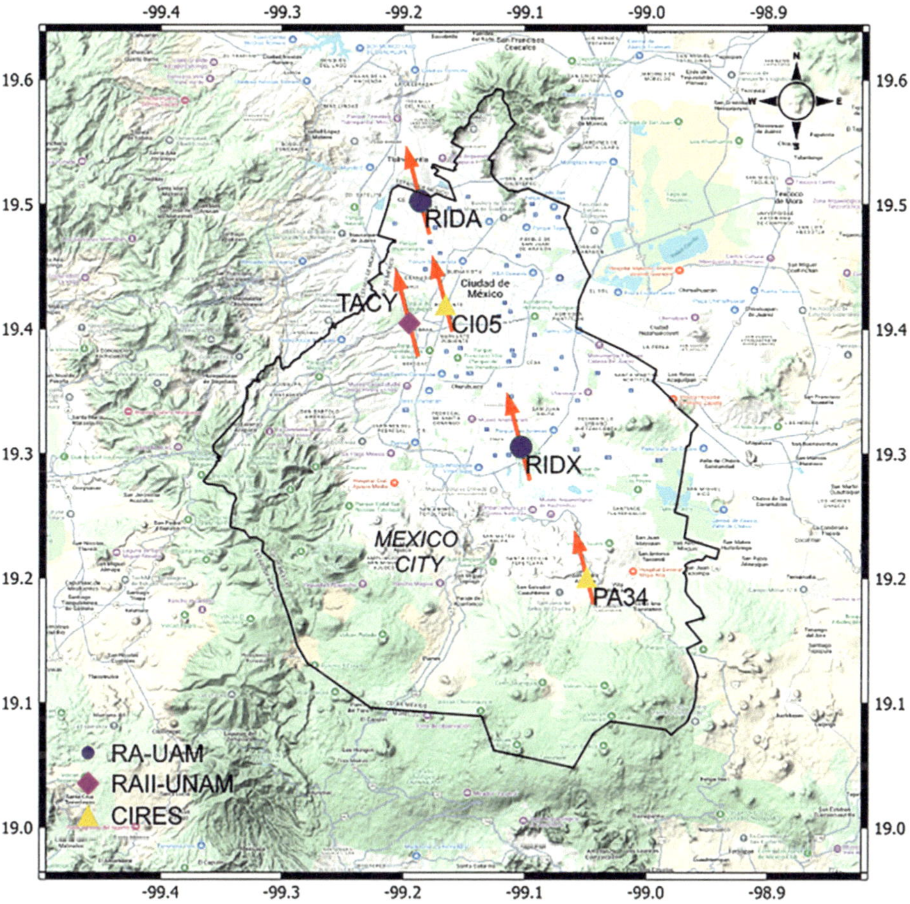

Figure 10.
Acceleration stations used in the waveforms identification of shear and surface waves, from the Puebla-Morelos earthquake, M7.1, with its epicenter roughly 130 km to the SE of Mexico City Basin.

stations RIDA, TACY, CI05, RIDX, and PA34 (panel a, c, d); additionally, **Figure 11** also shows the accelerations filtered at 0.2–0.7 Hz (panel b, e, f). **Table 1** comprises the ratios of the peak ground accelerations (PGA) relative to the firm soil site, TACY, for the unfiltered case, while **Table 2** presents the ratios for the filtered case. The PGA relative to the firm soil site is different when we consider shear wave arrivals separately from surface waves. While distinguishing surface waves from various body wave phases may often be difficult in practice, in the case of 2017 earthquake, PGA is often higher in the surface-wave window compared to the body-wave window, despite the attenuation of high frequency surface waves at this distance. Division between shear-wave arrivals and surface-wave arrivals is based on identification of vertical displacements phase of the filtered case.

When comparing unfiltered waveforms, shear and surface waves are amplified in a way similar relative to the firm site (**Table 1** and **Figure 11a, c, d**). But when comparing filtered waveforms at frequencies of 0.2–0.7 Hz, surface waves also are amplified in a way similar, relative to the firm soil site (**Table 2** and **Figure 11b, e, f**). However, when at each site the ratios of the accelerations of surface wave between shear wave accelerations are estimated, values between 2 and 4 times are obtained in

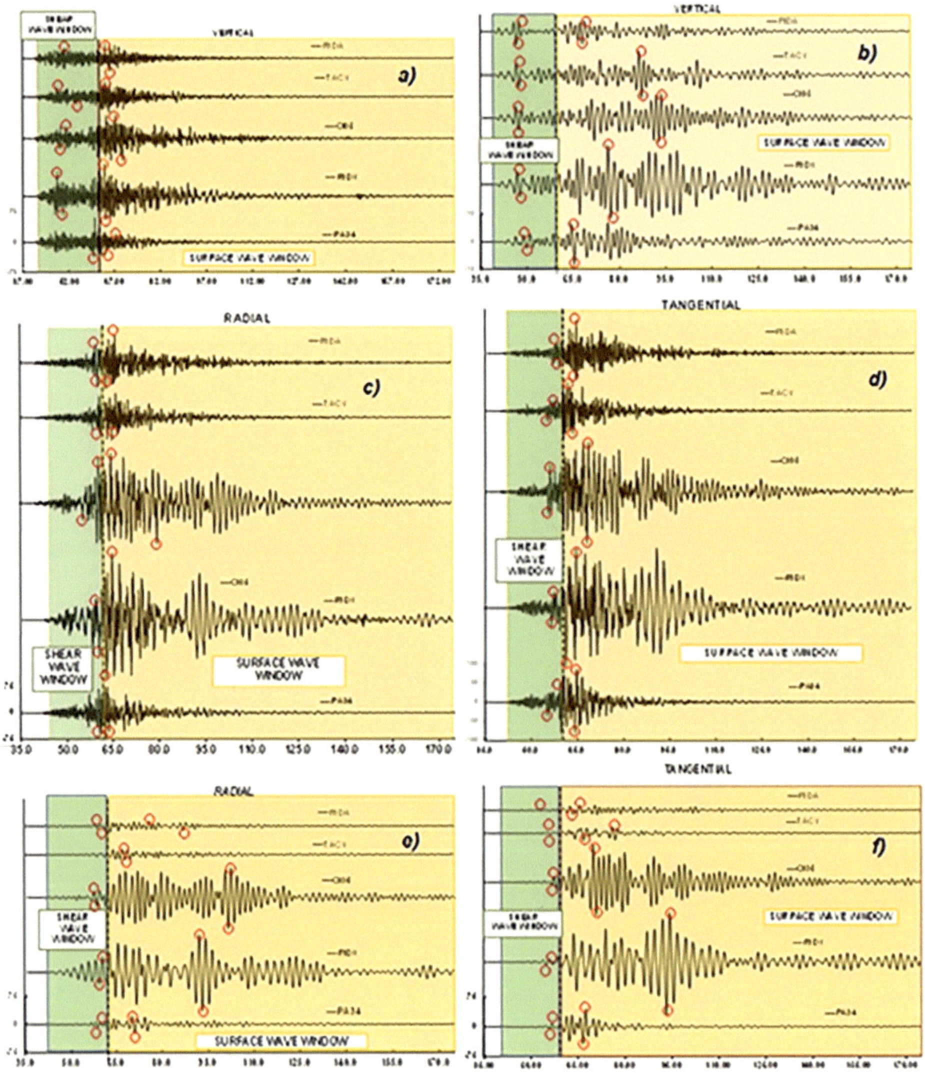

Figure 11.
Acceleration waveforms comparing shear and surface waves, from the P-M earthquake, M7.1. (a, c, d) Unfiltered accelerations for vertical, radial, and tangential components, respectively. (b, e, f) Filtered at 0.2–0.7 Hz. In each panel, PGA is identified for shear waves and surface waves separately.

the unfiltered case, but these differences are even more severe for the filtered case, especially in the tangential component.

4.1 Quantitative characterization of surface waves

After surface waves under investigation have been identified, in some cases it is necessary quantify basin effects through the use of some parameters, in addition to the coefficients presented in **Tables 1** and **2**. For example, Meza-Fajardo et al. [9], use the central frequency f_0, which is defined as the frequency associated with the maximum amplitude of the S-Transform of the extracted waves and an amplification coefficient A. They estimate amplification of the prograde Rayleigh and the Love

	Vertical		Radial		Tangential		Vertical		Radial		Tangential	
	Shear waves						Surface waves					
	AC1	**AC2**	**AC1**	**AC2**	**AC1**	**AC2**	**AC1**	**AC2**	**AC1**	**AC2**	**AC1**	**AC2**
TACY	21.8	−17.1	48.2	−38.7	25.5	−23.1	32	−29.9	54.8	−50.7	66	−66.7
RIDA	22.5	−22.5	46.7	−63.6	31.2	−32.5	32.3	−29.7	71.8	−69.2	88.3	−68.5
CI05	24.6	−21.8	98	−88.6	60	−49.2	51.6	−49.4	111.48	−93	112.7	−118.2
RIDX	49	−37	59	−68.5	30.5	−36.4	64.1	−47.1	133.3	−116.2	121.2	−117.7
PA34	59.9	−35.6	69	−58.9	25.2	−36.4	32.1	−32.8	76	−78.1	87.1	−81.5
	Ratio of the PGA to the firm soil site TACY						Ratio of the PGA to the firm soil site TACY					
TACY	1.00	1.00	1.00	1.00	1.00	1.00	1.00	1.00	1.00	1.00	1.00	1.00
RIDA	1.03	1.32	0.97	1.64	1.22	1.41	1.01	0.99	1.31	1.36	1.34	1.03
CI05	1.13	1.27	2.03	2.29	2.35	2.13	1.61	1.65	2.03	1.83	1.71	1.77
RIDX	2.25	2.16	1.22	1.77	1.20	1.58	2.00	1.58	2.43	2.29	1.84	1.76
PA34	2.75	2.08	1.43	1.52	0.99	1.58	1.00	1.10	1.39	1.54	1.32	1.22

Table 1.
Maximum Acceleration values of time-histories of wave trains of **Figure 11** *for unfiltered accelerations, and ratios of this PGA values in each station respect to TACY station.*

	Vertical		Radial		Transversal		Vertical		Radial		Transversal	
	Shear waves						Surface waves					
	AC1	AC2	AC1	AC2	AC1	AC2	AC1	AC2	AC1	AC2	AC1	AC2
TACY	3.3	−3.5	2.2	−4	2.5	−1.9	7.3	−6.9	12.4	−11.7	14	−16.2
RIDA	3.1	−3.9	3.4	−4.9	2	−2.5	2.8	−3.8	16.4	−13.7	12.2	−9.9
CI05	3.6	−3.6	15	−18.5	13	−7	7.4	−7.3	66.7	−70.3	82.3	−72.3
RIDX	3.51	−3.1	29.5	−22.6	5.5	−4.9	10.5	−9.2	75.3	−78.2	98.1	−105.7
PA34	2.88	−2.18	6.13	−6.9	5.6	−7	6.9	−8.5	21.7	−24.7	44.7	−44.3
	Ratio of the PGA to the firm soil site TACY						Ratio of the PGA to the firm soil site TACY					
TACY	1.00	1.00	1.00	1.00	1.00	1.00	1.00	1.00	1.00	1.00	1.00	1.00
RIDA	0.94	1.11	1.55	1.23	0.80	1.32	0.38	0.55	1.32	1.17	0.87	0.61
CI05	1.09	1.03	6.82	4.63	5.20	3.68	1.01	1.06	5.38	6.01	5.88	4.46
RIDX	1.06	0.89	13.41	5.65	2.20	2.58	1.44	1.33	6.07	6.68	7.01	6.52
PA34	0.87	0.62	2.79	1.73	2.24	3.68	0.95	1.23	1.75	2.11	3.19	2.73

Table 2.
Maximum Acceleration values of time-histories of wave trains of **Figure 11** *for filtered accelerations (at 0.2–0.7 Hz), and ratios of this PGA values in each station respect to TACY station.*

waves extracted at the Kanto basin in the frequency range (0.1–0.5 Hz) during the 2011 Tohoku earthquake. In another investigation, Bowden and Tai [15], propose the ratio:

$$\frac{A_n}{A_n^R} = \frac{u_n(0)}{u_n^R(0)} \left(\frac{UI_o}{U^R I_o^R} \right)^{-1/2} \tag{1}$$

This equation defines a frequency-dependent transfer function by which observations of ground motion at a reference site can be transformed to any other site, each described by a 1-D profile. In Eq. 3, $u_n(0)$ is the displacement eigenfunction measured at the surface (depth = 0) conforming to the type of wave measured by A_n (n = 1, 2, or 3 for radial, vertical, and tangential component of motion, respectively), U is group velocity, Io is an integral over the eigenfunctions and density, and superscript R refers to measurements at a reference site. In Eq. 3, Io are integrals usually used to describe the kinetic energy of surface waves; these integrals are defined differently for Rayleigh and Love waves as.

$$I_o = \int_0^{\infty} \rho(z)\left(u_1(z)^2 + u_2(z)^2\right) dz \quad I_o = \int_o^{\infty} \rho(z)\left(u_1(z)^2 + u_1(z)^2\right) dz \tag{2}$$

$$I_o = \int_o^{\infty} \rho(z)\left(u_3(z)^2\right) dz \tag{3}$$

The nature of local site effects can be estimated by measurements of ground surface motions from sites with different substratum, or by theoretical ground response analyses; in this way, amplification factors were obtained by Gómez-Bernal and Saragoni [16]. The amplification factors of surface waves were determined assuming that plane harmonic waves propagate in elastic media parallel to the free surface soil. The rock is treated as an elastic half space. Transfer functions between contiguous layers were obtained, and the amplification factors were obtained according to the following relations:

$$A_h^{'}(\omega) = \frac{U_s}{U_r^o} e^{i(k'-k)x} \tag{4}$$

$$A_v^{'}(\omega) = \frac{W_s}{W_r^o} e^{i(k'-k)x} \tag{5}$$

$$A_h^{''}(\omega) = \frac{U_s}{U_r} \tag{6}$$

$$A_v^{''}(\omega) = \frac{W_s}{W_r} \tag{7}$$

where Us, W_s, U_r°, W_r°, U_r, and W_r are the displacement components at the free surface of the layer, free rock surface, and interface between the rock and the layer, respectively.

Another procedure to estimate the amplification between two sites would be to extract the surface waves in the frequency range of interest and then apply an intensity parameter to establish the relationship between the two sites using, for example, the Arias intensity or the power spectrum.

5. Microearthquakes in the valley of Mexico basin

The Basin of Mexico is located in the central part of the Trans-Mexican Volcanic Belt (TMVB) and is surrounded by mountains and volcanic structures such as Popocatépetl (currently active), Iztaccihuatl, Ajusco, and Nevado de Toluca. The basin has approximate dimensions of 110 km in the NS direction and 80 km in its minor axis with an EW orientation.

The seismicity of the Mexico Basin can be considered constant with an annual average between 15 and 20 microseisms and has been concentrated south of Mexico City, with epicenters in the Sierra de las Cruces and Ajusco. A characteristic of the seismicity in the Valley of Mexico is that sequences of earthquakes occur in a short period of time; for example, Prince [17] reported a total of 17 events south of Mexico City during the month of July 1974; another sequence of 20 earthquakes occurred between July 12 and

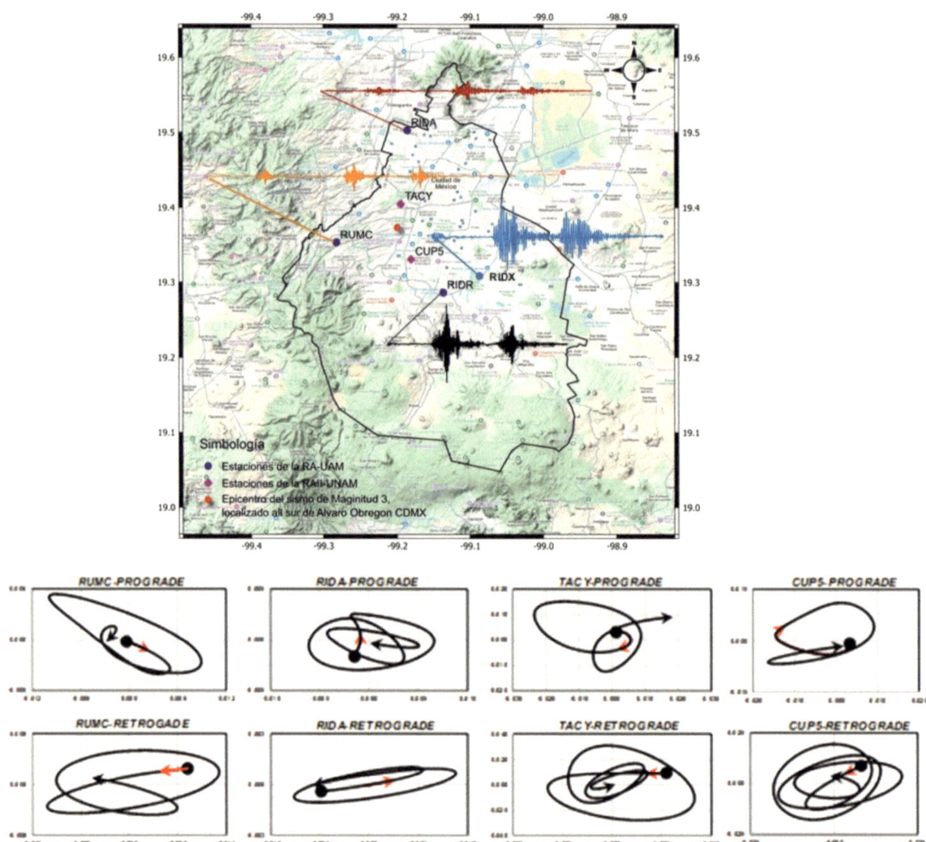

Figure 12.
Top: *Velocities waveforms, from the Microearthquake, M3.0, in the radial components in four station of* ***RSCDMX***. Bottom: *Particle motion of extracted Rayleigh waves in the R-Z plane during the Microearthquake for 0.09–1.0 Hz. In the first line, ellipse of retrograde Rayleigh wave at stations RUMC, RIDA, TACY, and CUP5 during 3 s, in the second line, ellipses of prograde Rayleigh wave during 4 s.*

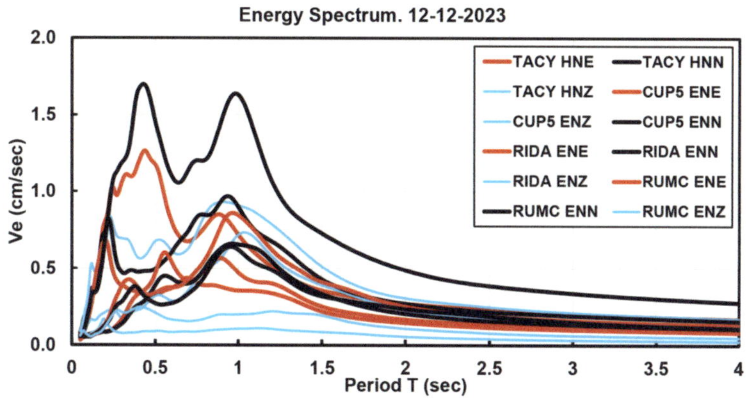

Figure 13.
Energy spectrum for the December 12, 2023, microearthquake (M3.0), in four stations on firm soil indicating the central period around the energy is concentrated in this local events.

18, 2019. The epicenters are located within or near densely populated urban areas and with depths between 1 and 7 km, and the magnitudes range between 1 and 4.4, normal and combined mechanisms type have been reported, indicating the existence of shallow faults (see for example Arce et al. [18]), and Ramirez-Guzmán et al. [19].

5.1 The microearthquakes and the surface waves in the higher frequency range (0.09: 1.0 Hz)

In December 2023, the National Seismological Service, SSN [20] reported a total of 23 microseisms that occurred in Mexico City (in the municipality of Álvaro Obregón) with epicenters very close to each other and with depths of approximately 1 km. Of this sequence of earthquakes, the most intense occurred on December 12 and 14 with M of 3.0 and 3.2, respectively (epicenter 19.363°N, 99.200°W). The first three events on the 12th occurred in a time span of less than 250 sec, and between the first and second earthquake, there was a time lapse of 85 sec, and between the second and third, it was 60 seconds. These earthquakes caused minor damage to some buildings near the epicenter in Mexico City. This sequence of earthquakes was recorded at four stations of the Accelerograph Network of the Autonomous Metropolitan University, RA-UAM, as well as other seismic networks. The ground velocity records of the first three consecutive events on December 12 are shown in the map of **Figure 12**.

Microseisms generated within the Mexico City basin have been observed to have a higher frequency content than events with subduction origin. As shown by the energy spectrum of stations RIDA, RUMC, TACY, and CUP5, the energy is concentrated around the period of 1.2 s during the December 12, 2023, earthquake. In addition, a train of Rayleigh waves in the range of 0.09–1.0 Hz was identified propagating through the city at stations RIDA, RUMC, TACY, and CUP5. First, short-period prograde Rayleigh waves were detected, followed immediately by a train of retrograde waves. This finding is illustrated in **Figure 12**, where we plot the elliptical motions of the particles in the vertical plane oriented in the propagation direction at the five stations. **Figure 12a** clearly shows that the first Rayleigh wave train identified at all five stations is prograde in the propagation direction (given by the positive sense of R), then switches to a retrograde motion as the rotation sense is opposite to the propagation direction. The detected period of the waves propagating in the valley is approximately 1.2 s (0.8 Hz), as also illustrated by the energy spectra of the December 12, 2023, microearthquake in **Figure 13**.

6. Conclusions

Surface waves in the Mexico Basin, recorded during the two earthquakes of September 19, Michoacán (year 1985) and Puebla-Morelos (year 2017), have been identified and analyzed, and a local microearthquake with epicenter within Mexico City is included. Waveforms of the different types of surface waves (retrograde Rayleigh waves, prograde Rayleigh waves, and Love waves) have been extracted and their polarization/propagation direction has been quantified. These two seismic events that caused severe damage in Mexico City allow comparisons of two very intense events and other smaller ones that excite the basin from different azimuths and with different focal mechanisms. Acceleration records, from of the Mexican network in operation on September 19, 1985, includes stations in the source zone, in the path, and in Mexico City, were studied simultaneously.

The findings of this study can be summarized as follows:

- In the analysis of the 0.06–0.15 Hz frequency range, at the beginning of the records in Mexico City, a strong arrival in the tangential direction was detected in the R-T plane, with a period of 8 sec, without radial or vertical motion, which is characteristic of Love wave. This wave was also detected at stations close to the source and those along the path, indicating that they were generated in the source.

- In the R-V plane of all the sites studied (source, path and Valley of Mexico), ground motions with a retrograde elliptical trajectory were identified, typical of Rayleigh waves, and with a long period of 12 sec. These 12-sec Rayleigh wave and the other 8-sec tangential wave are detected at the same time at source stations; however, in Mexico City, they are recorded with a difference of 9 sec between them, with the 8-sec surface tangential wave arriving first. In the 0.15–0.25 Hz band, a surface wave was detected in the RV plane. At some stations such as ZACA, PAPN, and OCTT, the motion is prograde, while at the rest, it is retrograde.

- In the 0.25–0.50 Hz range (2.0 to 5.0 sec), in the R-V plane, the trajectory of the 2.7 sec wave is prograde in Mexico City. It is observed that this wave starts at the same time as the long period wave (line A). At the path stations (AZIH, PAPN, and OCCT), the motion is also prograde, but at ZACA and UNIO, the trajectory is retrograde, as is the TEAC station.

From the above paragraphs, we conclude that in the earthquake of September 19, 1985, groups of surface waves were generated at the source that were guided along the continental crust; at the beginning of the records of the intense part, a Love wave of 8 sec with tangential motion stands out, as well as a group of Rayleigh waves, three of them with periods of 12, 5, and 2.7 sec, respectively. However, it should be noted that the arrivals of surface waves in the Valley of Mexico were constant for more than 100 sec, as shown by the sites on firm ground.

When a thick sedimentary layer exists, it happens an investment of the sign of the vertical eigenfunction near the surface, giving by result a prograde motion of

Rayleigh wave in the surface. This situation is demonstrated, with models of the Valley of Mexico, a retrograde motion happens in the surface within the frequency band between 0.05 and 0.15 Hz but changes the way to prograde motion in the frequency band of 0.15–0.5 Hz. The reason for the reversal in the motion from retrograde to prograde is attributed to elastic effects, mainly caused by extremely slow seismic superficial velocities. However, from the observations of the microseisms that occurred within the valley, this reversal effect was detected but with Rayleigh waves of higher frequency (0.09–1.0 Hz).

Using information of the September 19, 2017, earthquake and with a simple analysis, it was possible to predict the amount of surface-wave amplification separately from the body waves, amplification that is more commonly computed. Therefore, in the hazard analysis models, ground motion prediction equations should include terms for the presence of surface waves in valleys, especially to include long-period signals.

Acknowledgements

The acceleration information used in this work was provided by the following seismic networks: Accelerograph Network of the Autonomous Metropolitan University, RA-UAM; Accelerograph Network of the UNAM Engineering Institute, RA-UNAM; and the Center for Seismic Instrumentation and Recording, CIRES.

Author details

Alonso Gómez-Bernal
Universidad Autónoma Metropolitana, Azcapotzalco, México City, Mexico

*Address all correspondence to: agb@azc.uam.mx

References

[1] Swanger HJ, Boore DM. Simulation of strong motion displacements using surface-wave modal superposition. The Bulletin of the Seismological Society of America. 1978;**68**(4):247-263

[2] Joyner WB. Strong motion from surface waves in deep sedimentary basins. The Bulletin of the Seismological Society of America. 2000;**90**(6):95-112. DOI: 10.1785/0120000505

[3] Gómez-Bernal A, Saragoni R. Interpretation of dynamic soil effects on México City valley using the dense accelerograph network. In: Proceedings of the Fifth Int. Conf. On Seismic Zonation. Vol. 1. Nice, France: Ouest Éditions, Presses Académiques; 1995. pp. 747-754

[4] Gómez BA. Interpretación de los efectos del suelo en el valle de México usando la red acelerográfica de alta densidad (in Spanish) [thesis]. UNAM, México: Facultad de Ingeniería; 2002

[5] Ewing WM, Jardetzky WS. Press F: Elastic Waves in Layered Media. New York: McGraw Hill; 1957

[6] Bard PY, Bouchon M. The seismic response of sediment-filled valleys. Part 2. The case of incident P and SV waves. The Bulletin of the Seismological Society of America. 1980;**70**:1921-1941

[7] Tanaka T, Yoshizawa S, Osawa Y. Characteristics of strong earthquake ground motion in the period range from 1 to 15 sec. Proceedings of the 7th World Conference on Earthquake Engineering. 1980;**2**:609-616

[8] Pei D, Papageorgiou AS. Locally generated surface; waves in Santa Clara Valley: Analysis of observations and numerical simulation. Earthquake Engineering and Structural Dynamics. 1996;**25**:47-63

[9] Meza-Fajardo K, Aochi H, Papageorgiou A. Comparative analysis of Rayleigh and love waves detected propagating in the Nobi and Kanto basins during the 2004, 2007 Chuetsu and 2011 Tohoku earthquakes. Soil Dynamics and Earthquake Engineering. 2021;**143**. DOI: 106606.0.1016/j.soildyn.2021.106606.hal-03747350

[10] Mirwald A, Cruz-Atienza VM, Díaz-Mojica J, Iglesias A, Singh SK, Villafuerte C, et al. The 19 September 2017 (Mw7.1) intermediate-depth Mexican earthquake: A slow and energetically inefficient deadly shock. Geophysical Research Letters. 2019;**46**(4):2054-2064 DOI: 10.1029/2018GL080904

[11] Anderson JG, Bodin P, Brune JN, Prince J, Singh SK, Quaas R, et al. Strong ground motion from the Michoacán Mexico earthquake. Science. 1986;**233**:1043-1049

[12] Bard PY, Campillo M, Chávez-García J, Sanchez-Sesma FJ. A theoretical investigation of large and small scale amplification effects in the Mexico City Valley. Earthquake Spectra. 1988;**4**(3):609-633

[13] Campillo G, Sanchez-Sesma A. Destructive strong ground motion in México City: Source, path and site effects during great Michoacán earthquake. The Bulletin of the Seismological Society of America. 1989;**79**(6):1718-1735. DOI: 10.1785/BSSA0790061718

[14] Pérez Cruz GA. Estudio sismológico de reflexión del subsuelo de la Ciudad

de México (in Spanish) [thesis]. UNAM, México: Facultad de Ingeniería; 1988

[15] Bowden DC, Tsai VC. Earthquake ground motion amplification for surface waves. Geophysical Research Letters. 2017;**44**:121-127. DOI: 10.1002/2016GL071885

[16] Gómez-Bernal A, Saragoni R. Radiation and dispersion of surface waves in Mexico from the 1995 earthquake (mw = 7.3). In: Proceedings 12WCEE, 12th World Conference on Earthquake Engineering. New Zealand: Society for Earth-quake Engineering, 2000. p. 2303

[17] Prince J. Serie de Investigación IPS-1. In: Report of the Engineering Institute. Mexico: National Autonomous University, UNAM; 1974

[18] Arce JL, Layer PW, Macías JL, Morales-Casique E, García-Palomo A, Jiménez-Domínguez FJ, et al. Geology and stratigraphy of the Mexico Basin (Mexico City), central Trans-Mexican Volcanic Belt. Journal of Maps. 2019;**15**(2):320-332. DOI: 10.1080/17445647.2019.1593251

[19] Ramírez-Guzmán LM, Contreras Ruiz M, Aguirre González J, Aldama-Sánchez BD, Roy Burman A, Carrillo Lucia MA, et al. El rompecabezas de la sismicidad al poniente de la Ciudad de México. In: Report of the Engineering Institute. Mexico: National Autonomous University, UNAM; 2024

[20] Servicio Sismológico Nacional. 2024. Available from: http://www.ssn.unam.mx/sismicidad/reportes-especiales/

Chapter 2

Seismic Resistant Building

Abdallah Azizi

Abstract

Earthquakes are the leading cause of structural destruction, resulting in significant human and economic losses. Consequently, ongoing efforts have focused on developing solutions, which have evolved with advancements in architecture and construction sciences. These methods were first proposed by strengthening structures and then with the view of increasing ductility and design based on performance, and in parallel with that, structural control methods were proposed. In this chapter, methods of strengthening structures against earthquakes have been reviewed. Different viewpoints have been proposed and methods of structure control have been expressed. It has been shown that the factor of destruction of structural components is their strain energy reaching the ultimate capacity. Finally, the method of controlling the strain energy of the components in all methods was stated.

Keywords: seismic resistant, structures control, ductility, damper, isolation, TMD, TLD

1. Introduction

This chapter aims to introduce the primary methods for enhancing the resilience of structures in the face of earthquakes. Earthquakes are a significant source of structural damage, leading not only to tragic loss of life but also to substantial economic repercussions. In response to this ongoing challenge, a variety of strategies have been developed, each demonstrating varying degrees of effectiveness in mitigating earthquake-induced damage.

Before delving into these methodologies, we will first explore the mechanisms behind structural damage during seismic events, specifically examining the issue from an energy perspective. Understanding how energy is transferred and dissipated during an earthquake is crucial for developing effective strengthening techniques. Following this foundational overview, we will discuss each method of structural reinforcement, elucidating the underlying philosophy and principles that guide their design and implementation.

From the dawn of humanity, our ancestors sought refuge in naturally occurring shelters, such as caves, as a means of protection from natural disasters. As civilization advanced, so too did our architectural practices. Early human structures were primarily designed to support vertical loads—the weight of roofs and occupants. This focus on vertical load-bearing remained prevalent for centuries, with most constructions adhering to this simplistic approach.

Figure 1.
Tomb of Cyrus the Great [1].

While many of these early structures possessed some inherent capability to withstand lateral forces, such as those generated by seismic activity, this resilience was often insufficient for modern earthquake demands. Many historical edifices, such as ancient castles, have survived to the present day, primarily because they were either constructed in low seismic-risk areas or were built according to the rudimentary principles of structural engineering, albeit without the advanced knowledge we possess today.

A notable example of such enduring architecture is the Tomb of Cyrus the Great in Iran (see **Figure 1**). This remarkable structure has withstood the test of time for over 2500 years, largely due to its innovative construction techniques. The lower layers of the tomb utilize smooth stones that allow for sliding movements during seismic events, effectively accommodating the dynamic forces generated by earthquakes.

As we progress through this chapter, we will categorize and analyze various strengthening techniques, such as base isolation, energy dissipation systems, and retrofitting strategies. Each method will be discussed in terms of its energy management capabilities, detailing how it alters the energy transfer dynamics during an earthquake to protect the integrity of the structure.

For instance, base isolation involves decoupling a building from ground motion, allowing it to move independently and reducing the forces transmitted through its structure. Energy dissipation devices, on the other hand, are designed to absorb and dissipate seismic energy, transforming it into less harmful forms. Retrofitting approaches may involve reinforcing existing structures with modern materials to enhance their capacity to withstand seismic forces.

In conclusion, this chapter will provide a comprehensive overview of the methods available for strengthening structures against earthquakes, underpinned by a solid understanding of the energy dynamics involved. By learning from the past and applying modern engineering principles, we can significantly improve the resilience of our built environment against these natural disasters.

2. Strain energy

From a physical perspective, energy is the essence of any moving object. Every particle strives to reach the lowest possible energy state, a principle that is manifest in various natural phenomena. During interactions, particles transfer portions of their momentum to neighboring particles, effectively reducing the energy density per unit volume. However, the mechanism by which energy is received from a source by adjacent components varies, contingent upon the structural characteristics of those components. These characteristics are primarily related to mass, stiffness, and damping within the structural members. Consequently, different structures may exhibit diverse behaviors in their conversion of input energy, as illustrated by the following examples:

Example 1: When a metal ball falls onto a surface made of jelly, the impact generates concentric circular waves that propagate outward. The jelly's softness allows for significant deformation, which dissipates energy as it transforms into kinetic and strain energy.

Example 2: Conversely, if the ball lands on a paste-like layer, it will merely sink into it without generating substantial waves. In this scenario, the energy is quickly converted into kinetic energy and subsequently dissipated as internal or damped energy, resulting in minimal outward motion.

Example 3: If the ball strikes a hard foam surface floating on water, the foam will oscillate vertically without significant distortion. If the foam is thin, it may even break under the impact. Here, the majority of the ball's energy transforms into kinetic energy and potential energy, causing minimal structural damage to the foam unless it exceeds its capacity.

In the first example, most of the ball's energy is transformed into kinetic and strain energy. In the second, the energy is rapidly converted into kinetic and then damped energy. The third example highlights how a significant portion of the energy is retained as kinetic and potential energy, with the foam merely oscillating in response to the impact.

Among these energy types, strain energy is the primary factor that can compromise structural integrity. Unlike strain energy, non-strain potential energy—such as that arising from gravitational forces—does not inherently lead to destruction.

From these observations, three fundamental principles can be derived:

1. *Energy transfer*: Energy is transferred from the energy source to adjacent components, aiming to reduce the energy level of the source.

2. *Variability in energy reception*: The manner in which energy is absorbed from the source varies based on the structural characteristics of the neighboring components.

3. *Strain energy and structural integrity*: In structures, exceeding the strain energy capacity of members leads to structural failure.

During seismic events, the energy imparted to structures is primarily stored in the forms of kinetic energy, strain energy, and internal energy, which results from damping mechanisms.

It is also important to note that strain energy can be categorized into two distinct forms: volumetric and deviatoric strain energy. The latter, deviatoric strain energy, is particularly critical; it is the primary contributor to structural failure. Deviatoric strain energy results from distortional changes within the material, which can exceed the material's yield point, leading to irreversible deformation or rupture.

In conclusion, comprehending how energy is transferred and dissipated within structures is essential for developing effective designs that enhance resilience against seismic activities. By recognizing the nuances of energy behavior, engineers can implement strategies to mitigate the effects of strain energy, ultimately ensuring the structural integrity and safety of buildings in earthquake-prone regions.

This limitation arises from the inherent properties of materials, which possess a finite capacity to withstand strain and store potential energy. When the applied stress exceeds this threshold, materials can undergo permanent deformation, commonly referred to as plasticity. This phenomenon is effectively described by various behavioral models, such as the Von Mises and Tresca criteria, which predict the conditions under which materials yield. While these models provide valuable insights into material behavior under stress, a detailed exploration of their mechanics and applications is beyond the scope of this chapter. Instead, we will focus on the implications of these principles for material selection and design in engineering applications, emphasizing the importance of understanding these limitations to ensure structural integrity and performance.

3. Lateral load-bearing system

As structural engineering has advanced, the imperative to address lateral loads has become increasingly integral to the design process. Engineers have recognized the need to incorporate specialized components such as braces, shear walls, and bending frames—illustrated in **Figure 2**—collectively referred to as lateral load-bearing systems. These systems are vital for enhancing a structure's capacity to withstand lateral forces arising from environmental factors, such as wind, seismic activity, and other dynamic loads.

The incorporation of these lateral load-bearing systems fundamentally alters how structures respond to external forces. By strategically distributing loads, these components enable a significant portion of the energy input into the structure to manifest as kinetic energy. This dynamic response is essential because it ensures that the strain energy within the structural elements remains well below critical thresholds, thereby minimizing the risk of catastrophic failure.

However, while these systems considerably amplify a structure's strength, they also lead to an increase in lateral stiffness. This heightened stiffness can inadvertently result in a greater transfer of loads during seismic events, which poses challenges for

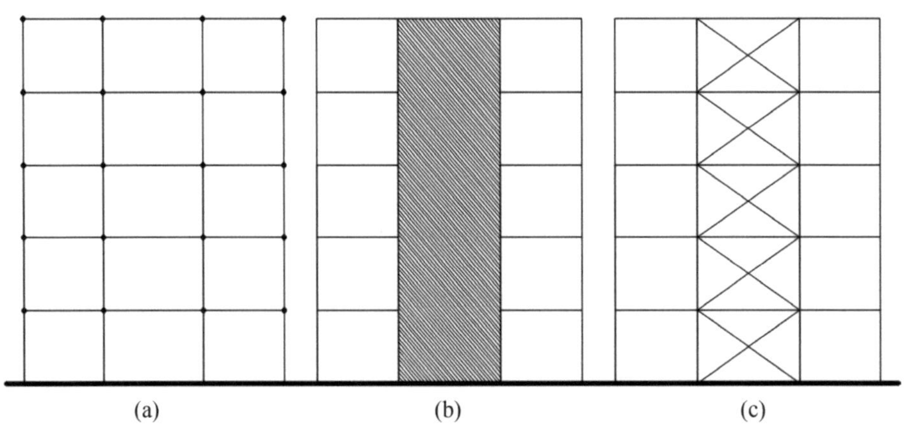

Figure 2.
Lateral load-bearing system. (a) Moment resisting frames. (b) Shear wall frames. (c) Braced frames.

structural integrity. As more load is directed to the building during an earthquake, it can trigger localized damage in both structural and non-structural components. For instance, critical connections between beams and columns may endure excessive stress, increasing the risk of joint failures. Non-structural elements, such as partitions, facades, and mechanical systems, may also suffer from amplified forces, leading to issues such as cracking, separation, or even collapse.

The design and implementation of lateral load-bearing systems require a comprehensive understanding of various factors, including material properties, connection detailing, and the dynamic behavior of the structure under varying loading conditions. Engineers must carefully balance the benefits of increased strength and stiffness with the potential for localized damage. This necessitates rigorous analysis, modeling, and empirical testing to accurately predict how different components will interact during extreme events, ensuring the safety and functionality of the structure.

4. Ductility

Comprehensive strengthening of structures to withstand seismic events has often proven to be both impractical and economically unfeasible. Consequently, some degree of damage to structural components during an earthquake is inevitable. Such damage typically signifies that the structure has exited the elastic zone, leading to the plasticization of certain members. This transition is critical because, if left unmanaged, the plasticization of structural components can escalate into severe structural failure.

In response to the challenges posed by potential plasticization, the concept of "ductility" emerged as a vital criterion in structural design. Ductility refers to a structure's ability to undergo significant deformation without failing, allowing it to absorb and dissipate energy during seismic events. The greater the energy absorption capacity of a structure, the higher its ductility. Essentially, a ductile structure can endure considerable stress and strain without collapsing, thereby enhancing its resilience during earthquakes.

To ensure that structures can effectively utilize ductility, building codes have been established with specific guidelines that facilitate elastic design. These regulations outline particular rules and criteria tailored to various types of structures, assigning a behavior coefficient that reflects the anticipated ductility of each design. This does not imply that a structure will remain entirely elastic during an earthquake; rather, the intent is to ensure that, even if some components yield plastically, the overall stability of the structure is maintained.

Furthermore, the implementation of ductility criteria in design serves several purposes. It allows engineers to create structures that can absorb seismic energy while controlling the extent and location of damage. By designing for ductility, engineers can predict where and how a structure will fail, allowing for strategic planning that minimizes risks to occupants and surrounding infrastructure. This approach not only enhances the safety of the building but also facilitates easier and more cost-effective repairs after an event.

5. Performance base design

The design of structures informed by performance-based methodologies has become increasingly vital in modern engineering, particularly as building codes evolve to address the complexities of seismic safety and resilience. These codes emphasize the necessity for structures to be designed with adequate ductility, which is crucial for absorbing and dissipating energy during dynamic events, such as earthquakes. Ductility enables structures to undergo significant deformation without experiencing catastrophic failure, thereby enhancing the overall safety of occupants.

In a performance-based design framework, structures are categorized into specific performance levels tailored to the anticipated behavior during various loading scenarios (illustrated in **Figure 3**).

Each level defines a range of acceptable damage and functionality, guiding engineers in creating designs that not only meet safety standards but also maintain usability after seismic events. This approach allows for a more nuanced understanding of how buildings will react under different conditions, ultimately leading to more resilient structures.

A critical aspect of performance-based design is the consideration of nonlinear effects, particularly in the analysis and modeling of joints and connections within the structural framework. These joints are pivotal areas where stresses can concentrate, and their performance under load can significantly influence the overall stability and integrity of the structure. By defining and analyzing the behavior of joints through nonlinear modeling techniques, engineers can better predict how the structure will respond to extreme loads, allowing for targeted design improvements that enhance ductility.

Moreover, performance-based design methodologies encourage innovation in materials and construction practices. Engineers can utilize advanced computational tools, such as finite element analysis, to simulate various scenarios and assess how different design configurations will perform under dynamic forces. This capability not only aids in optimizing ductility but also allows for the exploration of novel materials and assemblies that can contribute to enhanced structural performance.

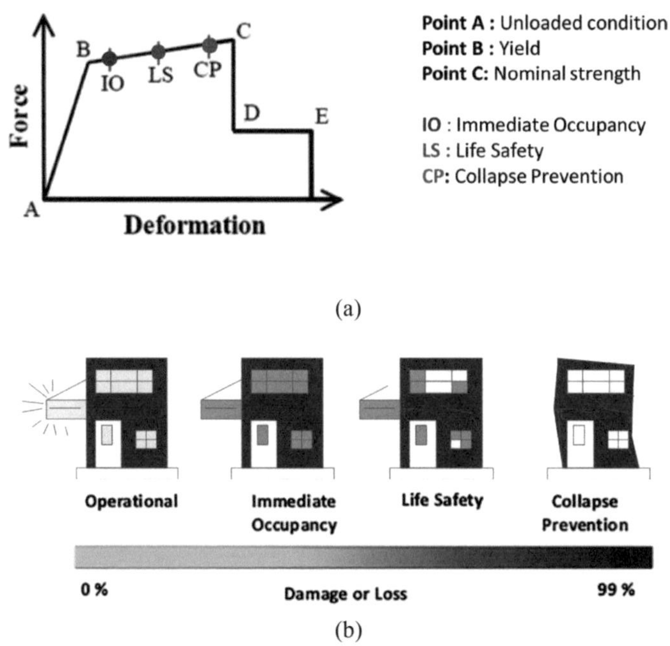

Figure 3.
(a) Schematic diagram of lateral load-deflection [2]. (b) Schematic of design levels on performance base method.

Important performance levels are usually defined as follows.

a. Operational

b. Immediate occupancy

c. Life safety

d. Collapse prevention or near collapse

Reinforcement strategies that prioritize ductility, along with performance-based design principles, allow engineers to effectively manage strain energy within various components of a structure. This approach ensures that strain energy is concentrated in designated areas, where it can reach critical levels without compromising the overall stability of the structure.

This mechanism not only enhances structural resilience but also helps prevent the accumulation of critical strain energy in more vulnerable sections of the building. As a result, potentially hazardous components are safeguarded from excessive stress, reducing the risk of failure. This targeted reinforcement enhances the overall performance of the structure.

6. Structural control systems

In addition to the conventional approach of reinforcing structures with lateral load-bearing components, there is an alternative perspective known as structural control science. This discipline shifts the focus from simply maximizing structural strength to effectively managing the energy input to the structure. The aim is to regulate external loads in a manner that minimizes potential damage to the structural components. Structural control encompasses various strategies that actively respond to dynamic forces, such as earthquakes and high winds. This approach is categorized into several key components, as illustrated in **Figure 4**. By employing advanced control systems—such as dampers, isolators, and active control mechanisms—engineers can significantly reduce the impact of external forces, thereby enhancing the overall resilience and safety of the building.

All the measures discussed in this chapter are designed with a specific goal: to minimize structural damage. This overarching objective can be further articulated in scientific terms.

The integration of specialized systems into structures to control and dissipate forces, thereby safeguarding the integrity of the structure and preventing potential failure. By actively mitigating the impact of external loads, these technologies enhance the resilience and longevity of the building.

6.1 Passive seismic control

Passive structural control systems are typically categorized into three distinct types:

Damping systems facilitate the entry of energy into structural components and subsequently dissipate that energy. Commonly referred to as dampers, their primary function is to convert the kinetic energy generated by dynamic loads into internal energy.

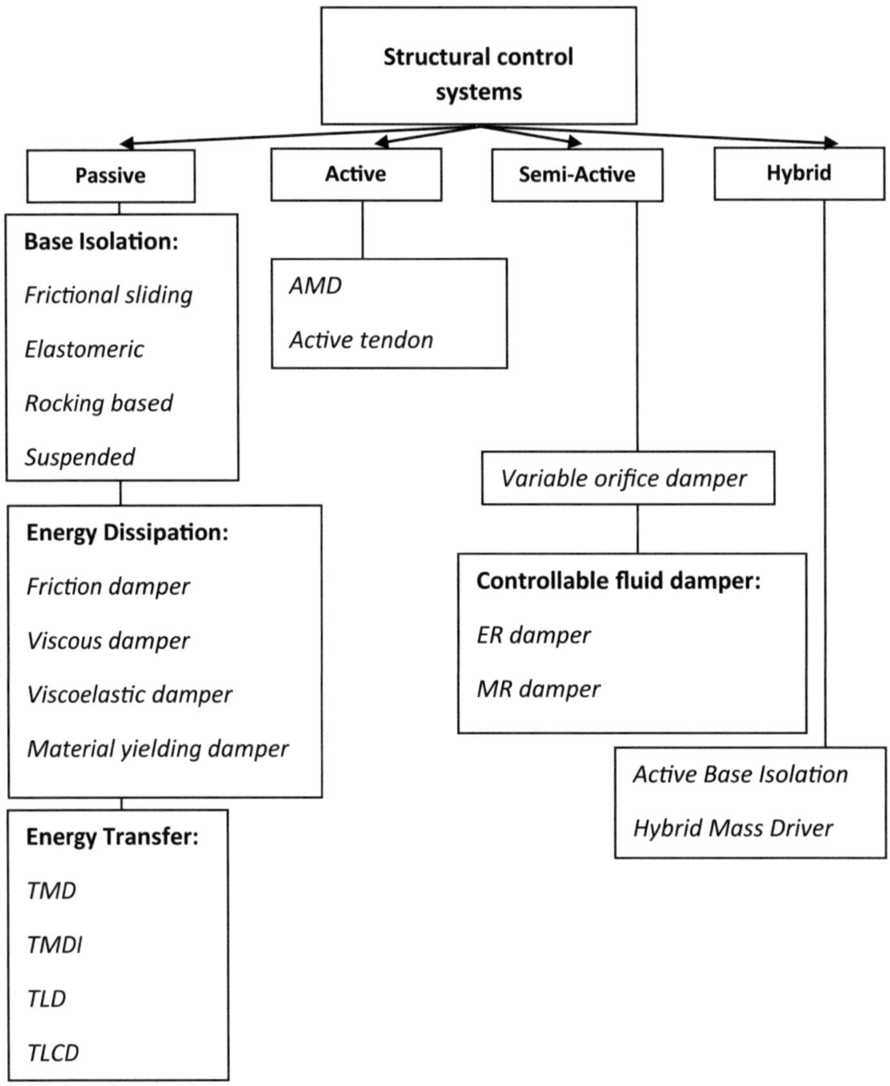

Figure 4.
Structures control methods classification chart.

This internal energy is eventually released as heat, effectively reducing the overall energy transferred to the structure.

Isolation systems create a separation between the structure and the ground, thereby minimizing the transmission of seismic or vibrational forces. By isolating the structure, these systems significantly reduce the impact of external forces, enhancing the building's resilience.

Energy transfer systems are designed to channel energy within the structure itself and dissipate it. By effectively managing the flow of energy, they help to mitigate the effects of dynamic loading, further protecting the structural integrity.

In summary, these passive structural control systems play a vital role in enhancing the safety and durability of structures by effectively managing energy transfer and dissipation during dynamic events.

6.1.1 Friction damper

Displacement-dependent dampers operate on the principle of Coulomb friction, which transforms seismic energy generated by earthquakes into thermal energy through friction. This process effectively increases the vibration period of the system.

One significant advantage of these dampers is their ability to resist fatigue under service loads, ensuring long-term reliability. Furthermore, their performance is unaffected by variations in loading velocity and ambient temperature. Typically, these dampers are installed parallel to braces, as depicted in **Figure 5**.

By efficiently dissipating energy during dynamic events, displacement-dependent dampers significantly contribute to the structural resilience and durability of buildings.

6.1.2 Material-yielding damper

These dampers rely on displacement to dissipate seismic energy through the yielding, nonlinear behavior, and inelastic characteristics of their components, which may include materials such as steel, lead, and others. Depending on their configuration and the specific structural application, these dampers can deform and dissipate energy in various modes, including shear, bending, and axial deformation.

Examples of these dampers are showcased in **Figure 6**, highlighting their adaptability and effectiveness in enhancing structural resilience during earthquakes.

6.1.3 Viscose damper

The performance of these dampers is influenced by the velocity of the fluid within them. They dissipate energy through the movement of a viscous liquid in a cylinder-like chamber, with the generated force being contingent upon the shape and size of the openings as well as the fluid velocity. One notable advantage of these dampers is their ability to produce maximum damping force out of phase with the maximum displacement of the structure. This phase difference arises from the relationship between force and deformation within the damper, which is dependent on the piston movement and fluid velocity.

Due to their ease of installation, variety of dimensions, and compatibility with other structural elements, these dampers have become popular choices for both new designs and retrofitting projects. They can be effectively implemented across a wide range of structures. Two types of these dampers are illustrated in **Figure 7**.

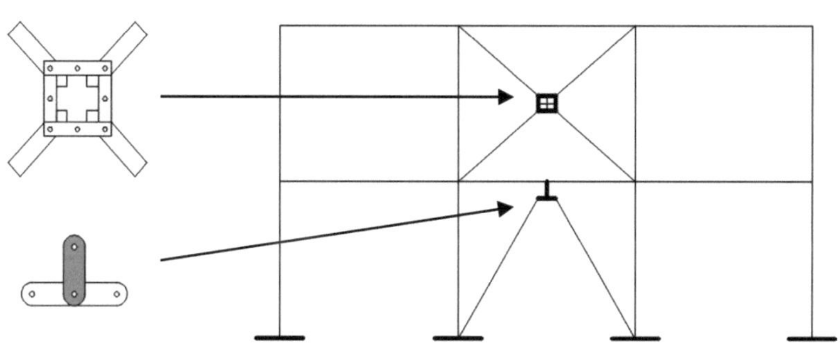

Figure 5.
Example of friction damper in the structure.

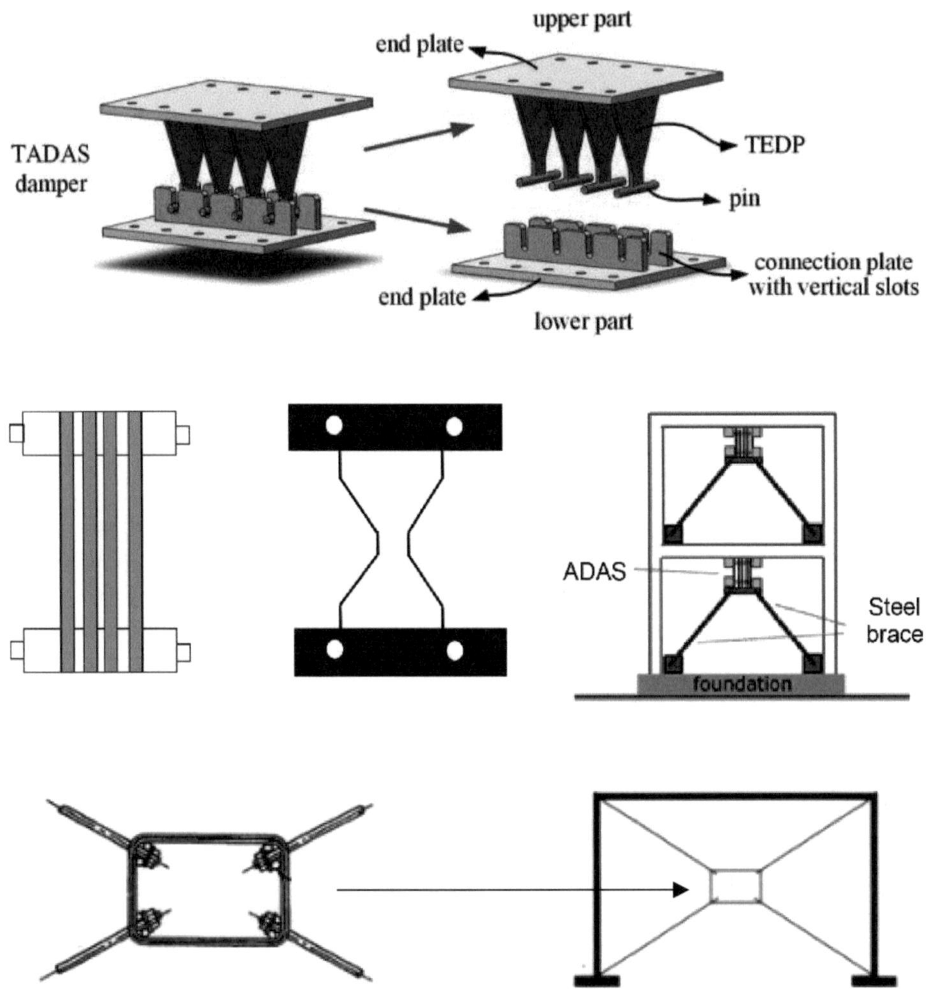

Figure 6.
Some examples of material-yielding dampers [3, 4].

6.1.4 Viscoelastic damper

These dampers, depicted in **Figure 8**, function based on the mechanism of viscoelastic shear deformations. This mechanism combines both viscous and elastic properties, meaning that the relationship between stress and strain is time-dependent and can revert to its original state upon unloading. They utilize materials such as polymers, which exhibit these properties across a wide temperature range.

The performance characteristics of these dampers are influenced by both temperature and loading frequency. The damping coefficient of the viscoelastic material is determined by its dimensions, the loss modulus, and the frequency of the applied load. This enables the dampers to be effectively tailored to respond to various dynamic conditions, enhancing their overall effectiveness in energy dissipation.

Dampers serve as plasticized components within a structure. By converting incoming energy, they prevent its accumulation as strain energy in structural

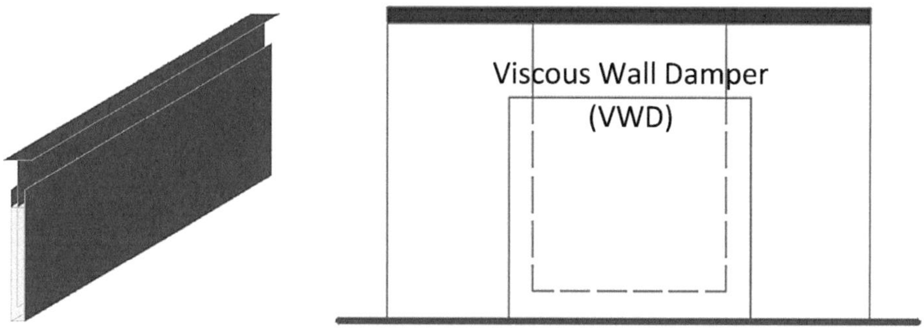

(a)

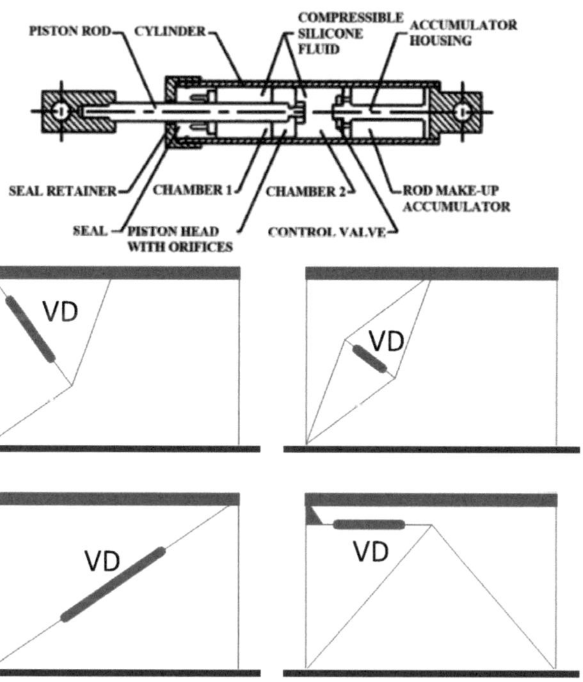

(b)

Figure 7.
Example of viscous dampers in the structure. (a) Viscous Wall Damper. (b) Viscous Damper [4].

elements, instead transforming it into internal energy. This process reduces the risk of damage and enhances the overall stability and resilience of the structure.

6.1.5 Energy transfer systems

Another type of structural control system involves mechanisms that enable energy to pass through the structure with minimal impact, effectively redirecting it into specialized damping systems. A primary example of this approach is represented by

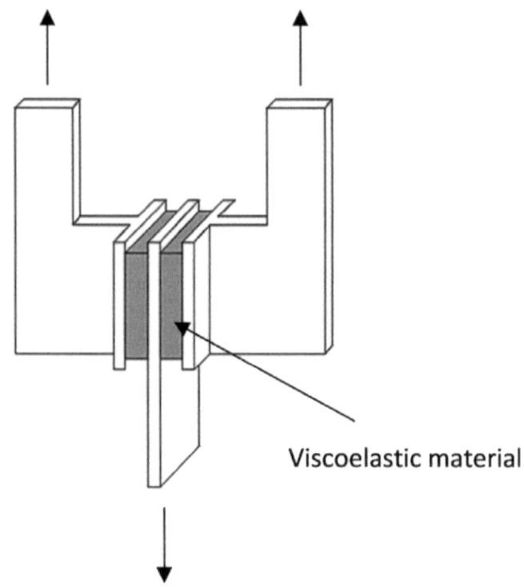

Figure 8.
Example of a viscoelastic damper.

Tuned Mass Dampers (TMD) and Tuned Liquid Dampers (TLD), as illustrated in **Figure 9**. These systems enhance structural integrity by absorbing vibrations, thereby ensuring the structure remains in a safe and stable condition.

A Tuned Mass Damper (TMD) consists of a mass connected to the main structure via a spring. This arrangement allows the mass to move in opposition to the vibrations of the structure, effectively dissipating kinetic energy and reducing oscillatory motion. The tuning of the mass and spring system is crucial, as it must be calibrated to resonate at the same frequency as the vibrations it aims to mitigate.

In contrast, a Tuned Liquid Damper (TLD) utilizes liquid—primarily water—as its damping medium instead of a solid mass. This system capitalizes on the fluid's inherent properties to absorb and dissipate vibrational energy. The mass and movement of the liquid allow for effective energy transfer.

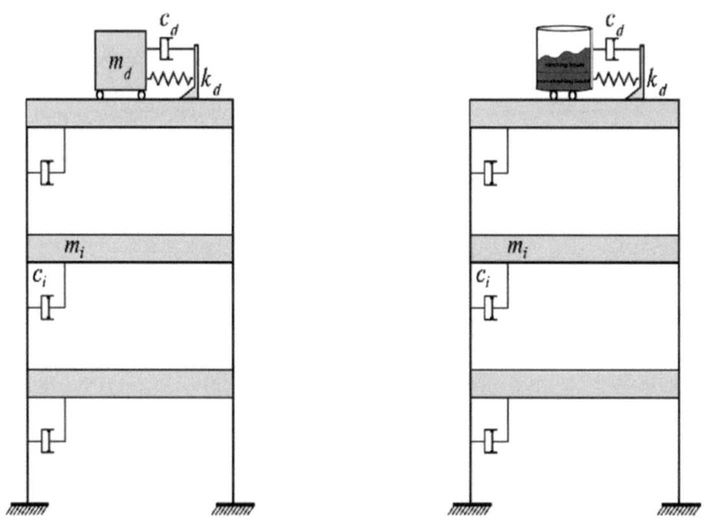

Figure 9.
Schematic of TMD and TLD in the structure.

Another notable variant is the Tuned Liquid Column Damper (TLCD), which relies on the inertia of a liquid column contained within a U-tube. As external forces act on the structure, the inertia of the liquid provides a counteracting force, effectively damping vibrations. TLCDs are particularly effective in structures subjected to periodic loads, such as seismic activity or wind-induced vibrations.

Additionally, the Tuned Mass Damper Inerter (TMDI), as shown in **Figure 10**, combines the principles of a TMD with an inerter mechanism. This innovative design enhances the system's performance by providing inertial resisting forces through relatively small masses. The inerter allows for greater flexibility in design.

These systems optimize their characteristics to align with those of the structure, facilitating the transfer of input energy into the damping mechanisms. By effectively absorbing and dissipating this energy, they minimize the strain energy accumulated within the structure. This process enhances structural resilience and reduces the potential for damage during dynamic events.

As noted, another type of structural control system operates directly at the interface between the structure and external stimuli, transforming earthquake vibrations from destructive to non-destructive.

6.1.6 Elastomer-based isolators

Elastomeric-based methods are among the most effective seismic isolation techniques employed in modern engineering to enhance the resilience of structures against seismic forces. These methods utilize specialized isolators that play a crucial role in mitigating the impact of ground motion during an earthquake. The standard design of these isolators consists of multiple layers of rubber, reinforced by thin steel plates or fiber layers. These components are meticulously compressed and interconnected through a unique manufacturing process, resulting in an isolator that possesses both flexibility and strength.

The various types of elastomeric isolators cater to different performance requirements and applications, each offering distinct advantages. The Low Damping Rubber

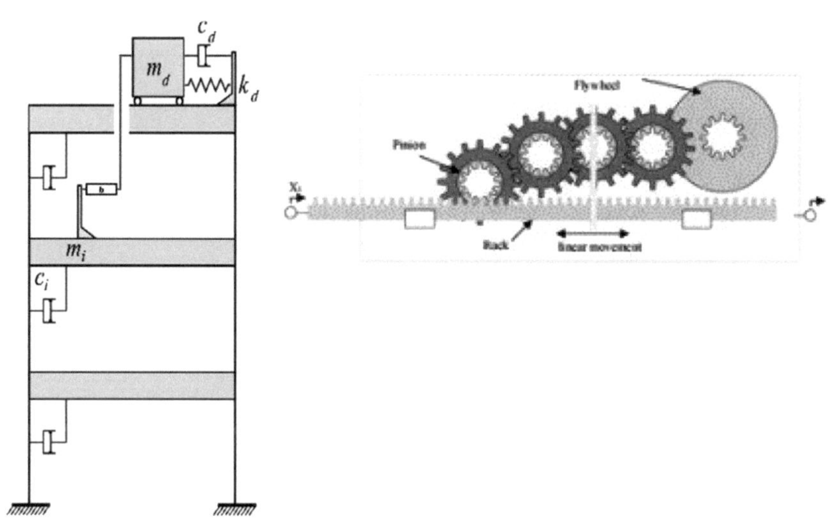

Figure 10.
Schematic of TMDI in the structure [5].

Bearing (LDRB) is engineered to provide minimal energy dissipation, which allows for increased flexibility and movement. This characteristic is particularly beneficial in structures that require a high degree of displacement during seismic events, as it enables them to sway without incurring significant damage.

Conversely, the High Damping Rubber Bearing (HDRB) incorporates specific additives that enhance energy dissipation capabilities. This feature effectively reduces the amount of seismic energy transmitted to the structure, making HDRBs suitable for buildings in high-risk earthquake zones where energy absorption and control are paramount.

The Lead-Rubber Bearing (LRB) is another notable type of isolator, which combines rubber and a lead core. This design allows for substantial energy absorption while maintaining a certain level of stiffness to control the overall movement of the structure. LRBs are particularly advantageous for tall buildings and bridges, where both flexibility and stability are critical.

Lastly, the Fiber-Reinforced Rubber Bearing (FRB) integrates fiber materials into the rubber matrix, enhancing the isolator's durability and load-bearing capacity. This innovation extends the lifespan of the isolator and improves its performance under varying load conditions, making FRBs suitable for a wide range of applications, from residential buildings to critical infrastructure.

Some instance of this type are shown in **Figures 11–13**.

6.1.7 Sliding-based isolators

In sliding isolators, the fundamental principle revolves around the relative motion between two surfaces during an earthquake. This sliding mechanism serves to decouple the structure from ground motion, thereby mitigating the transmission of seismic forces. The sliding process is activated when the intensity of the external exciting force surpasses the frictional force that resists this movement. This characteristic

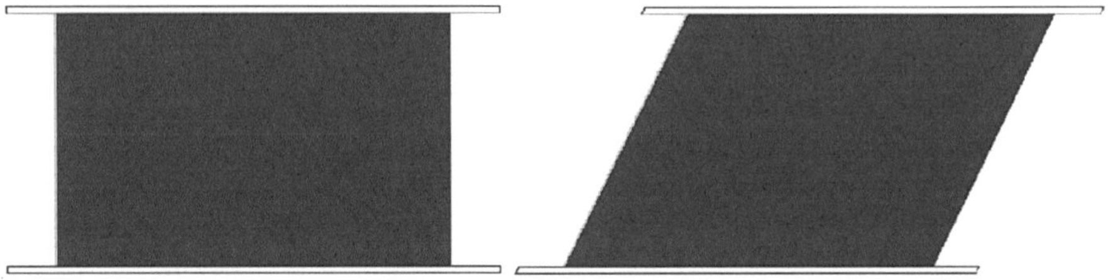

Figure 11.
Pure elastomeric seismic isolator.

(a) (b)

Figure 12.
(a) Elastomeric seismic isolator with steel layers. (b) Elastomeric seismic isolator reinforced with fibers.

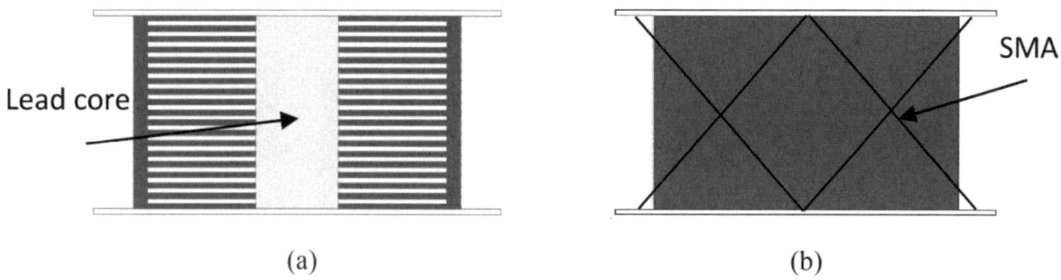

Figure 13.
(a) Elastomeric seismic isolator with lead core. (b) Elastomeric seismic isolator reinforced with SMA [6].

allows sliding isolators to effectively accommodate significant lateral displacements, which are common during seismic events.

Several types of sliding isolators have been developed, each with distinct operational features and advantages tailored to specific engineering needs.

Pure friction system (P-F) relies entirely on the friction generated between two surfaces to facilitate movement. When the applied seismic forces exceed the static friction threshold, the isolator begins to slide. While straightforward in design, the P-F system necessitates careful selection of materials and friction coefficients to ensure effective performance during seismic activity. It is often used in applications where simplicity and cost-effectiveness are prioritized.

Single friction pendulum system (SFPS) introduces a pendulum mechanism that allows the structure to move around a fixed point. This design not only enables sliding but also provides a restoring force that helps return the structure to its original position after displacement. The pendulum action enhances the isolator's ability to manage larger lateral movements, making it particularly effective for taller buildings or those in high-seismic zones.

Double friction pendulum system (DFPS), building on the principles of the SFPS, incorporates two friction surfaces arranged in a way that enhances energy dissipation and flexibility. This configuration allows the system to accommodate a wider range of seismic forces and displacements, making it suitable for structures that experience varying levels of seismic activity. The dual pendulum action provides improved control over the oscillations, ensuring greater stability during an earthquake.

Triple friction pendulum system (TFPS) further enhances the capabilities of the DFPS by introducing a third friction surface. This additional layer allows for even greater energy dissipation and control of structural movement. The TFPS is particularly effective in complex seismic scenarios where multi-directional forces are present, offering superior performance for critical infrastructure and high-value assets.

Resilient friction base isolation (RFBI) with a set of flat rings which can slide on each other and with a central rubber core and/or peripheral rubber cores, combines the principles of friction isolation with resilient materials, such as elastomers, to create a hybrid system that not only allows for sliding but also absorbs a significant portion of the seismic energy. This dual functionality enhances the overall effectiveness of the isolation system, making it suitable for a wide range of structures, from residential buildings to essential facilities like hospitals and emergency response centers.

The versatility of sliding isolators makes them a critical component in modern seismic design. By allowing controlled movement, these systems significantly reduce the risk of damage to structures during earthquakes, ultimately improving occupant safety and protecting valuable assets. As research and technology continue to advance, the design and application of sliding isolators are likely to evolve, further enhancing their effectiveness in mitigating seismic impacts. Some instances of the mentioned systems are shown in **Figures 14–16**.

6.1.8 Rocking-based isolators

In this type of isolation system, the primary components of the bearings include rolling rods, elliptical rods, spherical bearings, and curved surfaces. These elements usually are typically engineered with re-centering properties, allowing them to return to their original position after being displaced during an earthquake. **Figure 17** shows an instance of this type.

6.1.9 Suspended base isolation

The concept of suspended seismic isolation systems involves creating a flexible interface between the structure and its foundation. Rather than being directly connected, the building is connected to the foundation via pendant cables [7–9]. This design facilitates lateral movement of the structure during an earthquake, leading to a significant reduction in the acceleration transmitted to the building. The pendulum column—**Figure 18**—and multilayer isolator—**Figure 19**—are samples from this type that have been investigated and proposed recently [10].

The seismic isolators, usually work by changing the natural periodicity of the structure. Since the destructive frequencies of earthquakes are usually defined in a certain range, increasing the natural periodicity of the structure causes the structure to move away from the danger zone and the response of the structure to these vibrations is reduced.

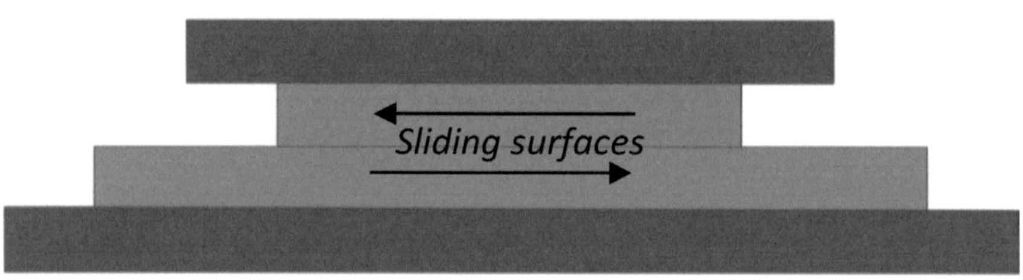

Figure 14.
Pure friction isolator.

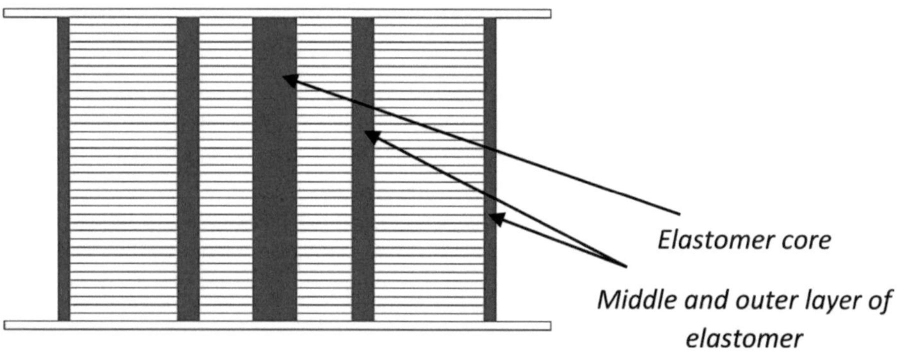

Figure 15.
Resilient friction isolator.

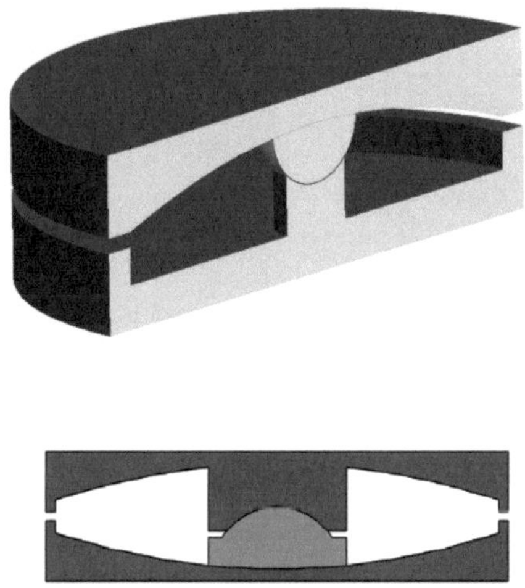

(a)

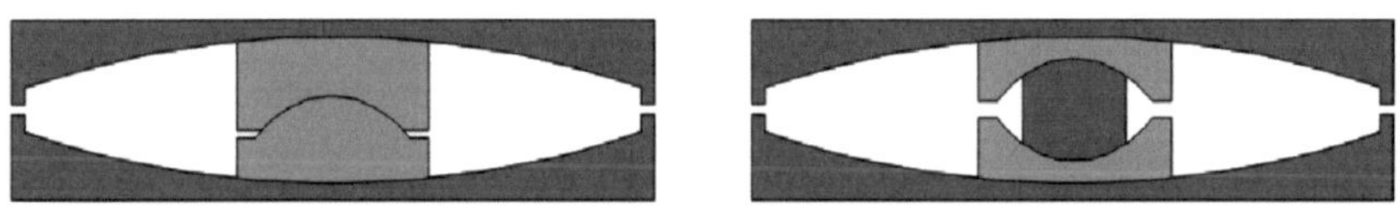

(b) (c)

Figure 16.
(a) Friction pendulum isolator. (b) Double friction pendulum isolator. (c) Triple friction pendulum isolator.

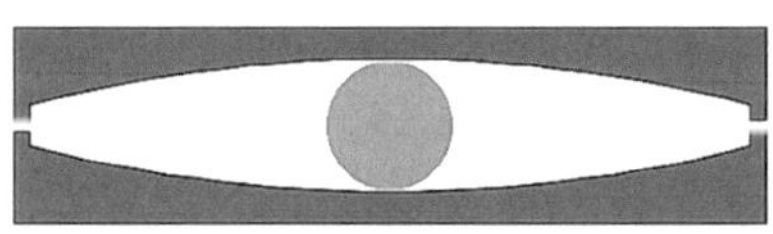

Figure 17.
Roller isolator.

By increasing the natural periodicity of the structure, the speed of wave transmission in the structure becomes lower than the speed of wave movement from the ground to the structure. Therefore, the structure gets a chance to spread the deformation created in the base uniformly in the components. In this way, the distortion and as a result the strain energy in the components of the structure is reduced.

6.2 Active seismic control

This system offers seismic protection by applying forces that counterbalance the forces induced by an earthquake. As an active system, it requires an energy source and computer-controlled actuators to operate specialized braces and Tuned Mass Dampers distributed throughout the building. While active systems can provide enhanced performance, they are more complex than passive systems due to their reliance on computer control, motion sensors, feedback mechanisms, and moving parts, all of which may necessitate regular maintenance and servicing. Additionally,

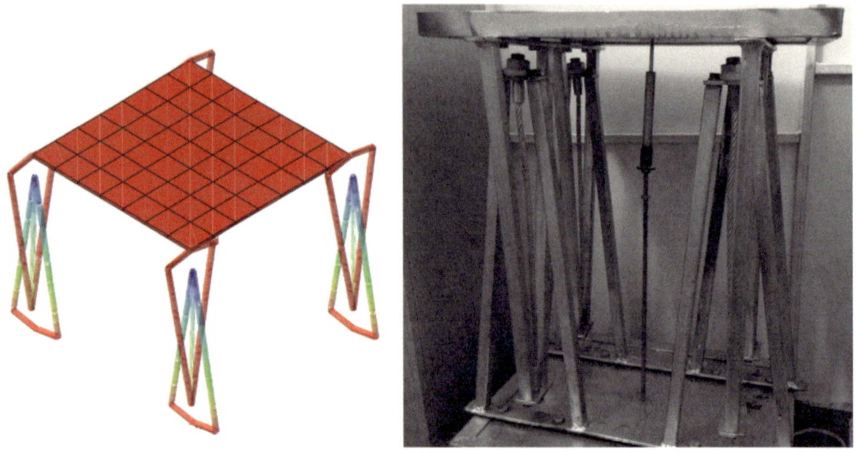

Figure 18.
Pendulum column.

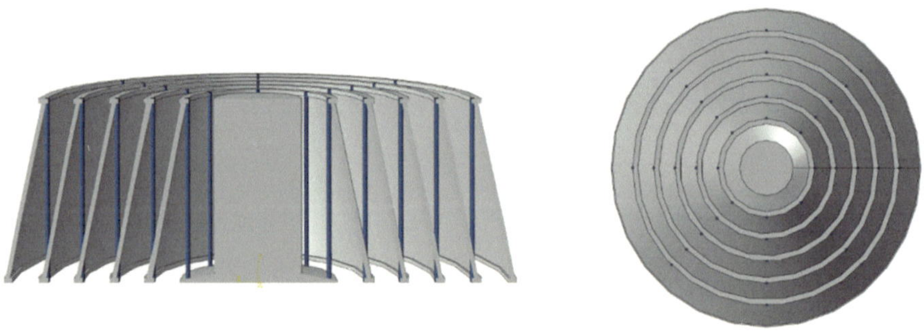

Figure 19.
Multilayer pendulum isolator.

an active seismic control system must have an emergency power source to ensure operability during a major earthquake and any immediate aftershocks.

In structural control systems, the terms "open" and "closed" refer to different control mechanisms used to regulate the behavior of a system. In an open-loop control system, the output is not directly measured or compared to the desired input. Instead, the control action is only based on the given input signal. This type of control does not pay attention to feedback, which means that no adjustment is made based on the response of the system output. Open systems are simpler and less complex, but have less accuracy and adaptability to changes in the system or external disturbances. In a closed-loop control system, the output is measured, compared with the desired input, and feedback is used to adjust the control action. This type of control system continuously monitors the output of the system and applies the necessary corrections to ensure that the output matches the desired input. Closed control systems are more advanced and offer better accuracy, stability and flexibility than open control systems. They are also able to respond to system changes or disturbances in real time.

In general, closed control systems are preferred because of their ability to provide more accurate and reliable control of structures or processes, especially in the presence of uncertainty or disturbances. On the other hand, open control systems are suitable for simpler applications that do not require feedback or adjustment based on the output response.

Active seismic control is one of the most recent advancements in the field of seismic mitigation systems, leveraging sensor technology for dynamic response. When an earthquake occurs, sensors embedded within the building detect the direction and intensity of seismic forces. This data is used to activate counterbalancing systems that move in the opposite direction to the earthquake-induced motion, thereby reducing structural stress.

Although this technology offers significant benefits in enhancing structural resilience, it is highly sophisticated and costly, which may limit its applicability for smaller projects in the foreseeable future. Prominent examples of active seismic control systems include the Active Mass Driver (AMD) and Active Tendon Cable (see **Figure 20**), both of which demonstrate the potential for cutting-edge solutions in earthquake protection.

6.3 Semi-active seismic control

The performance of semi-active control systems is akin to that of passive systems, yet these systems stand out due to their capability to modify their properties in real-time based on input from a controller. This adaptive functionality allows semi-active systems to respond to varying environmental conditions and dynamic loads with a degree of flexibility that passive systems lack.

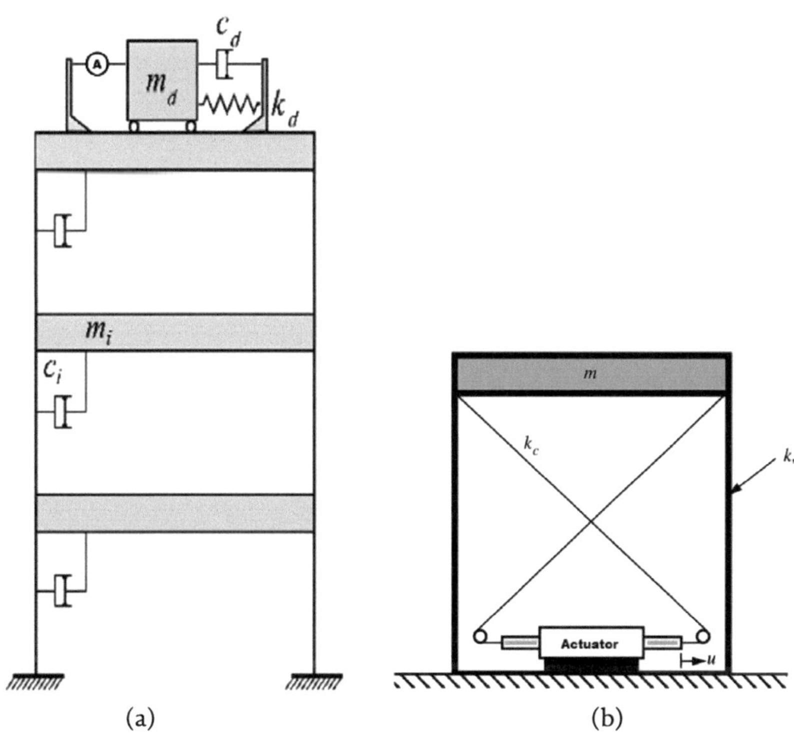

Figure 20.
(a) AMD schematic in the structure. (b) Schematic of active tendon cable.

Electrorheological (ER) and Magnetorheological (MR) dampers exemplify the potential of semi-active control systems. ER dampers utilize fluids that can change their viscosity in response to an applied electric field, enabling the system to adjust its damping characteristics dynamically. On the other hand, MR dampers operate similarly but rely on magnetic fields to alter the flow behavior of the fluid, allowing for rapid adjustments in resistance to motion.

One of the primary advantages of semi-active systems is their ability to provide enhanced performance without the complexity and energy demands associated with fully active systems. By leveraging the inherent properties of the materials they use, semi-active dampers can dissipate energy effectively while requiring minimal external power. This makes them suitable for a wide range of applications, from seismic protection in buildings to vibration control in vehicles and bridges.

6.4 Hybrid seismic control

This system seamlessly integrates the characteristics of both passive and active seismic control systems. It offers enhanced reliability and reduced costs when compared to fully active systems. By combining the benefits of passive elements, which provide inherent stability and damping, with active components that dynamically respond to seismic forces, this hybrid approach effectively minimizes the strain energy absorbed by the structure.

7. Conclusions

In this chapter, the methods of strengthening the structures were explained. From the most basic methods that work by increasing the lateral strength, to the view of structural engineering to increase the structure's plasticity and design based on performance, it was explained. Structural control systems that operate passive, active, semi-active, and hybrid were described and categorized. And finally, it was shown that all the methods work with the aim of reducing the level of strain energy.

Conflict of interest

The authors declare no conflict of interest.

DOI: http://dx.doi.org/10.5772/intechopen.1007347

Author details

Abdallah Azizi
University of Tabriz, Tabriz, Iran

*Address all correspondence to: a.azizi@tabrizu.ac.ir

References

[1] Wikipedia.org. Ancient City and Archaeological Site in Fars Province, Iranian National Heritage Site [Internet]. Wikimedia Foundation, Inc.; 2007. Available from: https://simple.wikipedia.org/wiki/Pasargadae

[2] Titirla MD. Using friction-yielding damper CAR1 to seismic retrofit a two-story RC building: Numerical application. Applied Sciences. 2023;**13**(3):1527. DOI: 10.3390/app13031527

[3] Li Z, Wang J, Li C, Cao J. Optimum arrangement of TADAS dampers for seismic drift control of buildings using accelerated iterative methods. Buildings. 2023;**13**(11):2720. DOI: 10.3390/buildings13112720

[4] Titirla MD. A state-of-the-art review of passive energy dissipation systems in steel braces. Buildings. 2023;**13**(4):851. Available from: https://www.mdpi.com/2075-5309/13/4/851. DOI: 10.3390/buildings13040851

[5] Liu X, Yang Y, Sun Y, Zhong Y, Zhou L. Inerter location-based vibration suppression study of a transmission line equipped with tuned-mass-damper-inerter (TMDI) under harmonic excitation. Buildings. 2022;**12**(5):657. DOI: 10.3390/buildings12050657

[6] Hosseini R, Rashidi M, Hedayati Dezfuli F, Karbasi Arani K, Samali B. Seismic assessment of a benchmark highway bridge equipped with optimized shape memory alloy wire-based isolators. Applied Sciences. 2019;**10**(1):141. DOI: 10.3390/app10010141

[7] Azizi A, Barghian M, Hadidi A, Yaghmaei-Sabegh S. Investigation of structures' seismic behavior when using the "pendulum column" as an isolator. Asian. Journal of Civil Engineering. 2023;**25**(3):2355-2366. DOI: 10.1007/s42107-023-00912-x

[8] Azizi A, Barghian M. Investigating the pendulum column isolator with flexible piers. Earthquakes and Structures. 2023;**24**(6):405-413. DOI: 10.12989/eas.2023.24.6.405

[9] Azizi A, Barghian M. Using the pendulum column as an isolator by reducing the gravity effect. Earthquakes and Structures. 2023;**25**(4):297-305. DOI: 10.12989/eas.2023.25.4.297

[10] Azizi A, Barghian M. Introducing a multi-layer pendulum isolator and investigating its effect on structures' responses during some earthquakes. Structure. 2023;**57**:105206. DOI: 10.1016/j.istruc.2023.105206

Chapter 3

Perspective Chapter: Research on Earthquake Disasters Based on Physical Constraints of Computational Gravimetry

Wei Li

Abstract

This chapter systematically explores the theoretical foundation of computational gravimetry and its application in the analysis of earthquake disasters. Firstly, the chapter introduces the historical development of computational gravimetry, the critical role of gravity in the earthquake process, and the significance of computational gravimetry in earthquake disaster research, as well as the limitations of traditional seismological methods. Subsequently, it outlines the characteristics and challenges of earthquake disasters and elaborates on the fundamental principles of computational gravimetry and the physical constraints in earthquake disaster research. Building on this foundation, the chapter analyzes the earthquake source mechanisms, seismic wave propagation characteristics, and their integration in model construction; examines the impact of crustal lithological structures on earthquake disasters; and discusses the relationship between temporal and spatial variations of the gravity field and earthquake disasters, as well as how to use gravity field changes for earthquake disaster prediction. The chapter also proposes methods for constructing and simulating earthquake disaster models and validates these models through case studies. The research indicates that computational gravimetry, as an emerging research method, has significant advantages in revealing the earthquake process, analyzing the causes of earthquake disasters, and formulating prevention and mitigation measures. However, to fully utilize its role in earthquake disaster research, continuous improvements in model construction, computational methods, and technological innovation are necessary. This chapter not only provides a theoretical basis and practical guidance for the prediction, prevention, and emergency response of earthquake disasters but also offers a new perspective and method for earthquake disaster prevention.

Keywords: computational gravimetry, physical constraints, earthquake disasters, regional research, model construction

1. Introduction

1.1 Background and significance

1.1.1 Historical development of computational gravimetry

Descriptions are provided of microgravimetric surveys conducted to verify the integrity of bedrock at a nuclear power plant's foundation and to delineate zones of minute voids and track grout placement at the base of a massive cooling tower. These examples are employed to elucidate the analytical process of the method.

It is emphasized that the distinction between conventional large-scale gravity surveys and microgravimetric surveys tailored for engineering purposes extends beyond mere reductions in grid spacing and precision thresholds. Beyond their smaller scale, such microgravimetric surveys necessitate the consideration of unique physical environments, stringent timing and budget constraints, and the imperative for definitive answers to precise inquiries. Consequently, meticulous adjustments are required in both field procedures and data reduction and interpretation methodologies.

When appropriately utilized, typically in conjunction with targeted drilling, microgravimetric surveys offer reliable, high-resolution subsurface information, albeit constrained by gravimeter noise. This approach can obviate the need for extensive and costly drilling operations. Across a broad range of applications, this method presents distinct advantages over alternative subsurface exploration techniques [1].

1.1.2 Historical development of computational gravimetry: Origins and evolution into computational methods

Gravimetry, the science of measuring gravitational acceleration, has undergone significant evolution from its inception to modern computational methods employed today. Below is a detailed examination of the historical development of gravimetry and its key milestones in geophysical research, particularly focusing on its evolution into computational gravimetry.

Origins of gravimetry – early developments: The measurement of gravity has a long history, dating back to ancient civilizations. However, it was not until the seventeenth century that Isaac Newton formulated his laws of universal gravitation, providing a theoretical foundation for gravimetry. Absolute Gravimeters: In the ninteenth and early twentieth centuries, various types of absolute gravimeters were developed, such as the pendulum gravimeter and the free-fall gravimeter. These instruments allowed for more precise measurements of gravitational acceleration.

Evolution into computational methods – Advancements in Instrumentation: With the advent of technology, gravimeters became more sophisticated and accurate. In particular, the development of laser interferometry in the twentieth century led to the creation of high-precision optical interferometric gravimeters. Introduction of Computational Techniques: As data collection became more efficient and voluminous, the need for computational methods to analyze and interpret these data arose. Computational gravimetry emerged as a field that leverages computer algorithms and numerical methods to process and analyze gravitational data.

Key milestones in the application of computational gravimetry in geophysical research – quantum gravimeters: In recent decades, quantum gravimeters have emerged as a new type of absolute gravimeter. These instruments use quantum mechanical principles, such as atom interference, to measure gravitational acceleration with high precision. Quantum gravimeters have been instrumental in various geophysical research applications, including volcano monitoring, oceanic and aerial gravity mapping, and earthquake studies.

1. Volcano monitoring: Continuous gravity observations using quantum gravimeters can analyze small changes in the gravity field caused by volcanic activity, providing valuable insights into the movement of magma and other geological processes.

2. Earthquake studies: Quantum gravimeters can detect changes in the gravity field caused by earthquakes, often earlier than seismic waves. This capability is crucial for real-time earthquake source studies and early warning systems.

3. Gravity mapping: Quantum gravimeters have been used to create high-resolution gravity maps of the Earth's surface, which are essential for understanding geological structures and tectonic processes.

Advancements in data analysis: Alongside the development of new gravimeters, significant progress has been made in data analysis techniques. Computational methods, such as numerical modeling and inversion algorithms, have been developed to interpret gravitational data and extract meaningful information about the Earth's interior.

Integration with other geophysical data: Computational gravimetry is often used in conjunction with other geophysical data, such as seismic, magnetic, and electromagnetic data, to provide a comprehensive understanding of geological processes and structures.

The historical development of gravimetry has seen significant advancements from early absolute gravimeters to modern computational methods. The evolution of quantum gravimeters and advancements in data analysis techniques have revolutionized geophysical research, enabling scientists to better understand the Earth's interior and geological processes.

1.1.3 The role of gravity in earthquake processes

Gravity, as a fundamental force in geophysics, exerts an influence on all objects within the Earth's system. In the realm of seismology, the relationship between gravitational forces and seismic activity represents a complex and fascinating area of research. While the precise mechanisms and degrees of influence remain subjects of ongoing investigation, it is recognized that gravitational changes can potentially impact earthquake processes in various ways.

Changes in the Earth's gravity field can serve as indicators of impending earthquakes, offering valuable insights into the underlying processes leading to seismic events. These changes may manifest through various phenomena, including the following:

1. Mass redistributions: Prior to an earthquake, there may be detectable shifts in the distribution of mass within the Earth's crust. These shifts can alter the local gravity field, potentially signaling the accumulation of stresses that may eventually lead to rock failure and seismic activity.

2. Crustal deformation: Earthquakes often precede crustal deformation, such as fault movements or the uplifting or subsidence of terrain. These deformations can cause subtle changes in the gravity field, which can be measured using sensitive gravimeters. These measurements can provide early warnings or help scientists better understand the mechanics of earthquake initiation.

3. Gravity gradients: Changes in gravity gradients, which represent the rate of change in gravitational acceleration over distance, can also be indicative of impending earthquakes. These gradients are influenced by variations in the density and structure of the Earth's crust, and their changes may reflect the ongoing tectonic processes leading to seismic events.

4. Tidal stress interactions: Gravitational tides, caused by the moon and sun, exert periodic stresses on the Earth's crust. These stresses can interact with pre-existing faults and weaknesses in the rock, potentially triggering or modulating seismic activity. Monitoring changes in the gravity field related to tidal stresses can provide insights into these interactions and their potential role in earthquake triggering (as shown in **Figure 1**).

In summary, changes in the Earth's gravity field can offer clues about impending earthquakes by reflecting underlying crustal deformations, mass redistributions, and interactions with tidal stresses. These changes, measurable using advanced gravimeters and computational techniques, are crucial for advancing our understanding of earthquake processes and potentially developing early warning systems.

1.1.4 Significance of computational gravimetry in earthquake disaster research

Improving the accuracy of earthquake prediction and risk assessment: Computational gravimetry plays a pivotal role in enhancing the accuracy of earthquake prediction and risk assessment. By leveraging advanced algorithms and high-resolution data, researchers can analyze subtle changes in the Earth's gravity field that may be indicative of impending seismic activity. These changes can provide valuable insights into the stress accumulation and release processes within the Earth's crust, ultimately leading to more precise earthquake forecasts and risk maps.

Enhancing our understanding of the physical mechanisms behind earthquakes: Computational gravimetry also contributes significantly to our understanding of the physical mechanisms behind earthquakes [4]. By modeling the interactions between gravitational forces, tectonic plates, and crustal deformations, researchers can gain deeper insights into the complex processes that lead to seismic events. This understanding is crucial for developing more effective mitigation strategies and improving our overall resilience to earthquake hazards.

Facilitating the development of early warning systems: Finally, computational gravimetry facilitates the development of early warning systems for earthquakes. By continuously monitoring the Earth's gravity field and analyzing real-time data, researchers can detect anomalies that may signify the imminent occurrence of a seismic event. These warnings can provide critical lead times for emergency response teams to initiate preparedness measures, evacuate vulnerable populations, and mitigate potential damage.

Computational gravimetry is a powerful tool for advancing earthquake disaster research. It improves the accuracy of earthquake predictions and risk assessments,

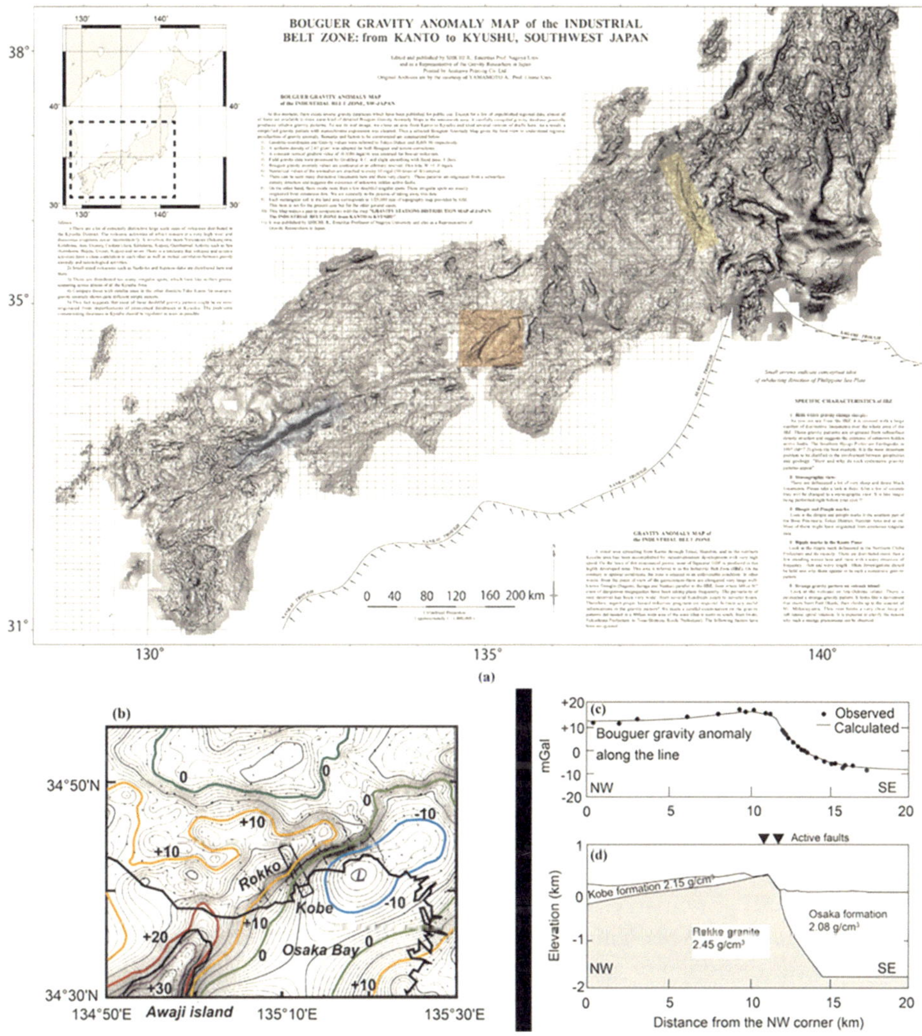

Figure 1.
(a) Gravity anomalies in the southwest to central parts of Japan. Contour lines are drawn every 1 mGal. Steep horizontal gradient zones are clearly observable along the Median Tectonic Line (blue), the Itoigawa-Shizuoka Tectonic Line (yellow), and the fault zone of the 1995 Kobe earthquake (red). Redrawing of the original figure after Shichi [2]. (b) Gravity anomalies around Kobe, where the M7 Kobe earthquake resulted in more than 6000 fatalities. The units are mGal, and the contours are drawn every 1 mGal. (c) Projection of the gravity anomalies along the rectangle in. (b) around Kobe. (d) Subsurface density model beneath the rectangle in (b) derived from the gravity anomalies [3].

enhances our understanding of the physical mechanisms behind seismic events, and facilitates the development of early warning systems. As technology continues to advance, the role of computational gravimetry in earthquake disaster preparedness and response will become increasingly critical.

1.1.5 Limitations of traditional seismological methods

The limitations of traditional seismological methods are as follows:
Inadequacies in capturing the full complexity of earthquake processes.

The need for complementary methods to provide a more comprehensive view.

Traditional seismological methods fall short in fully capturing the intricate and multifaceted nature of earthquake processes. These methods primarily rely on the analysis of seismic waves generated during earthquakes, which, while informative, often only provide a partial picture of the underlying physical and chemical interactions occurring within the Earth's crust.

Given these limitations, there is a pressing need for complementary approaches that can offer a broader and more holistic understanding of earthquake phenomena. These methods should aim to bridge the gaps left by traditional seismology, providing insights into the complex dynamics and mechanisms that drive earthquakes.

1.1.6 Advantages of computational gravimetry

The advantages of computational gravimetry are as follows:

Ability to process large datasets and detect subtle changes in the gravity field.

Integration of various data sources for a multi-disciplinary approach.

Potential for real-time monitoring and analysis.

Computational gravimetry excels in handling vast amounts of data, enabling scientists to detect even the slightest variations in the Earth's gravity field. This capability is crucial for monitoring subtle changes that may indicate underground activities such as magma movements, groundwater fluctuations, or even the subtle effects of climate change on the Earth's mass distribution.

Another significant advantage of computational gravimetry is its ability to integrate data from diverse sources, facilitating a multi-disciplinary approach to understanding and analyzing geophysical phenomena. By combining gravity data with information from seismic, electromagnetic, and other geophysical surveys, researchers can gain a more comprehensive and nuanced view of the Earth's internal processes and structures.

Computational gravimetry also holds the potential for real-time monitoring and analysis of gravity changes. This capability is invaluable for early warning systems, as it allows for the detection of anomalies that may signify impending hazards such as earthquakes, volcanic eruptions, or landslides. By providing timely information, computational gravimetry can contribute to the development of more effective disaster preparedness and response strategies [3].

1.1.7 Case studies: Successes in computational gravimetry for earthquake research

Examples of how computational gravimetry has contributed to earthquake disaster research:

1. Pre-earthquake gravity changes detection: Computational gravimetry has been instrumental in detecting subtle pre-earthquake gravity changes. For instance, studies have shown that changes in the gravity field can precede seismic activity by several days or even weeks. By analyzing these changes, researchers have been able to identify potential earthquake-prone areas, thus contributing to early warning systems.

2. Understanding fault mechanics: Computational gravimetry has also provided insights into the mechanics of faults, which are crucial for understanding how earthquakes occur. By modeling the gravity field around faults, scientists can

better predict how stresses accumulate and release, leading to more accurate assessments of earthquake hazards [5].

3. Post-earthquake assessment: Following an earthquake, computational gravimetry can be used to assess damage and understand the changes in the Earth's crust. By comparing pre- and post-earthquake gravity data, researchers can map out the areas most affected by the seismic event, which can guide rescue and recovery efforts (as shown in **Figure 2**).

The outcomes of computational gravimetry in earthquake research have had a significant impact on earthquake preparedness and response. By providing early warnings and insights into fault mechanics, computational gravimetry has contributed to the development of more effective earthquake hazard mitigation strategies. Furthermore, post-earthquake assessments using computational gravimetry have improved our understanding of seismic events and guided rescue and recovery operations.

1.1.8 Current trends and future perspectives

The latest advancements in computational gravimetry techniques.

Emerging technologies and their potential applications in earthquake disaster research.

The future role of computational gravimetry in hazard mitigation and disaster management.

The field of computational gravimetry has witnessed significant advancements in recent years, driven by innovations in sensor technology, data processing algorithms, and artificial intelligence (AI). One notable breakthrough is the development of the world's first low-altitude gravity measurement system, officially unveiled recently.

Figure 2.
Illustration of the building structure. Source: Eigenmann and Irfanoglu [5].

This system leverages advanced sensors and deep learning algorithms to provide unprecedented accuracy in gravitational data, marking a crucial milestone in both physical measurement technology and the application of AI in scientific research.

The core strengths of low-altitude gravity measurement technology lie in its high precision and efficiency, which are often difficult to achieve in traditional methods. By equipping specially designed sensors and incorporating machine learning techniques, the system can monitor and analyze the Earth's gravity field in real-time at relatively low flight altitudes. This technology enables scientists to map changes in the Earth's internal structure with greater clarity and detect precursors to natural phenomena such as earthquakes and volcanoes.

Emerging technologies, particularly in the realm of computational gravimetry and related fields, hold tremendous potential for advancing earthquake disaster research. Gravity exploration, a common geophysical exploration method, has been widely used in seismic fault studies due to its non-destructive, rapid, and high-resolution characteristics. By measuring changes in the Earth's surface gravity field, gravity exploration can infer the density distribution of underground structures, revealing the characteristics of underground formations.

In the context of earthquake fault research, gravity exploration can be applied to establish three-dimensional density models of seismic faults, identify fault zones, and assess fault activity levels. This information is crucial for understanding fault mechanics, predicting earthquake hazards, and developing effective mitigation strategies.

Moreover, the integration of satellite remote sensing, meteorological radar, and ground monitoring networksprovide comprehensive disaster monitoring system. These technologies can monitor a wide range of disaster indicators, including crustal deformations, weather conditions, and flood dynamics, providing timely and accurate information for disaster prediction and response.

The future role of computational gravimetry in hazard mitigation and disaster management is promising. As technology continues to mature, innovative measurement systems enhanced by AI are expected to bring more astonishing discoveries and potentially change our understanding of the Earth and the universe.

In hazard mitigation and disaster management, computational gravimetry can contribute to early warning systems by providing precise data on geophysical changes. These data can be used to develop more accurate models for predicting natural disasters such as earthquakes and floods. Additionally, real-time monitoring capabilities enabled by low-altitude gravity measurement systems can aid in rapid response and effective resource allocation during disasters.

Furthermore, the integration of computational gravimetry with other technologies, such as GIS, big data analytics, and drone imagery, can enhance disaster assessment and recovery efforts. By providing detailed information on damage extent and affected areas, these technologies can facilitate more efficient and targeted relief operations.

1.2 Overview

1.2.1 Earthquake disaster characteristics

1. Sudden onset and lack of prediction

    Earthquakes typically occur within a matter of seconds to minutes, often without significant precursory signs. This sudden onset makes it extremely difficult to predict and prepare for, leading to widespread devastation and loss of life.

2. Extensive destruction and widespread impact

   Earthquakes generate seismic waves that can cause extensive damage to buildings, infrastructure, and natural landscapes. When earthquakes occur in populated or economically developed areas, the destruction can be catastrophic, resulting in significant loss of life, injuries, and economic losses.

3. Far-reaching social and economic consequences

   The impact of earthquakes extends beyond the immediate physical destruction. They can disrupt social and economic activities, cause widespread panic and displacement, and have long-term psychological effects on survivors. The reconstruction and recovery process can also be lengthy and costly.

4. Secondary disasters

   Earthquakes often trigger secondary disasters such as fires, tsunamis, landslides, and chemical spills. These secondary disasters can exacerbate the initial damage and lead to further loss of life and property.

5. Long-lasting aftershocks

   After the main earthquake, a series of aftershocks may continue for days, weeks, or even months. These aftershocks can cause additional damage and disrupt the recovery process.

6. Periodicity

   Earthquakes tend to occur in cycles, with certain areas experiencing earthquakes more frequently than others. Understanding this periodicity can help in predicting and preparing for future earthquakes.

7. Interrelatedness with societal and individual preparedness

   The damage caused by earthquakes is closely related to societal and individual preparedness. Areas with better preparedness, such as those with stronger building codes and higher awareness of earthquake safety, tend to experience less damage and fewer casualties.

1.2.2 Challenges in earthquake disaster management

1. Prediction and early warning

   The lack of reliable earthquake prediction methods remains a significant challenge. While some warning systems can detect earthquakes shortly before they occur, the lead time is often too short to take effective protective actions.

2. Rapid response and rescue

   In the immediate aftermath of an earthquake, rapid response and rescue efforts are crucial. However, access to affected areas may be difficult due to damaged infrastructure, and rescue resources may be limited.

3. Disaster assessment and recovery

    Assessing the damage caused by an earthquake and planning for recovery can be complex and time-consuming. It requires coordination between multiple agencies and stakeholders, as well as access to accurate and timely information.

4. Mitigation and preparedness

    Mitigating the impact of earthquakes requires long-term investment in infrastructure, education, and preparedness. This includes building earthquake-resistant structures, conducting regular drills and exercises, and raising public awareness of earthquake safety.

5. Addressing secondary disasters

    Managing secondary disasters such as fires, tsunamis, and landslides requires specialized equipment and expertise. Ensuring that these resources are available and can be rapidly deployed is a significant challenge.

2. Physical constraints of computational gravimetry

2.1 Gravity data processing

2.1.1 Introduction

The unevenness of the Earth's surface and the uneven distribution of materials within the Earth are the main causes of gravity variations. The Earth's gravity field is crucial data for studying the shape of the Earth and the overall distribution of internal materials, while gravity anomalies are key data for investigating the flow of materials and density changes within the Earth. The gravity data obtained from surface measurements are the superimposed effects caused by the three-dimensional distribution of density at different scales and depths. It is extremely difficult to directly analyze all the density variation characteristics of the underground space through the acquired gravity data, and it is challenging to effectively obtain true and valid information. Therefore, it is necessary to analyze the approximate characteristics and attributes of the research object based on the understanding and knowledge of the target study area, utilize data processing techniques for gravity anomalies to extract the direct gravitational response of the target body, and finally conduct effective analysis and interpretation of the gravity anomalies.

2.1.2 Gravity and gravity anomalies

2.1.2.1 Gravity

Every object on Earth is subject to the force of gravity. All objects on the Earth's surface and in the nearby space are simultaneously subjected to two forces: one is the gravitational force F produced by the total mass of the Earth; the other is the inertial centrifugal force C caused by the Earth's rotation. The vector sum of these two forces acting on a certain object is called gravity P, where $P = F + C$, as shown in **Figure 3**.

The space around the Earth where gravitational forces exist is called the gravitational field. From the perspective of force, the properties of the gravitational field can

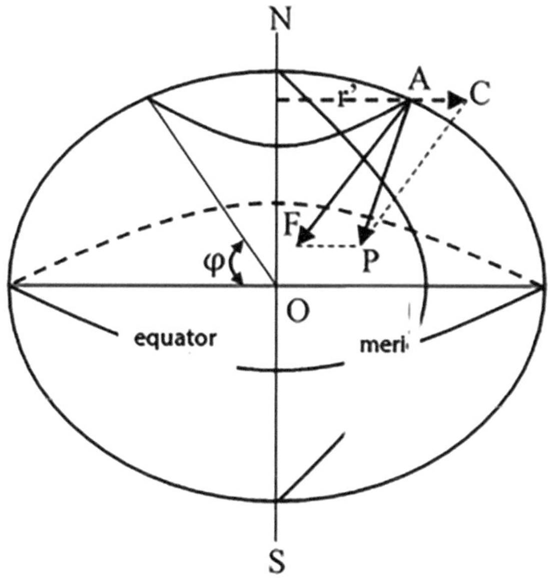

Figure 3.
The gravity experienced by a unit mass at any point outside the Earth.

be described by the intensity of the gravitational field. The intensity of the gravitational field at a point in the gravitational field is equal to the gravity experienced by a particle of unit mass at that point. According to Newton's second law, P = mg, where g represents the gravitational acceleration, then

$$\begin{aligned} P &= F + C = mg \\ F &= -G\frac{M_E \cdot m}{R^3}R \\ C &= m\omega^2 r' \\ g &= \frac{P}{m} = -G\frac{M_E}{R^2} + \omega^2 r'. \end{aligned} \tag{1}$$

In the equation, G is the universal gravitational constant (6.67×10^{-11} $m^3/kg{\cdot}s^2$); ME is the mass of the Earth; R is the position vector from the Earth's center to point m; the negative sign indicates that F is opposite to R; ω is the angular velocity of the Earth's rotation; r' is the perpendicular distance to the Earth's rotational axis. In studies of gravimetry, the gravitational acceleration or gravitational field intensity is often referred to simply as "gravity." Field gravity measurements actually determine the numerical value of the gravitational acceleration. Thus, the unit of gravity (i.e., gravitational acceleration) in the centimeter-gram-second system of units (CGS) is cm/s^2, and in the International System of Units (SI), the unit of gravity g is m/s^2 (units: 1 Gal = 10^3 mGal = 10^6 and μGal = 10^{-2} m/s^2).

From physics, we know that, in a conservative force field, the characteristics of the field source can be studied through potential functions. If V represents the gravitational potential, then the relationship between the gravitational field and the gravitational potential is as follows:

$$F = \text{grad}V = \nabla V \tag{2}$$

In the equation, the component forms of F are $F_x = \partial V/\partial x$, $F_y = \partial V/\partial y$, and $F_z = \partial V/\partial z$.

The gravitational potential of a particle is the work done by moving a unit mass of the particle from infinity to that point in the gravitational field, then

$$V = \int_{\infty}^{r} F dr = -G \int_{\infty}^{r} \frac{m}{r^2} dr = G \frac{m}{r} \tag{3}$$

Then, the gravitational potential outside the Earth's mass is as follows:

$$V = G \int_M \frac{dm}{r} \tag{4}$$

In the equation, r represents the distance from dm to the point of calculation. The component forms of the gravitational force outside the mass are as follows:

$$\left.\begin{aligned} F_x &= \frac{\partial V}{\partial x} = -G \int_M \frac{x - \xi}{r^3} dm \\ F_y &= \frac{\partial V}{\partial y} = -G \int_M \frac{y - \eta}{r^3} dm \\ F_z &= \frac{\partial V}{\partial z} = -G \int_M \frac{z - \zeta}{r^3} dm \end{aligned}\right\} \tag{5}$$

The centrifugal potential is defined as

$$Q = \int_0^{r'} C dr' = \int_0^{r'} \omega^2 r' dr' = \frac{1}{2} \omega^2 (x^2 + y^2) \tag{6}$$

Then, the components of the centrifugal force are

$$\begin{aligned} C_x &= \frac{\partial Q}{\partial x} = \omega^2 x \\ C_y &= \frac{\partial Q}{\partial y} = \omega^2 y \\ C_z &= \frac{\partial Q}{\partial z} = \omega^2 z = 0 \end{aligned} \tag{7}$$

The Earth's gravitational potential is equal to the sum of the gravitational potential and the centrifugal potential, so

$$W = V + Q = G \int_M \frac{dm}{r} + \frac{1}{2} \omega^2 (x^2 + y^2) \tag{8}$$

The component forms of gravity are

$$\left.\begin{aligned} g_x &= \frac{\partial W}{\partial x} = g\cos(g, x) = -G \int_M \frac{(x - \xi)}{r^3} dm + \omega^2 x \\ g_y &= \frac{\partial W}{\partial y} = g\cos(g, \mathrm{y}) = -G \int_M \frac{(y - \eta)}{r^3} dm + \omega^2 y \\ g_z &= \frac{\partial W}{\partial z} = g\cos(g, \mathrm{z}) = -G \int_M \frac{(z - \zeta)}{r^3} dm \end{aligned}\right\} \tag{9}$$

According to the definition of directional derivative, the component of gravity in any direction l is

$$g_l = \frac{\partial W}{\partial l} = g\cos(g \cdot l) \tag{10}$$

In the equations, the SI units for gravitational potential, gravitational potential due to mass, and centrifugal potential are all m^2/s^2.

According to potential field theory, the gravitational potential satisfies the Laplace equation $\nabla^2 V = 0$ outside the field source, and satisfies the Poisson equation $\nabla^2 V = -4\pi G\rho$ inside the field source. The centrifugal potential, satisfies $\nabla^2 Q = 2\omega^2$ both inside and outside the Earth (i.e., inside and outside the field source).

From this, we can see that the gravitational potential should satisfy the following differential equation:

$$\begin{aligned} &\nabla^2 W = 2\omega^2 = -4\pi G\rho \ \ (\text{inner}) \\ &\nabla^2 W = 2\omega^2 \end{aligned} \tag{11}$$

In mathematics, a function that satisfies the Laplace equation is called a harmonic function. Therefore, the gravitational potential in the external space of the Earth is a harmonic function, while the centrifugal potential and the gravitational potential (considering both gravitational and centrifugal effects) are not harmonic functions.

2.1.2.2 Gravity data processing

Gravity measurements can be divided into absolute gravity measurements and relative gravity measurements. The average absolute gravity value on the Earth's surface ranges from 9.78 to 9.832 m/s^2. The current measurement accuracy can reach the microgal (μGal) level or even higher. Relative gravity measurements determine the difference in gravity between various points relative to a certain gravity base point. They are easier and more accurate than absolute measurements, with an accuracy of 0.01 to 0.1 μGal. When the absolute gravity value of the base point is known, the absolute gravity values at various points can also be obtained through relative gravity measurements. Relative gravity measurements are the main form of modern gravity measurements.

Methods for measuring gravity can be divided into dynamic and static methods. The dynamic method observes the motion of an object under the action of gravity, and the directly measured quantities are time and distance, which are used for absolute gravity measurements. The static method is the basic method for relative gravity measurements, and the measured quantities are the angular displacement and linear displacement of the object's equilibrium position due to gravity changes, which are used to calculate the gravity changes between two points. The forms of gravity measurements can be divided into line measurements, profile measurements, and area measurements. Area measurements are the basic form of gravity measurements.

After correcting the results of field measurements using a gravimeter for zero drift, the reading differences of each measurement point relative to the base point are converted into gravity differences. These gravity differences cannot be directly considered as gravity anomaly values. Since ground gravity measurements are conducted on the actual Earth's surface, the unevenness of the Earth's surface causes these gravity differences to include various interference factors, and the degree of interference varies with the measurement point. In order to establish a common standard for the gravity differences at each measurement point, it is necessary to organize the observed data to obtain true gravity anomaly values, so that the gravity anomalies at various measurement points can be compared under consistent external conditions. The organization of gravity data mainly includes latitude correction, topographic correction, intermediate layer correction, and height correction.

2.1.2.3 Latitude correction

Latitude correction is also known as normal field correction. The Earth's normal gravity field is a function of latitude φ, gradually increasing from the equator to the poles. Even if the underground geological conditions are the same at measurement points of different latitudes, the gravity values at each measurement point will be different. Therefore, the purpose of this correction is to eliminate the influence of changes in latitude on the gravity values at the measurement points.

When conducting small-scale gravity measurements over a large area, the observed gravity values are directly subtracted from the normal gravity values (calculated using the normal gravity formula), which represents the theoretical gravity values on the sea surface where the potential function W is expressed as a function of latitude φ. The currently internationally adopted formula for normal gravity values is

$$g = 9.7801385\left(1 + 0.00530233\sin^2\varphi - 0.0000589\sin^2 2\varphi\right) \tag{12}$$

When conducting gravity measurements over a small area with a larger scale, the measurement range is limited, with north-south distances ranging from tens of meters to a few hundred meters, and up to a few kilometers at most. In this case, the latitude correction value can be calculated using the following formula:

$$\Delta g_{\mathrm{Lat}} = -8.14\sin 2\varphi \cdot \mathrm{D} \tag{13}$$

In the formula, φ represents the latitude of the overall base point or the average latitude of the measurement area and D represents the latitudinal distance between the measurement point and the overall base point.

2.1.2.4 Topographic correction

The undulations of the terrain often cause the materials surrounding the gravity measurement points not to be on the same horizontal plane, so it is necessary to eliminate the influence of the materials surrounding the observation points. The purpose of topographic correction is to eliminate the influence of the terrain undulations surrounding the measurement point on the gravity value at the observation point. The datum plane for topographic correction is the plane of the measurement point. The correction method involves removing the excess material above the plane of the measurement point and filling in the voids below the plane, as shown in **Figure 4**. For the positive terrain (1) above the plane of measurement point A, the excess material generates a vertically upward gravitational component f', causing the instrument reading to decrease, that is, the effect value is negative. The negative terrains (voids) (2) and (3) lack material relative to the plane of the measurement point, equivalent to a deficiency of gravitational pull at that point, which also causes the instrument reading to decrease, with a negative effect value. Therefore, regardless of whether it is positive or negative terrain, the topographic correction value is always positive. The process of topographic correction can be simply described as removing excess height and filling in lows relative to the plane of the measurement point.

The radius for topographic correction is specified as 90′ (166.7 km), and the correction density is taken within the range of 2.0 to 2.67 g/cm^3. When conducting measurements over a small area with a large scale, the correction radius can be reduced according to needs, generally to 5′ (7 to 10 km). If the terrain in the measurement area and adjacent areas is relatively flat, topographic correction may not be required under certain accuracy requirements.

The specific method for topographic correction is as follows: In **Figure 4**, a Cartesian coordinate system is taken with measurement point A as the origin, with the z-axis pointing vertically downwards. Let the coordinates of the underground mass element dm be (ξ, η, ζ). According to the law of universal gravitation, the vertical component of the gravitational force produced by dm on point A is $dg = G(dm/r^2)\cos\theta$, where G is the gravitational constant, r is the distance from dm to point A, and θ is the angle between r and the z-axis. When considering the influence of all the terrain surrounding point A that is either above or below it, the mass is integrated, resulting in

$$\Delta g = G\int \frac{dm}{r^2}\cos\theta \tag{14}$$

In this equation, $dm = \rho d\xi d\eta d\zeta$, $r^2 = \xi^2 + \eta^2 + \zeta2$, $\cos\theta = \zeta/r$, and ρ is the density of the material involved in the topographic correction. Substituting these into the equation gives

$$\Delta g = G\rho \iiint \frac{\zeta}{\left(\xi^2+\eta^2+\zeta^2\right)^{3/2}} d\xi d\eta d\zeta \tag{15}$$

For undulating terrain, the integral in Eq. (15) is difficult to compute. Therefore, an approximate method must be used to calculate the topographic correction using Eq. (15). The basic idea is to divide the undulating terrain into many small regions using a certain method and assume that, within each small region, the terrain undulation changes linearly. In this way, the influence of each small region is first calculated

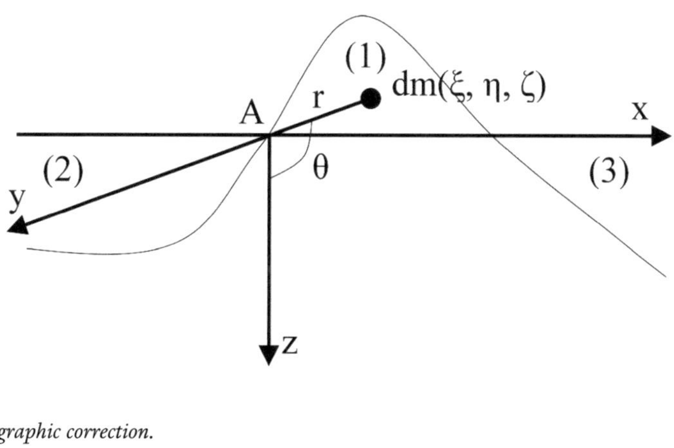

Figure 4.
Diagram of topographic correction.

using Eq. (15), and then all the small regions are combined to obtain the total gravitational influence of the entire terrain on the measurement point.

Currently, there is another method for topographic correction that combines the intermediate layer correction with the aforementioned topographic correction (i.e., the method of correction relative to the plane of the measurement point). Its purpose is to eliminate the influence of the mass (when the terrain surface is above the geoid) or the mass deficit (when the terrain surface is below the geoid) between the actual Earth's surface and the geoid on the gravity value at the measurement point. This correction can also be called generalized topographic correction. The datum plane for generalized topographic correction is the geoid, and the correction density is taken as 2.67 g/cm^3. However, for large lakes and oceans, a suitable density should be selected separately. The correction radius can be arbitrarily taken as 90′ (166.7 km), but the influence of the curvature of the Earth's surface on the topographic correction must still be considered when making corrections in the far zone. It should be noted that, due to errors in density selection and topographic measurement, topographic correction is inevitably imperfect, often leading to false anomalies related to the terrain, especially in mountainous areas.

2.1.2.5 Intermediate layer correction

After topographic correction, the surrounding area of the measurement point has become a plane. However, there still exists a horizontal material layer between the plane of the measurement point and the correction datum plane. Eliminating the influence of this material layer on the gravity value at the measurement point is called intermediate layer correction.

If the intermediate layer is treated as a uniform infinitely large horizontal material layer with a thickness of Δh (meters) and a density of ρ (g/cm^3), then for each increase of 1 meter in the thickness of this infinitely large material layer, the gravity value increases by approximately 0.0419ρ (mGal). Therefore, the formula for intermediate layer correction is

$$\Delta g_{\mathrm{I}} = -0.0419\rho\Delta h \tag{16}$$

When the measurement point is above the datum plane, Δh is taken as positive; otherwise, it is negative.

In actual measurements, variations in rock density within the measurement area and errors in determination will both lead to errors in intermediate layer correction. Additionally, since the radius of topographic correction is limited, while the intermediate layer correction is treated using an infinitely large horizontal layer, the mismatch between the two will inevitably result in errors in intermediate layer correction. During case study analysis, considering that part of the Himalayan orogenic belt within the Eastern Tectonic Syntaxis consists of Precambrian and Early Paleozoic strata, a rock density of 2.4 g/cm^3 is adopted for intermediate layer correction.

2.1.2.6 Height correction

After intermediate layer correction, only the influence of the material layer between the measurement point plane and the correction datum plane on the gravity value at the measurement point is eliminated. However, the influence of the distance of the measurement point from the Earth's center has not yet been eliminated.

Therefore, the purpose of height correction is to eliminate the influence of gravity variation with height at the measurement point and convert the gravity values of observation points at different heights to the same datum plane (typically the geoid). Height correction is also known as free-air correction or Faye correction.

If the Earth is treated as a spherical body with a concentric layered and uniform distribution of density, it can be derived that, for every increase of 1 meter in height above the ground, the gravity value decreases by approximately 0.3086 mGal. Thus, the formula for height correction for a spherical body is

$$\Delta g_H = 0.3086\Delta h \tag{17}$$

Where Δh is in meters. When the base point is above the datum plane, Δh takes a positive value; otherwise, it takes a negative value. It is worth noting that the height correction coefficient of 0.3086 is derived by treating the Earth as a spherical body with a concentric layered and uniform distribution of material density. However, the actual Earth is not an ideal sphere, and there are differences in the density distribution of its outer shell. Although this variation is slight, it must be taken into account in regional or local studies.

If the Earth is treated as an ellipsoidal body with a concentric layered and uniform distribution of density, a more accurate formula for height correction can be derived:

$$\Delta g_H = 0.3086(1 + 0.0007\cos 2\varphi)\Delta h - 7.2 \times 10^{-8}(\Delta h)^2 \tag{18}$$

Where Δh is in meters and φ is the geographic latitude of the measurement point.

Currently, regional gravity measurements use formula (18) for correction. When height correction and intermediate layer correction are combined, it is called Bouguer correction, with the formula:

$$\Delta g_H = 0.3086(1 + 0.0007\cos 2\varphi)\Delta h - 7.2 \times 10^{-8}(\Delta h)^2 - 0.0419\rho\Delta h. \tag{19}$$

During the process of Bouguer correction and topographic correction for gravity, the influences of height above the geoid, excess mass, and sea water depth and mass deficit on the gravity value at the measurement point have already been considered. Therefore, the obtained Bouguer anomaly should not have a significant correlation with topography and should be close to zero. In fact, the opposite of the above assumption is true. The Bouguer anomaly in mountainous areas is almost always negative, while large positive anomalies occur in oceanic areas, and the average value on the continental crust near sea level is close to zero. This phenomenon indicates that the distribution of matter inside the Earth, especially near the Earth's surface, is uneven. There is a mass deficit underground in mountainous areas, while there is a mass surplus under oceans and basins. This is an important example of the principle of underground mass compensating for Earth's surface morphology. When studying crustal structures, failing to consider this phenomenon will inevitably lead to certain false impressions in interpretation.

2.1.2.7 Isostatic correction

To study the isostatic state of the Earth's crust using gravity data, it is first necessary to apply an isostatic correction to the gravity observations in order to obtain the isostatic gravity anomaly. The calculation method for isostatic correction is related to the isostatic hypothesis adopted, but it generally consists of two steps.

The first step is to subtract the influence of the material (when the topographic surface is above the geoid) or the mass deficit (when the topographic surface is below the geoid) between the actual topographic relief of the Earth's surface and the geoid on the gravity value at the observation point. This step is called topographic correction. The second step is to fill all the mass involved in the topographic correction into the interior of the Earth (below sea level) to bring the Earth's crust to an isostatic state. At this point, a correction value is introduced either according to the isostatic density difference (Pratt hypothesis) or according to the isostatic depth (Airy hypothesis). This step is called compensation correction.

After the above two corrections, the gravity value at the measurement point corresponds to the gravity value when the natural surface of the Earth coincides with sea level and the material inside the Earth's crust is uniform. However, at this point, the measurement point is still "suspended" in the air, so a height correction is also introduced to place the measurement point on sea level (i.e., the geoid). Different isostatic hypotheses have different correction methods.

2.1.2.8 Isostatic correction based on Pratt hypothesis

The Pratt isostatic hypothesis in 1854 suggests that mountains rise upward like fermentation from underground. The better the fermentation, the higher the mountains and the smaller the density (the overall mass remains unchanged, while volume and density change inversely). This process occurs at a certain depth underground. The material above this depth exerts equal pressure on it everywhere, so this depth surface is also called an isobaric surface (or isostatic surface). The depth from the ocean surface to the isostatic surface is called the isostatic depth, which is generally taken as 100 or 113.7 km. According to this hypothesis, if the material above the isostatic surface is divided into many columns with the same cross-sectional area, then the mass of these columns should be equal, but their densities differ. In short, the higher the mountains, the smaller the density; conversely, the greater the density. The specific model is shown in **Figure 5**.

Topographic correction is also divided into inner and outer zones, with the correction methods detailed in the topographic correction section of this document, but the datum for correction is the geoid. The compensation correction for the inner zone is based on the isostatic density difference, as shown in **Figure 5**.

For land areas,

$$\begin{aligned} T\rho_0 &= (H+T)\rho' \\ \rho' &= \frac{T\rho_0}{H+T} \end{aligned} \tag{20}$$

where H is the height of the column, H_0 is the elevation of the observation point, T is the compensation depth, ρ_0 is the average density of the Earth's crust (i.e., 2.67 g/cm^3), and ρ' is the average density of any column. The compensation density is

$$\rho_{corr} = \rho_0 - \rho' \tag{21}$$

The formula for calculating the gravity anomaly caused by a land column with a density of ρ_{corr} at the observation point is

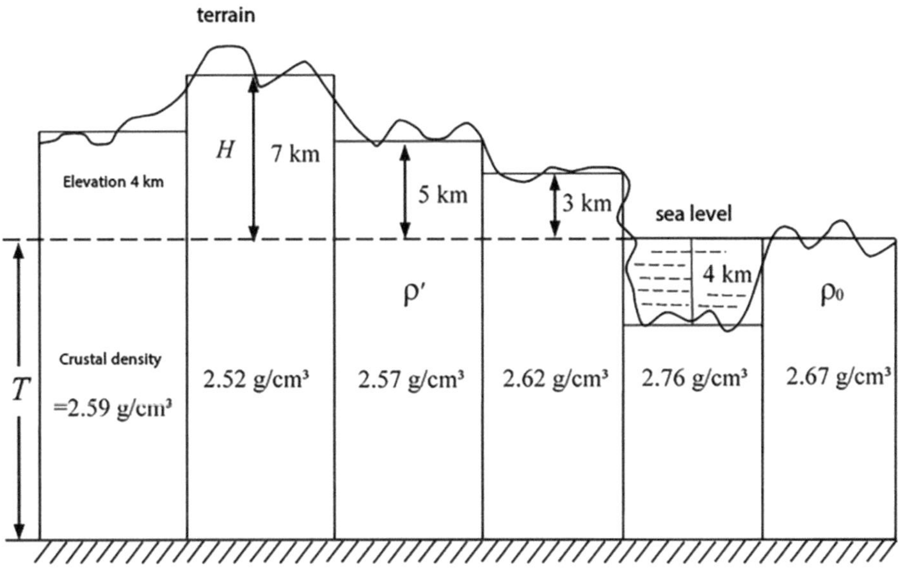

Figure 5.
Pratt Isostasy Model.

$$\Delta g = \frac{2\pi G\rho_{corr}}{n}\left[\sqrt{R_i^2 + (H_0 + T)^2} - \sqrt{R_{i+1}^2 + (H_0 + T)^2} - \sqrt{R_i^2 + H_0^2} + \sqrt{R_{i+1}^2 + H_0^2}\right] \tag{22}$$

For ocean areas,

$$\begin{aligned} T\rho_0 &= \rho''(T - h) + h\rho_O \\ \rho'' &= \frac{T\rho_0 - h\rho_O}{T - h} \end{aligned} \tag{23}$$

where ρ'' is the density within the ocean column, h is the depth of the seawater, and ρ_O is the density of the seawater. The compensation density is

$$\rho'_{corr} = \rho' - \rho_0 \tag{24}$$

The formula for calculating the gravity anomaly caused by an ocean column with a density of ρ'_{corr} at the observation point is

$$\Delta g = \frac{2\pi G\rho'_{corr}}{n}\left[\sqrt{R_i^2 + (H_0 + T)^2} - \sqrt{R_{i+1}^2 + (H_0 + T)^2} - \sqrt{R_i^2 + (H_0 + h)^2} + \sqrt{R_{i+1}^2 + (H_0 + h)^2}\right] \tag{25}$$

The compensation correction for the outer zone, which has been introduced in the topographic correction section of this document, is obtained by interpolation from the topographic compensation correction map, combined with the topographic correction.

2.1.3 Isostatic correction based on Airy's hypothesis

Airy's isostasy hypothesis in 1855 proposed that the Earth's crust is composed of many rock masses of varying thicknesses but uniform density, which float above a

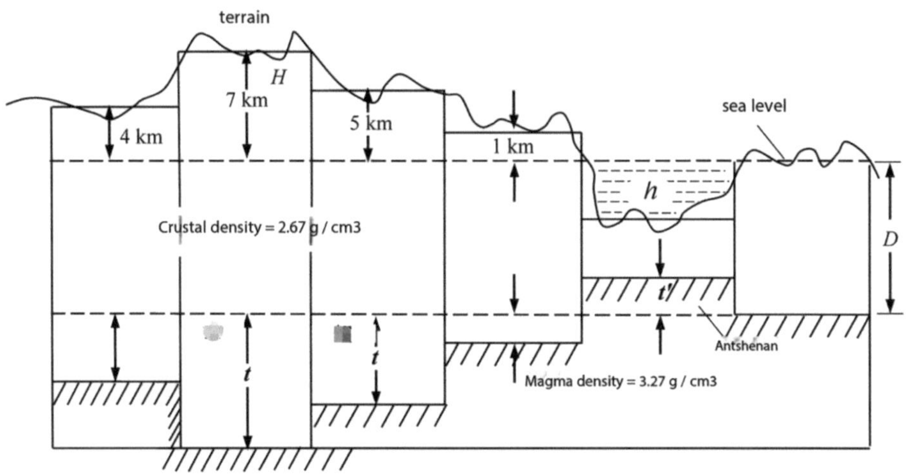

Figure 6.
Airy isostasy model.

plastic magma of higher density, akin to rafts on water. Thin rock masses intrude less deeply into the magma, while thicker ones intrude more deeply. This indicates that compensation is complete and occurs directly beneath such topographical features (i.e., compensation is local). The densities of the Earth's crust and magma were assumed to be 2.67 and 3.27 g/cm^3, respectively. The floating of rock masses above the magma follows the principle of Archimedes' buoyancy law. The distance between the sea level and magma surface is termed the normal crustal thickness, typically ranging from 30 to 60 km. The specific model is illustrated in **Figure 6**.

In research, the phenomenon where mountains are accompanied by thicker crust beneath them is called the "mountain root," while the thinning of the crust beneath the oceans is referred to as the "inverted mountain root." Thus, Airy's isostasy hypothesis is also known as the "mountain root" theory. In summary, crustal isostasy implies that, from a certain depth underground, the mass supported by the same area tends to be equal. Any increase or decrease in the mass over a large area on the surface must be compensated for underground.

The inner and outer zone topographic corrections are the same as those in Pratt's hypothesis. The compensation correction for the inner zone is illustrated in **Figure 6**. According to Archimedes' principle (a principle of hydrostatics), for land areas,

$$H\rho_0 = t(\rho - \rho_0) \tag{26}$$

where D is the normal thickness of the Earth's crust, H represents the height of the column, t is the mountain root, and ρ_0 is the average density of the Earth's crust. If ρ_0 is taken as 2.67 g/cm^3 and the density of magma as 3.27 g/cm^3, then

$$t = \frac{\rho_0}{\rho - \rho_0} H = 4.45H \tag{27}$$

This indicates that, for every 1 km, a mountain rises above the sea level, its subsidence into the magma increases by 4.45 km.

For ocean areas,

$$h(\rho_0 - 1.03) = t'(\rho - \rho_0) \tag{28}$$

where h represents the depth of the seawater and t' is the inverted mountain root, then

$$t' = \frac{\rho_0 - 1.03}{\rho - \rho_0} h = 2.73h \tag{29}$$

This shows that, in ocean areas, for every 1 km increase in seawater depth, the inverted mountain root increases by 2.73 km.

Using the same method, the proportional relationships for lakes in land areas and areas with thick sediments can be obtained, which are 2.78 and 1.12 km, respectively. The compensation correction for the inner zone in Airy's hypothesis involves removing the effects of the mountain root and inverted mountain root from the gravity value at the observation point, with the calculation formula being the same as Eq. (25). The compensation correction for the outer zone is carried out along with the topographic correction, using the same method as in Pratt's hypothesis.

2.1.4 Gravity anomalies

2.1.4.1 Free-air anomaly

In gravity surveying, the anomaly that has only undergone latitude correction and height correction is called the free-air anomaly. This anomaly is the simplest form of gravity anomaly because it makes no assumptions about the rock density above or below the sea level. However, this anomaly is also very meaningful.

In studying crustal structures, the Bouguer anomaly and the free-air anomaly are mainly used. Generally, in areas with gentle terrain, the free-air anomaly tends to be close to zero, and the average value over a large area ($1 \times 1°$) is also very low. The free-air anomaly is sensitive to the mass distribution at and near the surface, so in land areas, it has obvious topographic variation characteristics and is positively correlated with topographic relief. In ocean areas, the correlation is weak. Therefore, the free-air anomaly is widely used in oceans, where the normal gravity value at the corresponding point is subtracted from the observed gravity value at each measurement point.

2.1.4.2 Bouguer anomaly

The Bouguer anomaly is the anomaly obtained after latitude correction, Bouguer correction, and topographic correction. Since the Bouguer correction and topographic correction are equivalent to discarding the mass of material above the geoid, this naturally results in a deficiency of crustal mass. Therefore, the gravity anomalies corrected by the Bouguer method in mountainous or plateau areas are mostly negative anomalies. Furthermore, the Bouguer gravity anomaly mainly reflects the influence of abnormal mass inside the Earth on gravity measurement results. Specifically, geological heterogeneities from the surface to depths of tens of kilometers or even one or two hundred kilometers will cause Bouguer gravity anomalies as long as they have density differences. Generally speaking, the negative anomalies caused by changes in the thickness of sedimentary cover do not exceed $60 \sim 80$ mGal; anomalies caused by changes in the structure and composition of granite layers rarely exceed ± 50 mGal; anomalies within ± 100 mGal are related to changes in basalt; in addition, structures in sedimentary rocks and density heterogeneities such as metal ores can also cause small anomalies of a certain magnitude. Therefore, the local anomalies caused by density heterogeneities within the crust do not exceed ±200 mGal. The greatest role of the regional gravity field is to reflect the morphology of the upper mantle surface, that is, the depth of the Moho interface. Fluctuations in the Moho surface can cause anomalies with amplitudes within ±300 mGal over horizontal distances exceeding 100 km. Thus, large-scale variations in the Bouguer anomaly mainly reflect fluctuations in the Moho surface. This is an advantage in using gravity data to study crustal structures. The Bouguer gravity anomaly is also known as the "living fossil" for studying the structure of the lithosphere.

In addition to studying Moho surface fluctuations, the Bouguer gravity anomaly can also be used to delineate regional geological structural units, study fluctuations in sedimentary bases, delineate large igneous intrusions, and study regional deep and large faults.

2.1.4.3 Isostatic anomaly

The anomaly that has undergone latitude, height, and isostatic corrections is called the isostatic anomaly. The isostatic anomaly has unique significance in studying crustal movements and crustal structures and sometimes becomes the basic basis for explaining many geological phenomena. For example, gravity data have suggested that there are very deep mountain roots under mountain ranges, meaning that uplifts in surface topography correspond to depressions in the Moho interface. In contrast, deep tectonic depressions and deep ocean basins with thick and relatively new sediments correspond to inverted mountain roots, i.e., protrusions in the Moho interface. This relationship has been confirmed by seismic data and is generally correct. However, it is only meaningful over large areas (hundreds or thousands of square kilometers). There are three scenarios for the average value of isostatic anomalies:

$\overline{\Delta g_{Iso}} < 0$ indicates insufficient regional isostatic compensation, which is equivalent to a mass deficit in the crust;

$\overline{\Delta g_{Iso}} \approx 0$ indicates that the regional isostatic compensation is close to a balanced state;

$\overline{\Delta g_{Iso}} > 0$ indicates excessive regional isostatic compensation, which is equivalent to the existence of surplus mass in the crust.

These three scenarios can be illustrated using a floating object in water, as shown in **Figure** 7.

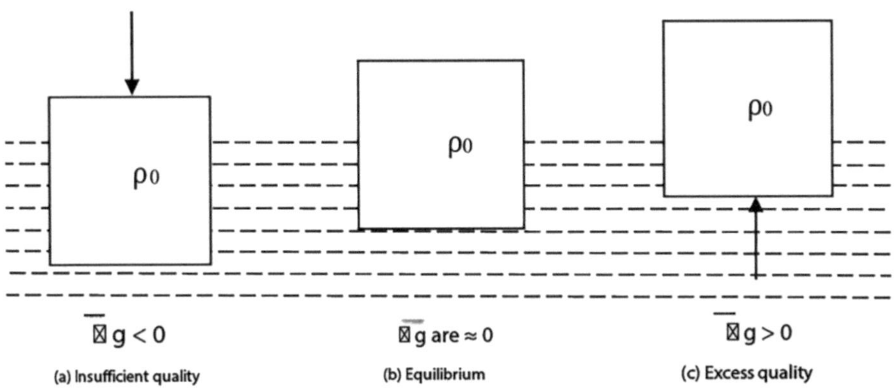

Figure 7.
Three scenarios of isostatic anomalies. (a) Insufficient quality; (b) Equilibrium; (c) Excess quality

DOI: http://dx.doi.org/10.5772/intechopen.1008859

Based on this principle, the isostatic anomalies found in modern tectonically active orogenic belts or deep-sea trenches (both accompanied by intense seismic activity) can be explained by the tectonic activity itself. According to plate tectonic theory, oceanic trenches are zones where the lithosphere subducts into the mantle. The lighter lithosphere is deeply inserted into the asthenosphere due to mantle convection, resulting in local negative anomalies. In contrast, orogenic belts are formed by upward compressive forces from deep within the crust (phase changes of hot subterranean materials and convection of the asthenosphere) that lift the lighter crust. During the crustal uplift, due to hydrostatic equilibrium, the mountain roots also thicken, but the thickening rate is less than the uplift rate of the crust, leading to a mass surplus and positive isostatic anomalies. Only those platforms and ancient land areas with weak or relatively stable tectonic movements in modern times can achieve an isostatic equilibrium state, with isostatic anomalies close to zero.

The phenomenon of crustal isostasy can also be observed through the distribution characteristics of Bouguer anomalies. For example, as the sea depth increases, positive Bouguer anomaly values become higher; whereas, as terrain height increases on the continent, negative Bouguer anomalies become lower. This phenomenon qualitatively demonstrates the existence of crustal isostasy. The issue of crustal isostasy is complex, and with crustal tectonic movements, glacier melting, and mountain destruction, crustal isostasy will continue to be disrupted. On the other hand, large-scale and long-term horizontal movement of deep crustal materials will gradually restore equilibrium in unbalanced areas. Therefore, to study the isostatic state of the Earth's surface using gravity data, it is not sufficient to rely solely on spatial distribution data; it is also necessary to study the laws of gravity changes over time.

Given the complexity of calculating gravity isostatic anomalies, which may even be impossible in some areas, free-air anomalies are sometimes used as substitutes. This is because, on a small scale, free-air anomalies can be approximated as isostatic anomalies within a range where the average terrain height does not exceed 2000 meters above the sea level and the average sea depth does not exceed 2000 meters below sea level. The gravity anomaly profile in **Figure 8** demonstrates this characteristic.

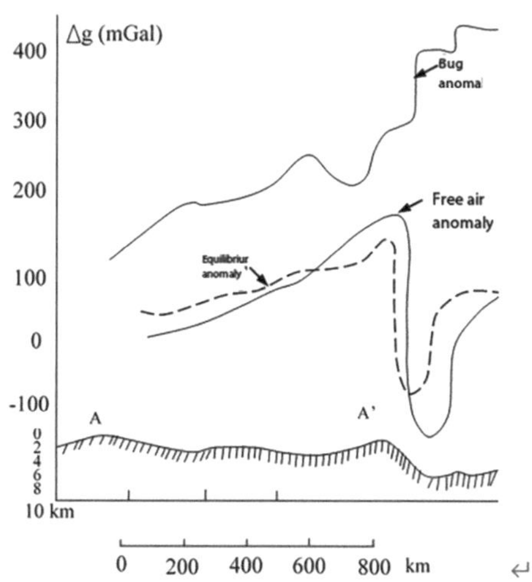

Figure 8.
Submarine topography and gravity anomaly profile AA' across an oceanic trench.

2.1.5 Gravity anomaly interpretation

2.1.5.1 Classification of gravity anomalies

2.1.5.1.1 Multiplicity of solutions in gravity anomaly interpretation

The multiplicity of solutions in gravity anomaly interpretation is determined by the complexity of gravity anomalies and the non-uniqueness of inverse problem interpretation.

- Complexity of Gravity Anomalies

  The complexity of gravity anomalies reflects various geological factors. As previously mentioned, any density difference from the surface to deep underground or even into the upper mantle can cause gravity anomalies. Therefore, although the observed values at any measuring point have undergone various corrections, they still represent the superimposed effects of the distribution of many substances from below the stratum, originating from different depths. Only by distinguishing the anomalous components from different depths using various methods can further interpretation be carried out.

- Non-uniqueness in inverse problem interpretation of gravity field

  In gravity interpretation, studying the characteristics and distribution range of anomalies caused by known geological bodies is called the forward problem in interpretation. Conversely, studying the occurrence of geological bodies based on the characteristics and variation patterns of anomalies is called the inverse problem in interpretation.

It is relatively easy to determine the gravity field generated by a known distribution of matter because the solution to the forward problem is unique. However, the solution to the inverse problem is more difficult and subject to multiplicity. This is determined by the equivalence of the gravity field, where geological bodies of different depths, shapes, and densities can cause the same gravity anomalies at the surface. The above situation poses certain difficulties for the interpretation of gravity anomalies. Therefore, in the interpretation of gravity data, it is essential to emphasize the comprehensive interpretation with geological and other geophysical data to reduce the multiplicity of interpretations and make the final interpretation more consistent with the actual situation.

2.1.5.2 Classification of gravity anomalies

Gravity anomalies are caused by density inhomogeneity in materials from the ground to tens of kilometers underground or even into the upper mantle. This indicates that they can be used for exploration purposes at different depths. However, it also highlights the complexity of anomalies, posing certain difficulties for underground structural research and exploration. Therefore, during the interpretation of gravity data, anomalies generally need to be classified. The regional background field caused by deep or large geological structures is called a regional anomaly, while anomalies related to local structures or ore bodies are called local anomalies. Local anomalies are the remaining part after subtracting the regional anomaly from the entire anomaly, so they are also called residual anomalies. For different research purposes, the anomaly components to be retained are also different. The classification of anomalies involves decomposing the anomalous field into two or more different parts and retaining the required parts for subsequent research and analysis.

- Graphic method

  The graphic method, also known as the freehand smoothing method, involves obtaining regional anomalies by smoothly averaging the observed anomalous field. Alternatively, when the regional anomaly changes regularly, the regional field can be directly obtained from the observed anomaly map, and then the local anomaly is obtained by subtracting the regional anomaly from the observed anomaly.

- Mathematical analysis method

  This method, also known as the averaging method of the gravity field, is widely used and effective. It employs a certain form of template to calculate the gravity average of several points evenly distributed on the edge of the template and considers it as the regional anomaly value at the center of the template. Then, the local anomaly is obtained by subtracting the regional anomaly from the central point anomaly. The drawback of this method is that it does not consider the influence of known geological factors.

- Gravity high-order derivative method

  Taking the first derivative ($\partial\Delta g/\partial z$ or Δgz) or second derivative ($\partial^2\Delta g/\partial z^2$ or Δgzz) of the gravity anomaly along the vertical direction can change the proportion of components contained in the anomaly, which is beneficial for anomaly classification. According to potential field theory, gravity derivatives of different orders reflect substances at different burial depths differently. Taking the maximum value of each order of gravity derivatives of a sphere with mass M and central burial depth h as an example,

$$\left.\begin{aligned}\Delta g_{\max} &= GM\frac{1}{h^2}\\ \Delta g_{z\max} &= 2GM\frac{1}{h^3}\\ \Delta g_{zz\max} &= 6GM\frac{1}{h^4}\end{aligned}\right\} \tag{30}$$

- Analytic continuation of the gravity field

  Gravity anomalies vary with the depth of the field source. When the depths of the field sources of superimposed anomalies are different, their rates of increase or decrease with changes in the observation plane's height also differ. Anomalies caused by shallow geological factors are highly sensitive to changes in the observation plane's height, while deep geological factors are relatively insensitive. Therefore, in anomaly classification, people propose using spatial conversion methods of anomalies to classify superimposed anomalies at different depths, which is called analytic continuation of anomalies. Commonly used analytic continuation methods include upward continuation and downward continuation. Upward continuation converts the actually measured anomalies at the ground surface to anomalies on another observation plane above the ground, while downward continuation obtains anomalies on an observation plane at a certain depth underground based on actually measured anomalies at the ground surface.

Generally speaking, upward continuation always provides a smoother anomaly map and is better at distinguishing anomalies caused by deeper field sources. It weakens the influence of shallow geological factors on superimposed anomalies and relatively enhances the influence of deep geological factors. In contrast, downward continuation relatively enhances the influence of shallow geological factors and weakens the influence of deep factors. However, when the depth of downward continuation is greater than or close to the depth of the field source, the continued field information will show dramatic fluctuations. In some cases, the horizontal plane where fluctuations begin may indicate the top depth of the anomalous object of the field source. As discussed above, analytic continuation is particularly useful for classifying anomalies from field sources at different depths.

As shown in **Figure 9**, Curve 1 represents the superimposed anomaly caused by two spheres with significant differences in mass and burial depth. Curve 2 is the anomaly obtained by converting this superimposed anomaly to a certain height above the ground surface. It can be seen in the figure that the local anomaly component (caused by the smaller sphere) has been eliminated, while the regional anomaly (caused by the larger sphere) changes little. Subtracting Curve 2 from Curve 1 yields Curve 3, in which the regional anomaly is almost eliminated, and the local anomaly becomes apparent. The dashed lines of the large and small spheres in the figure correspond to the positions (increased burial depth) of the observation plane being lifted. The positive and negative signs indicate the sign of the residual mass causing Curve 3. If Curve 2 in the figure is regarded as the actually measured anomaly and Curve 1 is seen as the anomaly continued to a certain depth underground from Curve 2, it is obvious that downward continuation highlights the local anomaly caused by the shallow sphere.

From the above discussion, it can be seen that upward continuation is equivalent to "low-pass filtering," which smooths anomalies. When the accuracy of the original anomalies is low, it has little impact on the results of upward continuation, and

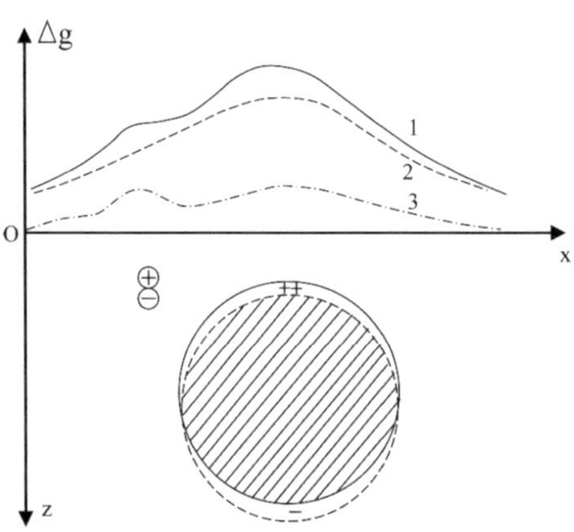

Figure 9.
Characteristics of analytic continuation of gravity anomalies.

relatively smooth anomalies can still be obtained. In contrast, downward continuation requires higher accuracy of the original anomalies because it is equivalent to "high-pass filtering," and individual errors can be "amplified" to cause drastic fluctuations in the continued anomalies. Therefore, when performing downward continuation of anomalies, the anomaly data should first be smoothed before the downward continuation is carried out.

2.1.6 Causes of gravity anomalies

2.1.6.1 Geotectonics

Geosynclines are structurally complex tectonic units of the Earth's crust, characterized by intense folding, metamorphism, and volcanism. In the final stages of their development, these areas typically feature standard mountainous terrain and generally have thicker crusts. The characteristic feature of the Bouguer gravity anomaly in geosynclines is their linear and parallel arrangement, extending for hundreds to thousands of kilometers, with anomalies reaching tens to hundreds of milligals. Moreover, the regional anomalies mirror the tectonic terrain. Due to intense folding in these areas, the morphology of gravity anomalies is also complex, generally manifesting as local small jumps against the backdrop of highly regular regional gravity anomaly gradient zones.

Platforms are relatively stable tectonic units, featuring flat terrain or gently rolling hills. The regional Bouguer gravity anomaly is characterized by smooth, stable changes with relatively small amplitude variations.

The transition zone between geosynclines and platforms is marked by prominent gravity gradient zones in Bouguer anomalies, primarily reflecting stepped faults or parallel faults. The gravity anomaly characteristics reflected between geosynclines, platforms, and transition zones are illustrated in **Figure 10**. In this figure, the anomaly characteristics of geosynclines and platforms are very clear. In the northern section of the profile, which is within the platform area, the gravity anomaly changes smoothly. In contrast, in the folded geosyncline area, the anomalies show jumping characteristics that mirror the terrain's undulations.

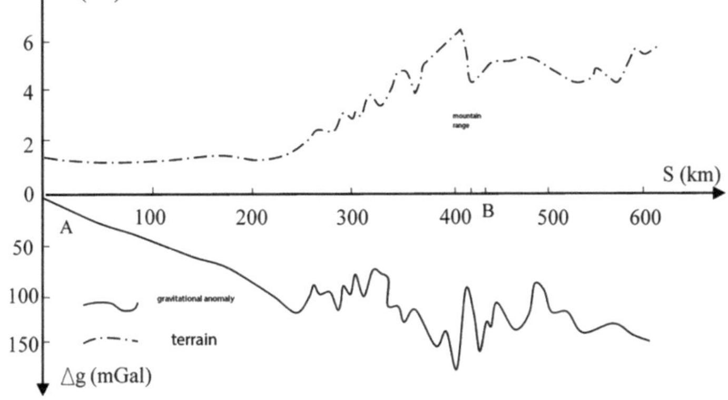

Figure 10.
Schematic diagram comparing Bouguer gravity anomalies and terrain in geosynclines and platforms.

2.1.6.2 Regional geology

Gravity anomalies serve as the fundamental basis for interpreting underground geological structures and mineral deposits. They arise from density inhomogeneities from the surface to deep underground. Collectively, the primary geological factors determining gravity anomalies include: variations in crustal thickness and density inhomogeneities within the upper mantle; internal structures and basement undulations of crystalline basement rocks; internal structures and compositional changes within sedimentary basins; mineral deposits; and near-surface density inhomogeneities.

Gravity anomalies manifest on the gravity observation surface with varying depths, scales, morphologies, and physical property characteristics. Meanwhile, the gravity field exhibits spatial superposition, which poses challenges in geological interpretation and analysis. To enhance the resolution of gravity anomalies and highlight more beneficial anomaly information, we have conducted research on data processing and interpretation theories for gravity anomalies, proposed a rapid gravity inversion algorithm, and established a data processing workflow. This provides theoretical and methodological support for the subsequent gravity data analysis and interpretation in this book.

2.2 Basic principles of computational gravimetry

From the geophysical point of view, most earthquakes can be reasonably modeled in terms of a dislocation embedded in an elastic medium [3]. Consequently, we can formulate the gravity changes as well as crustal deformations (i.e., displacements, tilt, and strains of the ground) in the framework of the dislocation theory [6, 7]. In the following subsections, we will provide an overview of the theoretical and observational results with respect to the gravity changes for several important cases.

2.2.1 Gravity changes excited by a large earthquake when magnitude M < 8

The terrestrial gravity field changes when a large earthquake occurs partly because the slip motion on the fault causes compression/dilation of the rocks around the fault and partly because the observation points are raised/lowered in space. The former necessarily causes a density change based on whether the rocks are compressed or dilated. The latter is the result of measuring the gravity at different points in space before and after an earthquake, provided that they are fixed on the deformable ground surface. Considering these two factors, Okubo [8] derived formulas for the gravity and potential changes (Δg and $\Delta \Psi$, respectively) due to faulting on a finite rectangular plane in a homogeneous half-space (**Figure 11**). These equations are useful for estimating Δg when the curvature of the Earth and the inhomogeneity of physical constants can be reasonably ignored. This is the case for large earthquakes having moment magnitudes of less than 8 because their fault dimensions are usually several tens of kilometers or less. Using the Cartesian coordinates (x_1, x_2, x_3), Δg at a point $(x_1, x_2, x_3 = 0)$ fixed on the ground can be expressed as

$$\begin{aligned}\Delta g(x_1,x_2) = \{\rho G\left[U_1 S_g(\xi,\eta) + U_2 D_g(\xi,\eta) + U_3 T_g(\xi,\eta)\right] + (\rho' - \rho) G U_3 C_g(\xi,\eta)\}\| \\ -\beta \Delta h(x_1,x_2),\end{aligned} \tag{31}$$

$$\Delta h(x_1,x_2) = [U_1 S_h(\xi,\eta) + U_2 D_h(\xi,\eta) + U_3 T_h(\xi,\eta)]\|, \tag{32}$$

where U_1 and U_2 are the strike-slip and dip-slip components of fault motion during an earthquake, respectively; U_3 is the component perpendicular to the fault, which usually vanishes for ordinary earthquakes but is not equal to 0 when magma or hydrothermal water intrudes along a thin sheet to form a tensile opening (*i.e.*, a dike or sill); ρ and ρ' are the densities of the half-space and intruding matter, respectively; Δh and $\beta \sim 0.3086$ (mGal/m) denote the uplift of the ground and the vertical gradient of the terrestrial gravity field, respectively. The symbol || is defined by Chinnery [9] as

$$f(\xi,\eta)\| \equiv f(x_1,p) - f(x_1,p-W) - f(x_1-L,p) + f(x_1-L,p-W), \tag{33}$$

$$p \equiv x_2 \cos\delta + d \sin\delta, \tag{34}$$

where L, W, and δ are the length, width, and dip angle of the fault, respectively (**Figure 11**); S_g, D_g, T_g, and C_g denote the contributions of strike-slip, dip-slip, tensile opening, and filling of the cavity sheet, respectively, and are given by Okubo [8] in an analytical form; and S_h, D_h, and T_h are the contributions of strike-slip, dip-slip, and tensile opening to the uplift, respectively, as given by Okada [6].

2.2.2 Gravity changes excited by a great earthquake (M > 8)

Estimation of the regional- to global-scale gravity changes excited by a great earthquake requires the consideration of the curvature and radial inhomogeneity of the Earth [10], because they have unusually large fault areas covering several thousand square kilometers [11]. For example, the fault length L and width W of the 2011 Tohoku earthquake were 500 and 200 km, respectively [12, 13]. A fault length greater than 100 km implies that significant displacement and gravity changes are imposed beyond the epicentral distance of a few hundred kilometers and that the Earth can no longer be approximated to be flat. In addition, the density and elastic moduli vary significantly with depth along a fault because it penetrates the boundary between the crust and mantle located at a depth of 10–30 km.

Sun and Okubo [14] was the first who succeeded in obtaining the theoretical coseismic gravity changes for a spherically symmetric, nonrotating, perfectly elastic, and isotropic (SNREI) Earth with self-gravitation. They presented a set of Green's functions

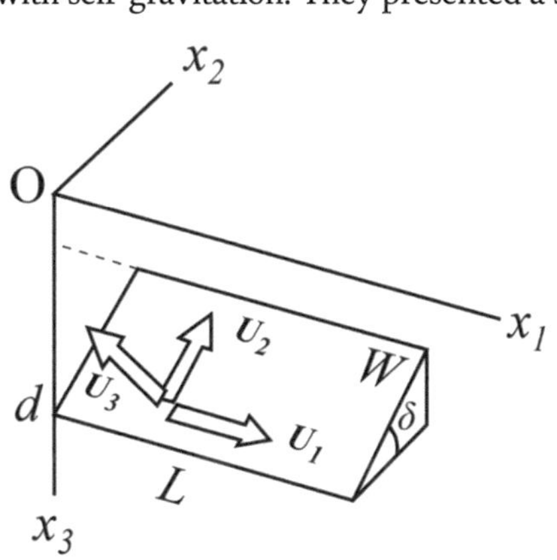

Figure 11.
A rectangular fault of length L *and width* W *embedded in a homogeneous half-space with a dip angle of δ.* U_1, U_2, *and* U_3 *denote the strike-slip, dip-slip, and tensile opening components, respectively Ref. [3].*

$\{G_{ij}(D_s, \theta, \lambda); i, j = 1, 2, 3\}$ describing the global gravity changes due to the dislocations of a unit seismic moment located on the polar axis (**Figure 12**). Once the set of Green's functions $\{G_{ij}; i, j = 1, 2, 3\}$ has been computed, the gravity change Δg due to an earthquake can be easily estimated through numerical integration over the fault surface S as follows [15]:

$$\Delta g(\theta, \lambda) = \iint_S G_{ij}(D_s, \Theta, \Lambda) \left[\mu u_i n_j\right] (\theta', \lambda') dS(\theta', \lambda'), \tag{35}$$

where (θ, λ) denotes the colatitude and longitude of the observation point; D_s and (θ', λ') denote the depth, colatitude, and longitude of the infinitesimal fault dS in the integral, respectively; u_i and n_j denote the Cartesian components of the dislocation and a vector normal to the fault, respectively; and μ denotes the rigidity. In addition, Θ and Λ are the epicentral distance (as an angle) and azimuth of the observation point measured from the source given by the following spherical trigonometric relations:

$$\cos\Theta = \cos\theta \cos\theta' + \sin\theta \sin\theta' \cos(\lambda' - \lambda), \tag{36}$$

$$\sin\Theta \sin\Lambda = \sin\theta' \sin(\lambda' - \lambda), \tag{37}$$

$$\sin\Theta \cos\Lambda \sin\theta' = \cos\theta - \cos\theta' \cos\Theta \tag{38}$$

2.3 Physical constraints in earthquake disaster research

Here we attempt to further classify these foreshock physical models into three categories [16]: elastic stress changes, aseismic slip, and fluid flows (**Figure 13** and **14**). Some physical models presented before (**Figure 15**) can be combined into one category. For example, the preslip model (e.g., Bouchon et al. [18]), the recently proposed migratory slow-slip model [19–21] and the after-slip model [22] can be all combined into the

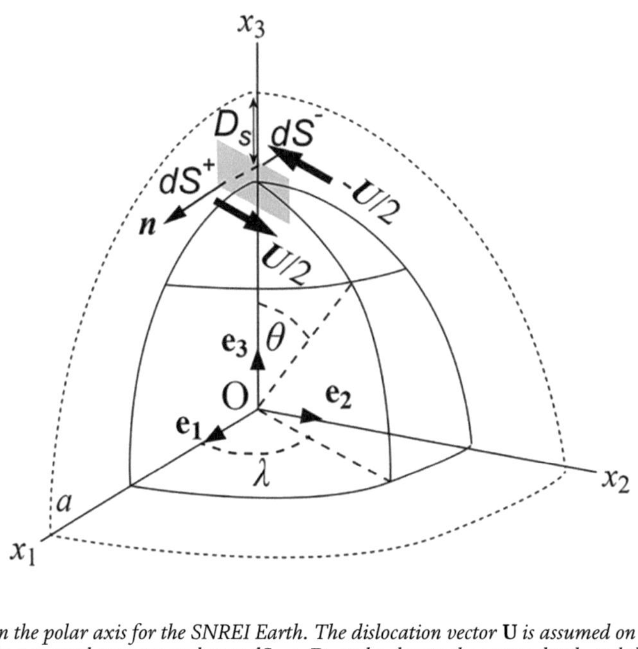

Figure 12.
Point dislocation on the polar axis for the SNREI Earth. The dislocation vector **U** *is assumed on an infinitely small rectangle defined by a normal vector* **n** *and area* dS 1. D_s *and* a *denote the source depth and the Earth's radius, respectively. Ref. [3].*

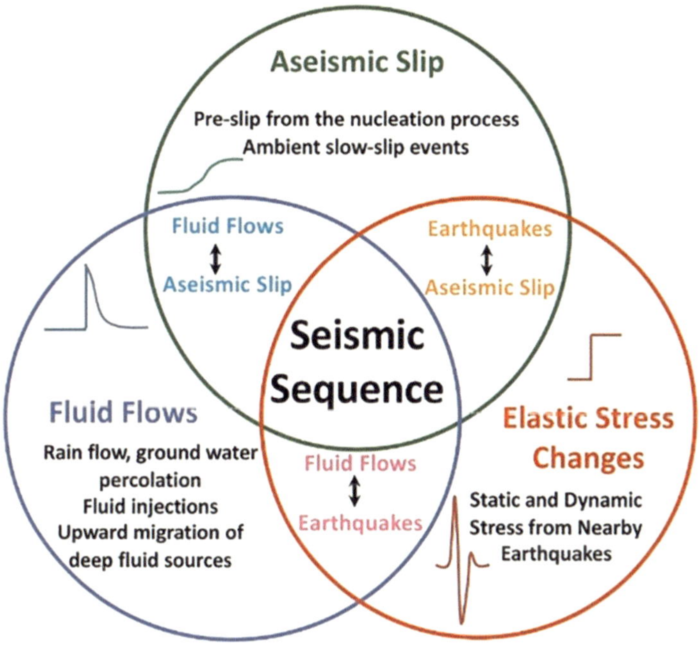

Figure 13.
An updated summary of three primary physical models that drive seismicity (foreshocks, aftershocks, and swarms) and potential interactions between them. Ref. [16].

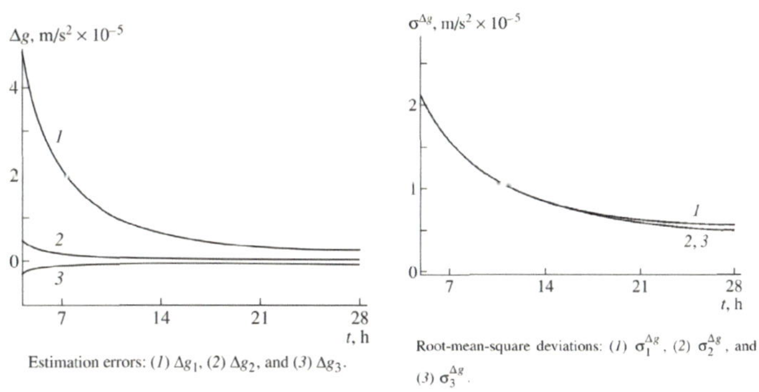

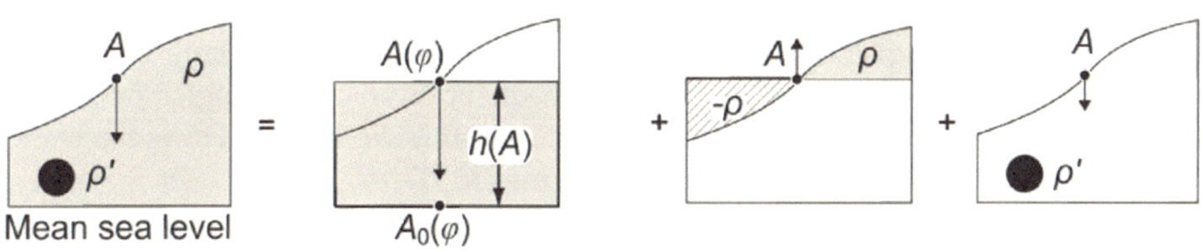

$$g_{obs}(A) = [\gamma(\varphi) - \gamma' h(A) + 2\pi G\rho h(A)] + [-T_c(A)] + \Delta g(A)$$

Figure 14.
The gravity anomaly Δg*. Terrestrial gravity* g_{obs} *at point* A *on the surface is decomposed into the following: (1)* γ *– the normal gravity at the mean sea level* A_o *at latitude* φ*, (2)* $\gamma' h$ *– the normal gravity difference between* A *and* A_o*, (3)* $2\pi G\rho h$ *– the attraction of an infinitely large flat plate of thickness* h *with density* ρ*, (4)* T_c *– the topographic correction, and (5)* Δg *– the attraction of a subsurface body with an anomalous density. Note that we may always take both* γ' *and* $T_c(A)$ *to be positive because we consider downward to be the positive direction for terrestrial gravity. Refs. [3, 4].*

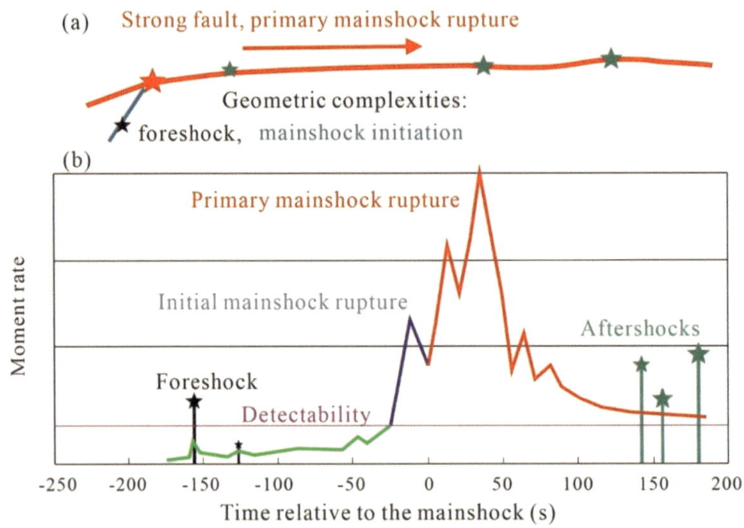

Figure 15.
(a) A schematic diagram showing a fault model where a large earthquake rupture on the main fault which is strong and is preceded by a foreshock. The initial mainshock rupture occurs on a subsidiary fault that has either a different fault orientation or different faulting style [17]. The mainshock ruptures unilaterally to the right and is followed by numerous aftershocks. (b) A normalized moment rate function showing the initial and primary mainshock ruptures, which is preceded by a foreshock and followed by many aftershocks [16].

aseismic slip category. Similarly, the elastic stress change category would include both static and dynamic stress changes from nearby earthquake ruptures (i.e., the original and alternative cascade models), and possibly dynamic stress changes from nearby earthquake ruptures [23–25] and large distant earthquakes [26, 27]. It is also obvious that the physical process in each category is not mutually exclusive and can 'trigger' each other. For example, an earthquake can trigger aseismic slip surrounding the mainshock rupture plane, which is termed after-slip. After-slip is known to drive aftershocks [28–31], and if the triggered aftershocks happen to be larger than the previous earthquake, we now have a case where after-slip (or more generally aseismic slip) from previous earthquake can drive foreshocks and the next larger mainshock [22]. Similarly, based on the physical mechanisms of dilatancy hardening and fault slip affecting its hydraulic diffusivity, there are also interactions between fluid flows and slow preslip that can either promote or inhibit each other [32, 33]. Thus, while fluid flows can drive seismicity, the occurrence of moderate to large earthquakes are also known to break seals within the fault zones, resulting in rapid changes of fluid flows and subsequent seismicity through a fault-valve mechanism [34, 35]. Finally, as mentioned before, aseismic slip and fluid flow can work in concert to drive foreshocks and earthquake swarms [36, 37].

2.4 Importance of physical constraints in model construction

The gravity model is a type of spatial interaction model used in various fields, including transportation, economics, and urban planning, to predict the flow of people, goods, or information between two locations. The model is based on the principle that the flow between two locations is proportional to their sizes and inversely proportional to the distance separating them. Here is why physical constraints are important in the construction of gravity models:

1. Maintaining realism

   Physical constraints ensure that the gravity model reflects the real-world limitations and characteristics of the system being modeled. For instance:

   Distance decay: The assumption that interaction decreases with distance is a fundamental physical constraint. The model must account for the fact that longer distances generally entail higher costs, lower speeds, or less convenience, leading to lower interaction.

2. Conservation of flow

   In a gravity model, the total flow from a set of origins to a set of destinations must be conserved. This means that the sum of all flows from a particular origin should equal the total supply at that origin, and the sum of all flows to a particular destination should equal the total demand at that destination.

3. Balancing attraction and repulsion

   The model must balance the attraction between locations (often based on population or economic activity) with the repulsion effect of distance. Physical constraints help in setting up the model parameters to achieve this balance.

4. Parameter estimation

   Physical constraints guide the estimation of model parameters, such as the ‘friction factor’ or ‘distance decay parameter’. These parameters should be chosen to reflect real-world conditions:

   Friction factor: This parameter adjusts the impact of distance on the flow. A physically constrained model will estimate this factor based on observed travel times, costs, or other relevant factors.

In conclusion, physical constraints are vital in the construction of gravity models to ensure that the models are realistic, reliable, and useful for decision-making. They help maintain the integrity of the model’s assumptions and its predictive capabilities, allowing for better-informed planning and policy decisions.

3. Earthquake focal mechanism and seismic wave propagation

3.1 Earthquake focal mechanism and its role in earthquake disasters

Recently, Stein and Bird [17] proposed an alternative cascade model where large continental earthquakes (such as the 2023 M7.8 Pazarcik, Türkiye, earthquake) likely nucleated on a branch or splay fault before jumping onto the major strike-slip fault (**Figure 3a**) [16]. The initial rupture on the branch fault could be considered as a foreshock, but depending on the time separation between the initial and the major ruptures, it can also be viewed as part of the mainshock rupture at teleseismic distances. For example, the 2024 M7.5 Noto Peninsula mainshock was preceded by a M5.9 foreshock 14 seconds earlier [37]. But this event was only listed in the JMA catalog, not by the USGS or other global catalogs. Ma et al. [38] showed that there was likely more than one event in these 14 seconds, and hence, they argued for a continuous initial slow rupture, rather than one single foreshock (**Figure 3b**). Ozacar and Beck [39] also showed that the initial ruptures of the 2001 M7.9 Kunlun fault and the 2002 M7.9 Denali fault earthquakes all started on fault structures with different faulting styles than the main strike-slip faults (**Figure 3a**). Regardless of the name of the branched fault rupture and its faulting style, it likely reflects an alternative view to consider how large ruptures on continental faults are initiated (as shown in **Figure 16**) [17].

The 2023 Kahramanmaraş earthquakes occurred on active faults that were known to be a high seismic hazard, yet the devastating impacts of these earthquakes show that the risk was not adequately considered. Vulnerabilities arising from exposure, corruption, and poverty led to a lack of seismic preparedness which amplified the earthquake risk into a tragic disaster [40].

3.2 Seismic wave propagation characteristics

Seismic wave propagation refers to the behavior of waves as they travel through the Earth due to earthquakes or other seismic sources. The characteristics of seismic wave propagation are crucial for understanding the structure of the Earth's interior, locating earthquake epicenters, and assessing seismic hazards. Here are some key characteristics:

Wave types:

Body waves: These waves travel through the interior of the Earth. They are further categorized into:

Primary (P) waves: Also known as compressional or longitudinal waves, they are the fastest seismic waves and can travel through solids, liquids, and gases.

Secondary (S) waves: Also known as shear or transverse waves, they are slower than P waves and can only travel through solids.

Surface waves: These waves travel along the surface of the Earth and are slower than body waves. They cause the most damage during earthquakes. There are two main types:

Love waves: These are transverse waves that cause horizontal ground motion perpendicular to the direction of wave travel.

Rayleigh waves: These are a type of elliptical motion that cause both vertical and horizontal ground motion.

Wave velocity:

The speed at which seismic waves travel depends on the medium they are passing through, including its density and elasticity. P waves generally travel at speeds between 5 and 8 kilometers per second, while S waves travel at speeds between 3 and 5 kilometers per second. Surface waves are slower than body waves, with speeds typically less than 3 kilometers per second.

Refraction and reflection:

As seismic waves pass from one medium to another (e.g., from crust to mantle), they can bend (refract) or bounce back (reflect) depending on the angle of incidence and the difference in wave speeds between the two media.

Diffraction:

Seismic waves can bend around obstacles and spread out after passing through small openings, a phenomenon known as diffraction.

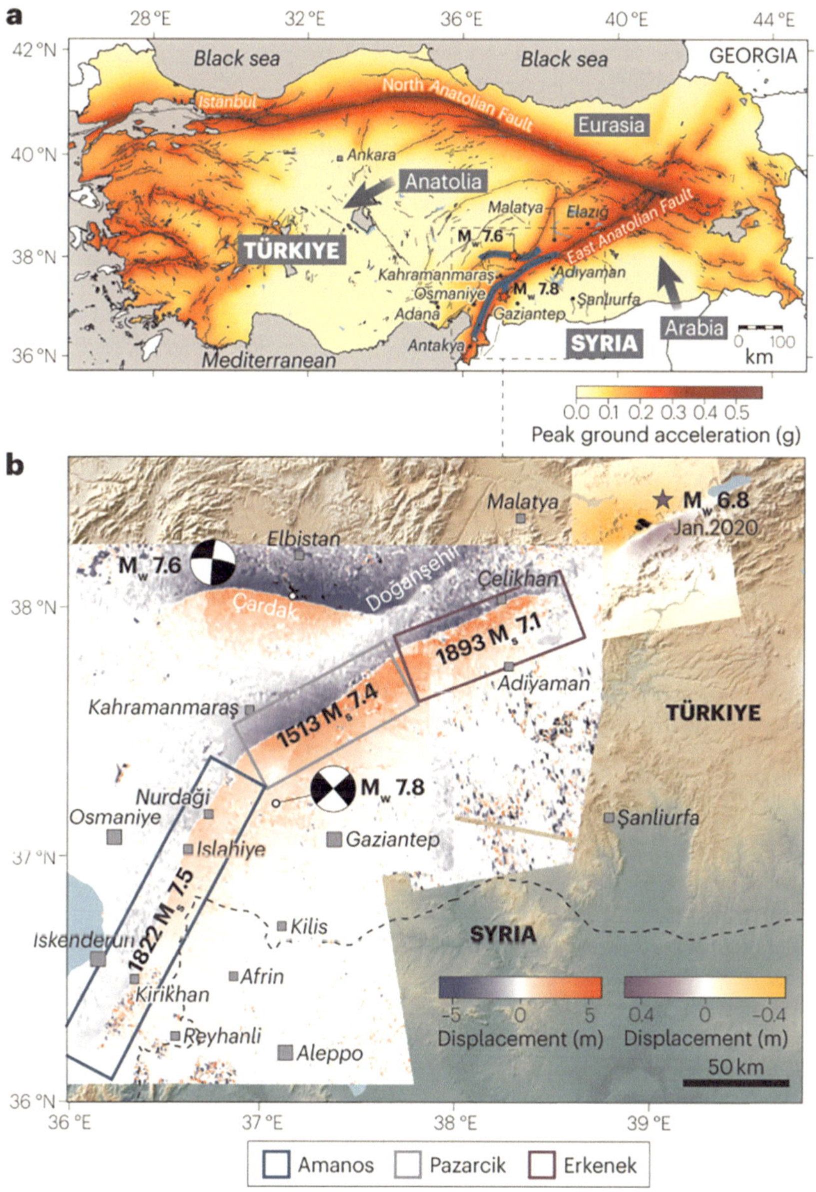

Figure 16.
Known seismic hazard and surface rupture maps of the 6 February 2023 earthquakes. ((a) Earthquake hazard map of Türkiye from the Turkish disaster and emergency management authority (AFAD) published in 2018. The 6 February 2023 earthquakes (blue lines) occurred on faults known to have high potential hazard. The dashed box is the area shown in panel. (b) Surface ruptures of the 6 February 2023 magnitude 7.8 and magnitude 7.6 earthquakes and historical events along the East Anatolian Fault. Red-blue colors show surface displacement due to the Kahramanmaraş earthquakes projected into the N79 °E direction (approximately parallel to the EAF), which are estimated from a 3D inversion of pixel offsets of Sentinel-1 radar and Sentinel-2 optical satellite images. Purple-orange colors show fault-parallel surface displacement due to the 2020 magnitude 6.8 earthquake measured using interferometry of Sentinel-1 satellite data. Focal mechanisms and surface rupture (black lines) are from the United States Geological Survey. Earthquake hazard map of Türkiye in panel a adapted with permission from AFAD). Ref. [40].

Attenuation:

The energy of seismic waves decreases as they propagate through the Earth. This is due to the conversion of wave energy into heat and the scattering of seismic energy.

Dispersion:

In some cases, the velocity of seismic waves depends on their frequency, leading to the spreading out of wave packets over time. This is known as dispersion and is more pronounced in surface waves.

Amplitude and frequency:

The amplitude of seismic waves determines the shaking intensity at a given location, which is an important factor in earthquake hazard assessment.

The frequency of seismic waves can also vary, with high-frequency waves often being more damaging.

Wavefronts and ray paths:

Seismic waves propagate as wavefronts, which can be approximated by rays in many cases. The paths that these rays take through the Earth can be used to infer the structure of the Earth's interior. Understanding these characteristics is essential for seismologists to interpret seismic data and for engineers to design structures that can withstand the forces exerted by seismic waves during earthquakes.

3.3 Integration of focal mechanism and seismic wave propagation in model construction

The integration of focal mechanism analysis with seismic wave propagation in model construction is a critical step in seismological research, particularly for understanding the source characteristics of earthquakes and the behavior of seismic waves through the Earth's crust and upper mantle. Here's how these two aspects are integrated:

Focal Mechanism Analysis:

The focal mechanism of an earthquake provides information about the orientation of the fault plane and the slip direction. It is determined from the first-motion polarities of P waves recorded by seismic stations around the event.

The focal mechanism can be represented by a beachball diagram, which shows the two nodal planes (the possible fault planes) and the strike, dip, and rake of the fault.

Seismic wave propagation:

This involves the study of how seismic waves generated by an earthquake travel through the Earth. It includes understanding the velocity structure, attenuation, and scattering of waves, as well as the refraction and reflection at different interfaces.

Integration Process:

1. Data collection and analysis

   Collect seismic data from a network of stations that recorded the earthquake, including the arrival times of P and S waves, amplitudes, and first-motion polarities. Perform a focal mechanism analysis to determine the style of faulting (strike-slip, normal, or thrust) and the orientation of the fault plane.

2. Inversion for earthquake source parameters

   Use the seismic wave arrival times to invert for the earthquake's source parameters, including the hypocenter location, depth, and origin time. Combine this with the focal mechanism to estimate the slip vector and the orientation of the fault.

3. Waveform modeling

   Model the observed seismic waveforms using synthetic seismograms generated from source, path, and site effect models. Adjust the models to match the observed waveforms, which involves refining the velocity structure and the earthquake source model.

4. Integration into earthquake source and wave propagation models

   Integrate the focal mechanism and source parameters into a comprehensive model that includes the Earth's velocity structure and the physical properties of the rocks. Use the integrated model to simulate how seismic waves propagate from the source to the recording stations.

5. Model validation and refinement

   Validate the model by comparing the synthetic waveforms with the observed data. If there are discrepancies, refine the model by adjusting parameters such as the velocity model, source depth, or focal mechanism details. Perform sensitivity analyses to understand how changes in different parameters affect the seismic waveforms.

6. Application in seismic hazard assessment

   Once the model is validated, it can be used for various applications, such as seismic hazard assessment, where the understanding of both the source characteristics and wave propagation is crucial for predicting ground motion in future earthquakes.

By integrating focal mechanism analysis with seismic wave propagation, seismologists can create more accurate and comprehensive models of earthquake sources and the Earth's subsurface. This integration is essential for improving our understanding of seismic processes and for applications such as earthquake early warning systems, seismic risk mitigation, and resource exploration.

4. Lithological structure of the crust and its impact on earthquake disasters

4.1 Overview of lithological structure and its importance

The lithological structure of the Earth's crust refers to the distribution, composition, and arrangement of the various rock types that make up the crust. This structure is complex and varies significantly from one region to another, influenced by geological processes such as sedimentation, magmatism, metamorphism, and tectonic activity. Understanding the lithological structure is crucial for several reasons, particularly in the context of earthquake disasters:

4.1.1 Importance of lithological structure

Seismic wave propagation: The type of rocks and their physical properties (e.g., density, porosity, elasticity, and anisotropy) affect how seismic waves travel through the crust. For example, soft sedimentary layers can amplify seismic waves, while hard igneous or metamorphic rocks can reflect or refract them differently.

Earthquake generation: The lithological structure can influence the occurrence of earthquakes. Faults often form along boundaries between different rock types due to differences in strength and deformation behavior. The interaction between the lithological layers and the stress fields can lead to the accumulation and release of strain, resulting in seismic activity.

Ground shaking: The lithological structure influences the intensity and duration of ground shaking during an earthquake. Soft soils and loose sediments can undergo liquefaction, leading to increased shaking and damage to structures built on them.

Site effects: Local geology, including the lithological structure, plays a significant role in site effects, which are the modifications of seismic waves as they pass through the uppermost layers of the crust. This can amplify or attenuate the ground motion at specific locations.

Landslide triggering: The lithological structure can affect the stability of slopes. Certain rock types and geological structures are more prone to landslides, which can be triggered by seismic activity.

Resource distribution: Lithological structures often control the distribution of natural resources, which can impact the development and construction practices in an area. This, in turn, can affect the vulnerability of structures to earthquake damage.

4.1.2 Characteristics of lithological structure

Layering: The crust is often layered, with different rock types stacked on top of each other. These layers can be of varying thickness and can extend over large areas or be confined to specific regions.

Heterogeneity: The lithological structure is heterogeneous, with variations occurring on multiple scales, from the macroscopic (e.g., between tectonic provinces) to the microscopic (e.g., within individual rock samples).

Faults and fractures: The lithological structure is often disrupted by faults and fractures, which can serve as pathways for seismic wave propagation and as potential sources of earthquakes.

Fluid content: The presence of fluids, such as water or hydrocarbons, within the lithological structure can significantly affect the mechanical properties of the rocks and the way seismic waves propagate through them.

4.2 Effects of lithological structure on earthquake disaster processes

1. Seismic wave amplification:

    Soft sedimentary layers: When seismic waves encounter soft sedimentary layers, such as sand, silt, or clay, the ground motion can be amplified. This is particularly true for surface waves, which can cause significant damage to structures at the surface.

    Resonance: Certain lithological structures can cause resonance, where the frequency of the seismic waves matches the natural frequency of the ground, leading to increased shaking.

2. Seismic wave attenuation:

   Hard rock layers: In contrast to soft sediments, hard rock layers can attenuate seismic waves, reducing their amplitude and potential for damage.

   Complex geometries: The intricate geometries of lithological structures can scatter and dissipate seismic energy, which can mitigate the impact of earthquakes.

3. Liquefaction:

   Loose sediments: In areas with loose, water-saturated sediments, seismic shaking can cause the soil to lose its strength and behave like a liquid, a process known as liquefaction. This can lead to the collapse of buildings, bridges, and other infrastructure, as well as the creation of sand boils and ground subsidence.

4. Landslides and rockfalls:

   Steep slopes: Lithological structures that include steep slopes or weakly consolidated rocks are prone to landslides and rockfalls during seismic events. The type and arrangement of rocks can determine the stability of slopes and the potential for mass movement.

5. Surface rupture and faulting:

   Fault zones: The presence of faults and fault zones within the lithological structure can lead to surface rupture during earthquakes. The type of rocks adjacent to the fault can influence the displacement and the potential for surface damage.

6. Tsunami generation:

   Subduction zones: In subduction zones, the interaction between the lithological structures of the overriding and subducting plates can result in significant vertical displacement of the seafloor, potentially generating tsunamis.

7. Building response:

   Foundation conditions: The lithological structure beneath a building affects its foundation conditions. Buildings on firm rock will respond differently to seismic waves than those on soft soil or sedimentary layers.

8. Water and utility systems:

   Pipe and cable networks: The lithological structure can impact the integrity of water, gas, and electrical utility systems. Earthquakes can cause pipes and cables to break or be displaced, leading to service disruptions and potential hazards.

9. Post-seismic effects:

   Aftershocks: The lithological structure can influence the occurrence and distribution of aftershocks, which can further damage already compromised structures and impede recovery efforts.

4.3 Incorporating lithological structure into earthquake disaster models

1. Geophysical and geological mapping:

   Subsurface imaging: Advanced geophysical techniques such as seismic refraction, reflection, ground-penetrating radar, and magnetotellurics can be used to map the subsurface lithological structure.

   Drilling and sampling: Geological drilling and the analysis of rock cores provide detailed information about the lithological composition and structure.

2. Site response analysis:

   1D, 2D, and 3D modeling: Models can simulate how seismic waves propagate through different lithological layers in one, two, or three dimensions. This helps in understanding site-specific amplification effects.

   Wave propagation studies: Numerical methods such as finite difference, finite element, or boundary element methods can be used to study the effects of lithological structure on wave propagation.

3. Seismic hazard assessment:

   Ground motion models: Lithological structure is a key parameter in ground motion prediction equations, which estimate the shaking intensity at various distances from the earthquake source.

   Probabilistic seismic hazard analysis (PSHA): PSHA incorporates the variability of lithological structures to estimate the likelihood of different ground motion levels over a given period.

4. Earthquake scenario modeling:

   Synthetic seismograms: These can be generated to simulate the ground motion that would result from an earthquake based on the local lithological structure.

   Disaster scenario simulations: Models can simulate the effects of an earthquake on buildings, infrastructure, and population based on the local lithological conditions.

5. Liquefaction potential assessment:

   Liquefaction susceptibility maps: These maps are created by analyzing the lithological structure to identify areas prone to liquefaction.

   Cyclic stress analysis: Models can simulate the cyclic stresses on soil layers during an earthquake to assess the potential for liquefaction.

6. Landslide and rockfall hazard mapping:

   Slope stability analysis: Incorporating lithological data into slope stability models helps identify areas at risk of landslides and rockfalls during seismic events.

DOI: http://dx.doi.org/10.5772/intechopen.1008859

7. Urban planning and building codes:

    Zoning regulations: Lithological structure can inform zoning regulations to prevent construction in high-risk areas or to enforce stricter building codes.

    Design parameters: Engineers use lithological data to design foundations and structures that can withstand the expected seismic forces based on the local geology.

8. Emergency response and recovery planning:

    Risk maps: Detailed risk maps that include lithological structure can guide emergency response efforts by highlighting areas likely to experience the most severe effects.

    Infrastructure resilience: Planning for the resilience of critical infrastructure, such as hospitals and emergency services, takes into account the local lithological conditions.

9. Education and public awareness:

    Outreach materials: Incorporating lithological information into educational materials and public awareness campaigns can help communities understand the risks they face.

4.4 Structural geology and deformation characteristics of the Tibetan Plateau and adjacent areas

4.4.1 Introduction

The research area of this paper spans from 20 to 45°N latitude and from 65 to 110° E longitude. It encompasses the Tibetan Plateau, which is the largest and highest plateau in China and globally (**Figure 17**). This chapter primarily introduces the tectonic and regional geological background of the Tibetan Plateau and adjacent areas;

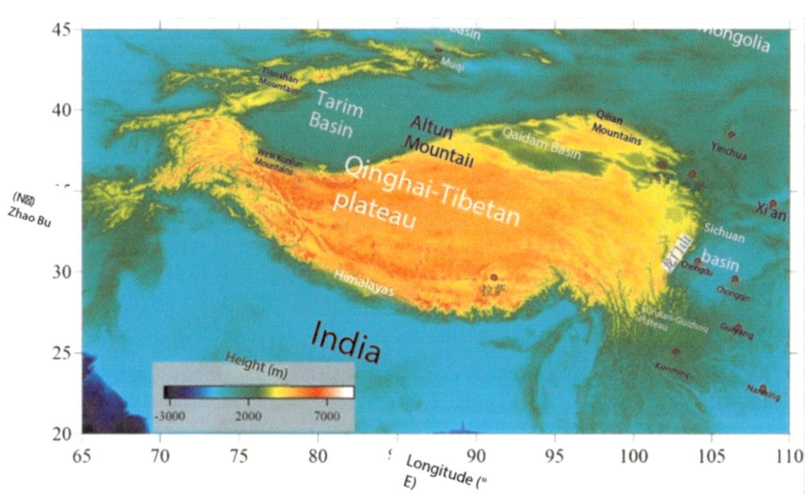

Figure 17.
Geographic feature distribution map of the Tibetan Plateau and adjacent areas.

conducts a statistical analysis of current seismic activity in the region; calculates and obtains the current strain field using existing deformation fields; and analyzes the characteristics of crustal movement and strain rate fields in the region.

Affected by the collision between the Eurasian Plate and the Indian Plate, the Tibetan Plateau and adjacent areas exhibit intense and large-scale crustal movements. The northern part of the Himalayas constitutes the Tibetan Plateau, with the Gangdise Mountains to the south and the Tanggula Mountains in between. The Kunlun Mountains form the most prominent mountain range to the north. The Tibetan Plateau stretches approximately 3000 km east-west, with two northwest-trending shallow fault valleys on its western side and two distinct northeast-trending long fault valleys in its eastern half. The intersection of the east-west Kunlun Mountains and the Tianshan Mountains with the northwest-west-trending Qilian Mountains and the northeast-east-trending Altun Mountains encloses the Tarim and Qaidam deserts, as well as the nearly triangular Junggar Basin north of the Tianshan Mountains. Due to long-term compression and tectonic influences, numerous mountains exceeding 6000 meters in elevation have formed within this region, with the peaks of the Himalayas being the most renowned, many of which exceed 8000 meters. Influenced by the plateau's topography and global climate, the region boasts a complex internal environment with a volatile climate. It is also the birthplace of many major rivers in East Asia, Southeast Asia, and South Asia, including the Yangtze River, Yellow River, Yarlung Zangbo River, Ganges River, Indus River, Nujiang River, Lancang River, and Tarim River. Additionally, frequent underground magma activity on the plateau results in multiple geothermal activity zones, forming numerous hot springs, thermal springs, and volcanoes, such as the famous Yangbajing geothermal field and Tengchong volcanoes (**Figure 17**).

4.4.2 Regional structural and geological background

Since the advent of the plate tectonic theory, numerous geophysicists and geologists from both domestic and international communities have recognized the Tibetan Plateau as a typical region of continent-continent plate collision, serving as a golden key to exploring the treasury of Earth sciences. The main research directions in the current study of the fine structure of plate tectonics in the Tibetan Plateau and adjacent areas focus on the delineation of major tectonic plates and boundaries. This is also a hot topic in the study of intraplate tectonics [41]. The Tibetan Plateau and adjacent areas exhibit complex lithospheric structures and excessively thick crusts. Based on the global tectonic evolution framework of arc-arc, arc-continent, and continent-continent collision zones, the basic structural framework of this region encompasses the Tianshan-Changning-Altun-Qilian-Qinling suture zone, collectively defining the Proto-Tethys tectonic system (Tarim Plate, Qaidam Basin). The Yarlung Zangbo suture zone and the South Kunlun subduction-collision zone constitute the Paleo-Tethys tectonic system (Tibetan Plateau). The Meso-Tethys tectonic system (Lhasa Block) is formed by the Yanshan-Himalayan tectonic belt and the Bangong-Nujiang suture zone, along with numerous active fault zones (**Figures 17** and **18**).

As can be seen from **Figure 18**, the interior of the Tibetan Plateau is mainly composed of three blocks, from south to north: the Lhasa Block, the Qiangtang Block, and the Songpan-Ganzi Block, which are separated by the Bangong-Nujiang suture zone and the Jinshajiang suture zone. Its exterior is also bounded by the Yarlung Zangbo suture zone and the East Kunlun suture zone, simultaneously delineating the

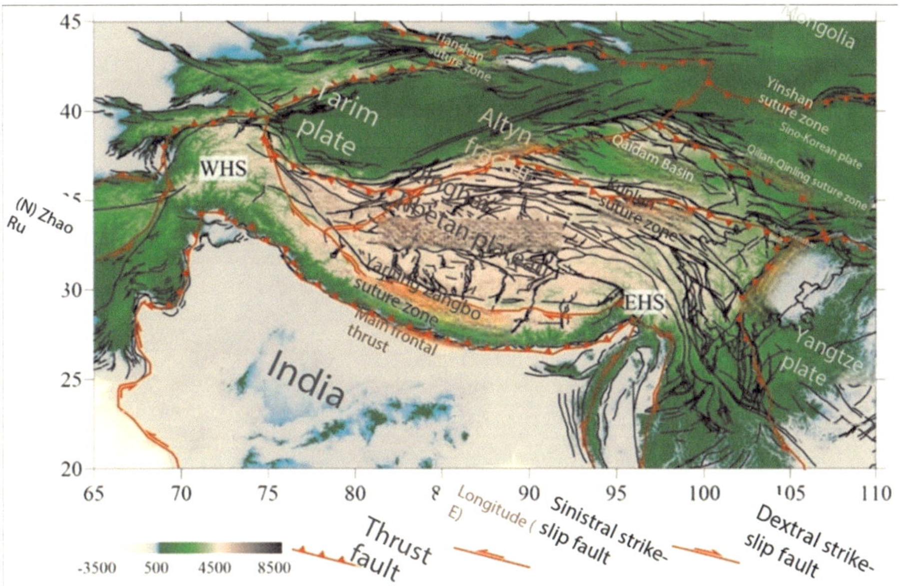

Figure 18.
Simplified diagram of major tectonic plates and boundaries in the Tibetan Plateau and adjacent areas (Modified from: Pan et al. [41], Wang et al. [42].

Himalayan Block and the Kunlun Block. Besides the widespread strike-slip shear faults that generally form around the plateau, there are extensive northwest-trending normal faults in Tibet. Meanwhile, India continues to move northward at a relatively uniform convergence rate.

Geological studies reveal that the Tibetan Plateau and adjacent areas are assembled from numerous continental crusts, mainly during the Paleozoic and Mesozoic eras, with the most recent amalgamation occurring during the collision between the Eurasian and Indian plates in the early Cenozoic (**Figure 19**) [44, 45].

Prior to the collision, the surface area of the Indian Plate was much larger, extending further north than the modern Himalayas [46]. Therefore, as the Indian Plate moved northward, the portion of its crust that disappeared is likely buried within a north-dipping detachment structure in the Himalayas and southern Tibet, such as the Main Central Thrust (**Figure 20**). This feature is also confirmed by the studies of regional seismic activity and strain fields. Meanwhile, normal faults in the eastern and western regions occur within the South Tibetan Detachment System, which played a crucial role in exposing the High Himalayan Metamorphic Core Complex. Thus, the Miocene evolution characteristics of the Himalayan orogeny are mainly characterized by synchronous crustal shortening along the horizontal deep structures of the thrust system, extending to the shallow horizontal horizons of the detachment system [48]. These tectonic activities may serve as an effective mechanism to explain crustal compression in southern Tibet [49]. Chung et al. [43], in their study of the geological and tectonic evolution models of the Tibetan Plateau and adjacent areas, investigated the systematic spatial and temporal variations of Cenozoic magmatic activity on the plateau. This spatiotemporal variation elaborates a geodynamic evolution model that describes when and how the Indian continental lithospheric mantle began to propel under Asia through rollback and rupture of the subducting Neo-Tethys plate and subsequent removal of the

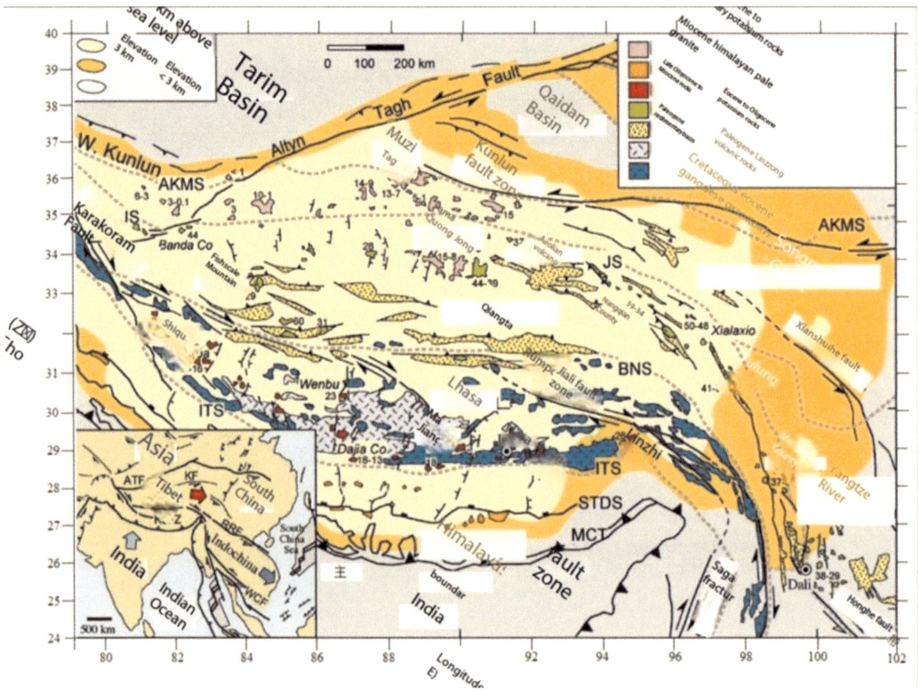

Figure 19.
Geological map of the Tibetan Plateau and adjacent areas. (Modified from: Chung et al. [43]).

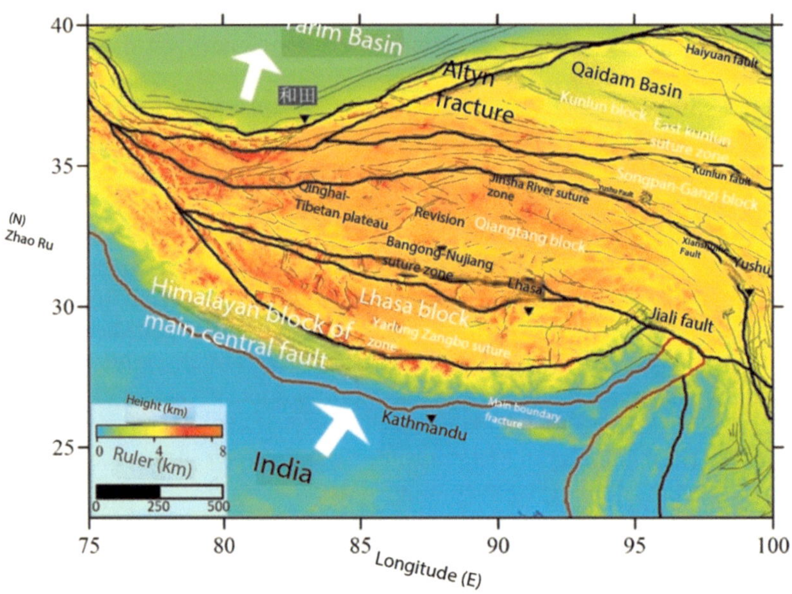

Figure 20.
Simplified structural diagram of the Tibetan Plateau. (Modified from: Bai et al [47]).

thickened Lhasa lithospheric root. The model suggests that it was only after the removal of the lithosphere at ~26 Ma that the Indian mantle lithosphere began to tilt northward, becoming a key controlling factor in the orogenesis of the Himalayan suture zone.

Geological data (**Figure 19**) indicate that normal faults also initiated in the central part of the Qiangtang Block in the northern Tibetan Plateau during the middle Miocene (~13.5 Ma). Extensive Paleogene sedimentary basins have developed in the northern Tibetan Plateau. Magmatism commonly occurred throughout the Qiangtang terrane and coincided with the formation of sedimentary basins around the Red River Fault Zone at the westernmost end of the Yangtze Craton or South China Block. Meanwhile, it can be observed that Miocene-Quaternary igneous rocks are only distributed in the northern Tibetan Plateau in a diffuse pattern, suggesting that this region may be one of the latest uplifted areas of the Tibetan Plateau and has continued to the present. The shape of this region aligns with the shape of the low-velocity lithospheric structure, the area of heat flow anomalies, the horizontal structure of the Moho discontinuity, and the surface velocity field [43, 50].

The continued interaction between the Eurasian and Indian plates during the early Cenozoic collision period led to the formation of the mighty Himalayas and the Tibetan Plateau and created the Yarlung Zangbo suture zone, which successfully separates the Himalayas from the Tibetan Plateau and also represents the northern boundary of the Tethyan Himalayas. Geographically, the Himalayas are located between the eastern and western Himalayas. The northern boundary of the Himalayas is considered to be the eastward-flowing Yarlung Zangbo River and the westward-flowing Indus River, while the southern boundary is the Main Frontal Thrust (MFT), which forms the northern boundary of the Indian Gangetic Plain [51]. Geologically, the Himalayan region is divided into the following: (1) outer or sub-Himalayas (tertiary layers); (2) Himalayas (low-metamorphic rocks without fossils); (3) Himalayas (crystalline complexes composed of granite); and (4) Tethyan Himalayas (marine, fossiliferous strata). There are four major structural units in the Himalayan region: (1) the Main Frontal Thrust (MFT) zone located between the Indian Gangetic Plain and the outer Himalayas; (2) the Main Boundary Thrust (MBT) zone located between the outer and lesser Himalayas; (3) the Main Central Thrust (MCT) zone located between the lesser and higher Himalayas; and (4) the South Tibetan Detachment (STD) zone located between the ultra-high Tethyan Himalayas (**Figure 21**).

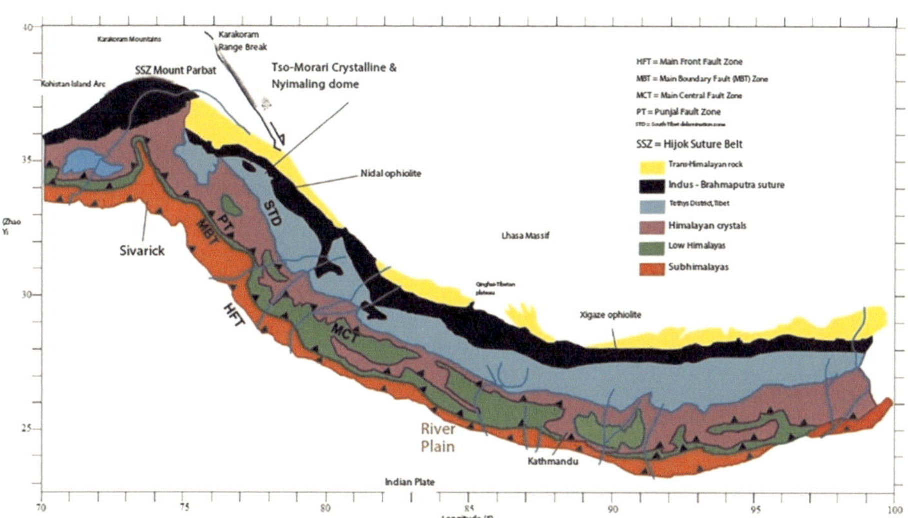

Figure 21.
Geological map of the southern margin (arc belt) of the Tibetan Plateau. (Modified from: Gupta et al. [52]).

4.4.3 Regional seismic activity

The Tibetan Plateau is one of the most seismically active continents in the world. Shallow seismic activity is widely distributed across the plateau, while intermediate-depth earthquakes dominate in the eastern and western Himalayas (**Figure 22**). **Figure 22** shows the random distribution of shallow events, which is inconsistent with the tectonic models of the evolutionary history of the Tibetan Plateau [50]. Statistical analysis of earthquakes with magnitudes Mw $\geq$ 5.0 in the Tibetan Plateau and its vicinity over the past 40 years (as shown in **Figure 23**) reveals the presence of approximately 1500 shallow earthquakes (H $\leq$ 50 km) and 700 intermediate-depth earthquakes (50 < H $\leq$ 300 km) in this region. As seen in Figure 2.6, since 1976, there

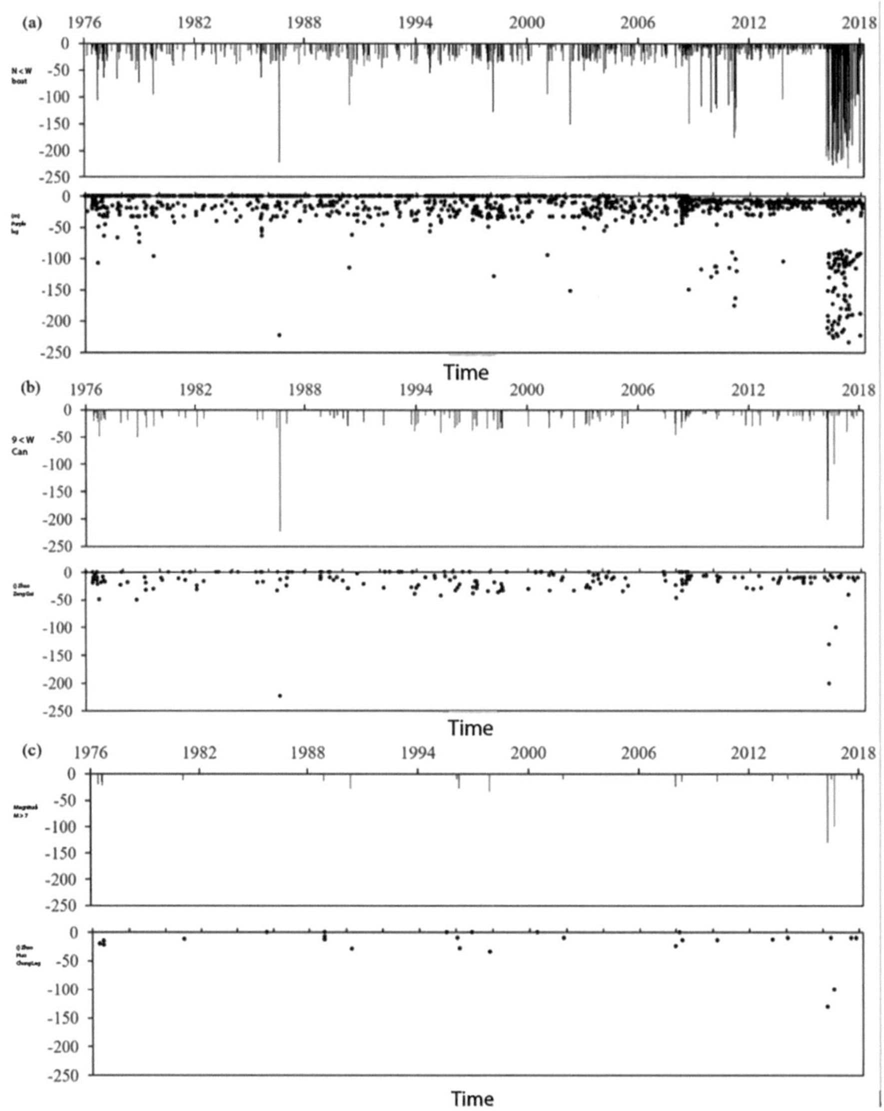

Figure 22.
Distribution of magnitude (M) and focal depth of main earthquakes in the Tibetan Plateau and adjacent regions (From 1976 to present, according to: USGS).

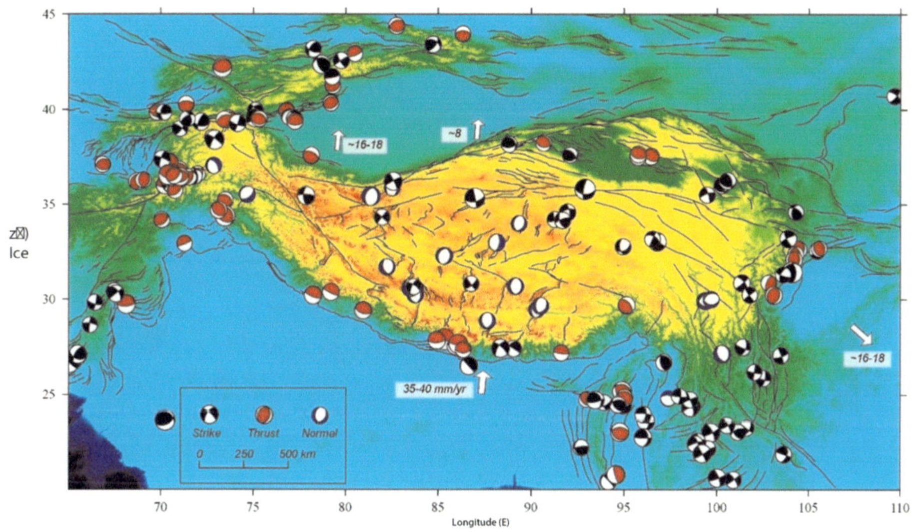

Figure 23.
Focal mechanism solutions for Mw ≥ 6.0 main earthquakes in the Tibetan Plateau and adjacent regions (From 1976 to present, according to: USGS).

have been eight earthquakes with Mw ≥ 8.0 and 52 earthquakes with 7.0 ≤ Mw < 8.0 [53]. Shallow earthquakes can occur throughout the entire Tibetan Plateau, while intermediate-depth seismic activity is concentrated in the eastern and western wedge-shaped turns of the Himalayan arc, specifically in the Eastern Himalayan Syntaxis, Yunnan, and Myanmar in the east, and the Pamir Plateau in the west, with earthquakes migrating back and forth between these two ends. The projection map of earthquake focal depths clearly shows that earthquakes in the region have shallow depths in the middle and deeper depths at both ends, which aligns with the tectonic characteristics of the region (**Figure 22**).

As can be seen from **Figure 23**, the diversity of focal mechanism solutions for Mw ≥ 6.0 main earthquakes in the Tibetan Plateau and adjacent regions indicates the complexity of the current stress field in the Himalayan-Tibetan region [47]. Further analysis of the focal mechanism solutions reveals that faults in the western Pamir Plateau are predominantly reverse-oblique, the central Himalayan arcuate region is dominated by reverse faults, and the eastern region is mainly characterized by strike-slip faults. A comprehensive analysis of the focal distribution and focal mechanism solutions in this region reveals that the focal distribution presents as an arcuate protrusion on both horizontal and vertical slices (**Figure 22**), with deeper foci in the east and west and shallower foci in the middle. This observation supports the hypothesis that there may be an asymmetric wedge-shaped body underlying the Tibetan Plateau, which slopes northward and is gently inclined in the west but steeply inclined in the east.

4.4.4 Regional deformation and stress field

With the development of space geodetic techniques, it has become feasible to monitor contemporary continental crustal movement and study its processes. Research results on large-scale crustal horizontal motion fields and the activity and deformation of major tectonic blocks have become increasingly abundant. Zheng et al.

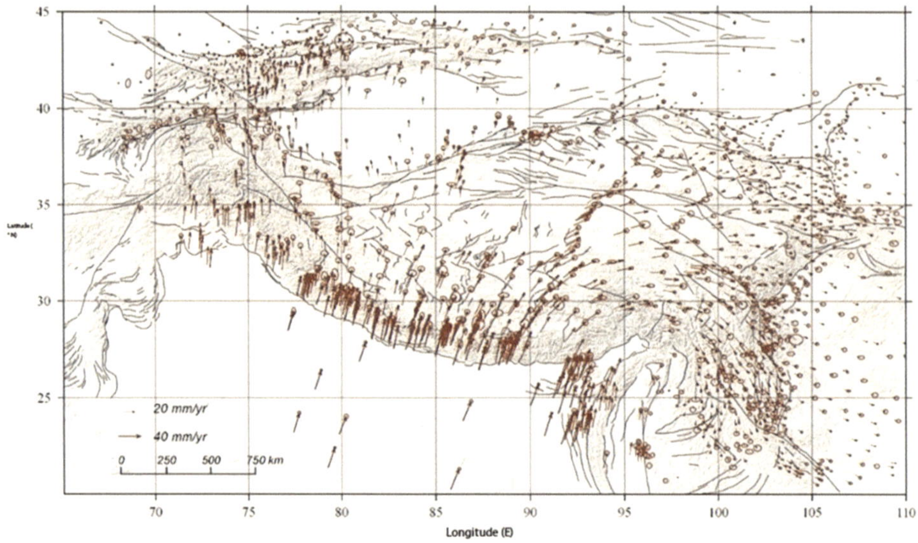

Figure 24.
Interseismic GPS velocity field in the Tibetan Plateau and adjacent regions from 1991 to 2015 (error ellipses represent 95% confidence intervals, according to Zheng et al. [54]).

[54], using GPS observation techniques, presented the most complete, precise, and advanced inter-seismic velocity field for the India-Eurasia continent to date, including 2576 velocity values. The velocity field reveals several large undeformed regions within the Tibetan Plateau and adjacent areas, with deformation concentrated along major faults, diffuse strain zones, and phenomena of plateau expansion (**Figure 24**). Analysis of the macroscopic characteristics shows that crustal movement and deformation intensity in the Tibetan Plateau and adjacent regions are stronger in the south and weaker in the north. The overall differential movement of the Tibetan block, caused by the push of the Indian Plate, is the most significant. Crustal deformation in the region presents as northeastward crustal shortening and northwestward crustal extension, with the Tibetan block as the main body. The crustal shortening rate in the NNE direction is approximately −28 mm/a, and the crustal extension rate in the NWW direction is approximately 25 mm/a (from the Tarim Basin to the Sichuan-Yunnan block). From the orderly distribution of crustal relative motion, two significant features are observed: first, the most prominent feature is the intense clockwise rotational motion around the Eastern Himalayan Syntaxis; second, there is a counterclockwise rotational motion in the northwestern part of the Tibetan block. Influenced by these two rotational motions, the near-east-west horizontal motion from the middle to the eastern edge of the Tibetan block gradually weakens, presenting a stress platform of crustal shortening, while its western edge exhibits a state of crustal extension deformation.

Although GPS data provide the most intuitive images of crustal movement, the velocity field, which is relative to a reference frame, cannot directly quantify tectonic deformation. However, further deriving the strain rate field based on the velocity field can potentially provide a quantitative reflection of tectonic deformation in the Tibetan Plateau and adjacent regions. Various parameters of the strain field can comprehensively express the different nature and intensity of deformation, and they are independent of the reference frame. Given the above research background and considering the uneven distribution of GPS stations in actual observations, this paper utilizes the least squares collocation method to solve for the strain rate field of crustal movement in the Tibetan Plateau and adjacent regions [55]. This method focuses on the strain rate field in the low-frequency domain, which exhibits relative stability.

Figure 25 shows the principal strain rates of the strain field in the Tibetan Plateau and adjacent regions, calculated from GPS observations by Zheng et al. [54], represented by two pairs of mutually perpendicular arrows at each grid point. From the distribution of principal strain rate vectors, those with larger lengths are mainly concentrated in the southern part of the Tibetan Plateau and the western Tianshan Mountains in Xinjiang. The dominant distribution of principal strains in the Tibetan block, where tectonic deformation is the strongest, is characterized by a northeast-oriented minimum principal strain rate and a northwest-oriented maximum principal strain rate. In contrast, the dominant distribution of principal strain rates in the Tianshan Mountains of Xinjiang is oriented north-south for the minimum principal strain rate and east-west for the maximum principal strain rate. Regarding the tectonic deformation patterns in different regions, especially near the boundary between the Indian and Eurasian plates, which has the highest values of principal compressive strain across the entire area, the direction of principal compressive strain is nearly perpendicular or slightly dextral to the main faults, and the principal tensile strain rate is much smaller than the principal compressive strain rate, indicating that the region is primarily a compressive deformation zone.

Within this context, the main body of the Tibetan Plateau is undergoing a strain state of north-south compression and east-west extension, but the Sichuan-Yunnan region in the eastern part of the Tibetan Plateau is the opposite, experiencing north-south extension and east-west compression. The orientation of the principal strain rates in the Tibetan Plateau and adjacent regions is generally consistent with the directions of the P-axis and T-axis in focal mechanism solutions. High values of the maximum principal compressive strain rate are distributed along the Main Himalayan

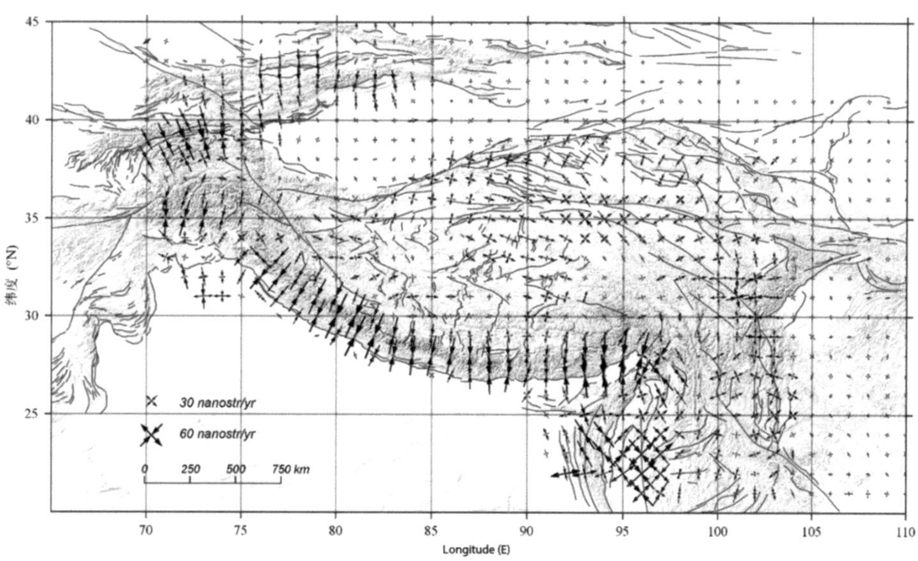

Figure 25.
Principal strain rate field in the Tibetan Plateau and adjacent regions.

Thrust and nearby areas, while within the plateau, the phenomenon of tensile strain rates exceeding compressive strain rates indicates that the interior of the plateau is in a state of tensile strain. Outside the Himalayan arc along the southern edge of the Tibetan Plateau, in the front and southern regions, the direction of the principal compressive stress of earthquakes is perpendicular to the arcuate structures, trending NNE-SSW at the western end and NNW-SSE at the eastern end [56]. The calculated strain rates in the Tibetan Plateau and adjacent regions also reveal a certain inheritance relationship between the current ground strain in this area and longer-term geological activities [57, 58]. These can be combined with seismic and geological data to analyze the seismic hazard in the region.

5. Temporal and spatial variations of the gravity field

5.1 Gravity field variations and their relationship with earthquake disasters

The theoretical coseismic gravity change was observationally tested by Tanaka *et al.* [59] against the gravity changes observed during the Shizukuishi earthquake on September 3, 1998 (M6.1) by fulfilling all the technical requirements to verify the theory. They showed that the observed spatiotemporal gravity variations agreed remarkably well with those expected from Eqs. (31)-(34) when the fault parameters were estimated from the displacement data alone without using the gravity changes (**Figure 26**). This was the first unequivocal detection of coseismic gravity changes and confirmed the validity of the theory. Currently, these expressions have been implemented in an earthquake simulator in California [60, 61].

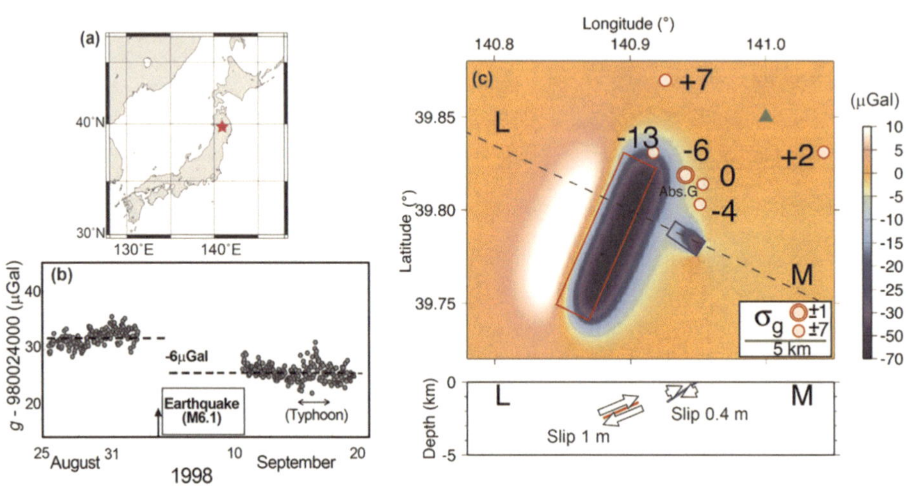

Figure 26.
Gravity changes during the Shizukuishi earthquake (redrawing of the original figures from Tanaka et al. *[59]). (a) Epicenter is denoted using a red star. (b) Absolute gravity change during the earthquake. The measurement error is approximately 1–2 μGal. (c) (top) Theoretical gravity changes (colors) and observed gravity changes (figures) during the M6.1 earthquake (the units are μGal), where σg denotes the measurement error. The blue and red rectangles are the faults projected on the surface. (Bottom) Cross-section along L-M viewed from the southwest. The slips on the larger and smaller faults are 1 and 0.4 m, respectively. Ref. [3].*

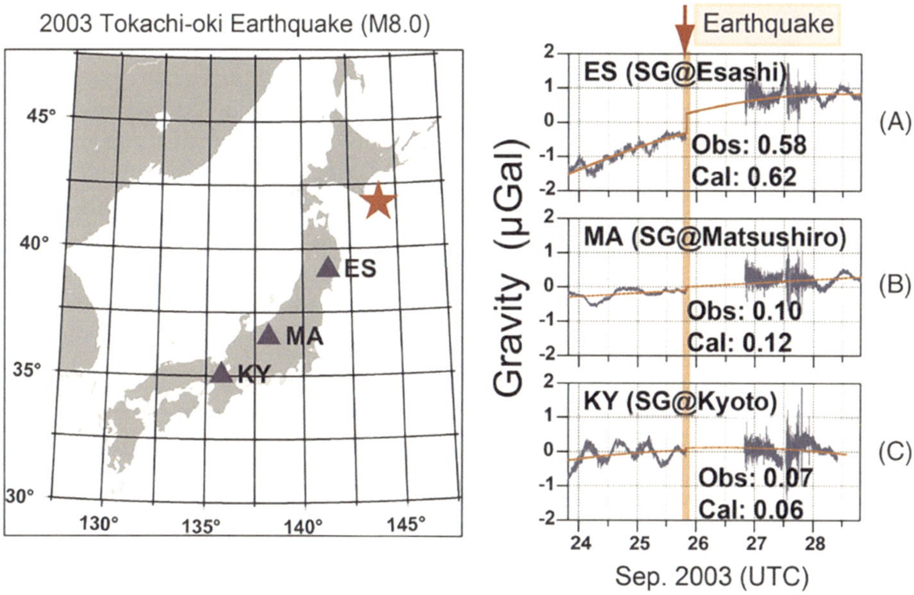

Figure 27.
Regional gravity changes detected with an array of SGs (redrawing of the original figures from Imanishi et al. *[62]). The units are μGal. (left) Epicenter (a red star) of the 2003 Tokachi-oki earthquake and the SG station locations (triangles). ES, Esashi; MA, Matsushiro; KY, Kyoto. (right) Gravity data (blue) recorded by SGs for the Tokachi-oki earthquake in 2003. The units are μGal. (A) Esashi, (B) Matsushiro, and (C) Kyoto. The red curves are the smoothed values obtained by considering the coseismic gravity steps and instrumental long-term drift. Ref. [3].*

5.2 Monitoring and analysis of gravity field changes

It took 10 years for the theory of Sun and Okubo to be verified by the observations of Imanishi *et al.* [62], who detected the regional gravity changes during the 2003 Tokachi-oki earthquake (M8.0) using an array of superconducting gravimeters (SGs) (**Figure 27**). In a series of studies, the author's research group predicted that great earthquakes cause gravity changes on regional to global scales exceeding the signal level of modern satellite gravity missions [63–65]. In fact, the GRACE (Gravity Recovery and Climate Experiment) satellite mission [66] detected significant gravity changes during the 2004 Indian Ocean earthquake off the coast of Sumatra, Indonesia (M9.1), the 2010 Maule earthquake in central Chile (M8.8), and the 2011 Tohoku earthquake in Japan (M9.1) consistent with the theoretical expectations according to Eq. (35). Refs. [67–69].

5.3 Utilizing gravity field variations in earthquake disaster prediction

The geological and geophysical work carried out in the Tibetan Plateau and its adjacent areas has provided important evidence for studying the deep structural changes, surface uplift, and related issues in the region. The Earth's gravity field is a direct reflection of the distribution of Earth's material density. The detailed distribution of gravity anomalies and the temporal and spatial dynamics of the gravity field can better reflect the differences in crustal thickness, changes in crustal density, and the migration of deep crustal materials, which are related to tectonic movement information. Based on geological and geophysical data, analyzing the geotectonic characteristics of the study area from the perspective of gravity studies will be of great significance for the research on geodynamics in the region.

This part utilizes actual gravity measurement data for regional gravity field modeling and obtains regional Bouguer gravity anomalies through gravity data processing methods. It conducts a qualitative analysis of the characteristics of the regional gravity field. Using seismic and geological data as a reference model for density inversion, density structure inversion is performed. Finally, the density anomaly information is quantitatively interpreted. It explores the corresponding earth science issues related to tectonic movements within the region.

5.3.1 Regional gravity data

1. Source of gravity data Through project collaboration, literature review, and personal communication, the gravity data collected for this study includes the following categories:

    Seismic gravity observation data provided by the National Gravity Network Center, starting from 1998 (National Network: 410 stations, Sichuan Network: 186 stations).

    Gravity survey network in the Sichuan Basin area conducted by Fu et al. [70] (302 stations).

    Gravity survey network in the eastern Himalayan structural node area conducted by Fu and She [71] (107 stations).

    Gravity survey network in the Yuxi Basin provided by Researcher Chen Shi of the Institute of Geophysics, China Earthquake Administration (641 stations) (personal communication).

    EGM2008 gravity field model (up to 2190 degrees).

2. Data description: The distribution of the data is shown in **Figure 28**. The region has a certain density of discrete surface gravity data and GPS-leveling elevation

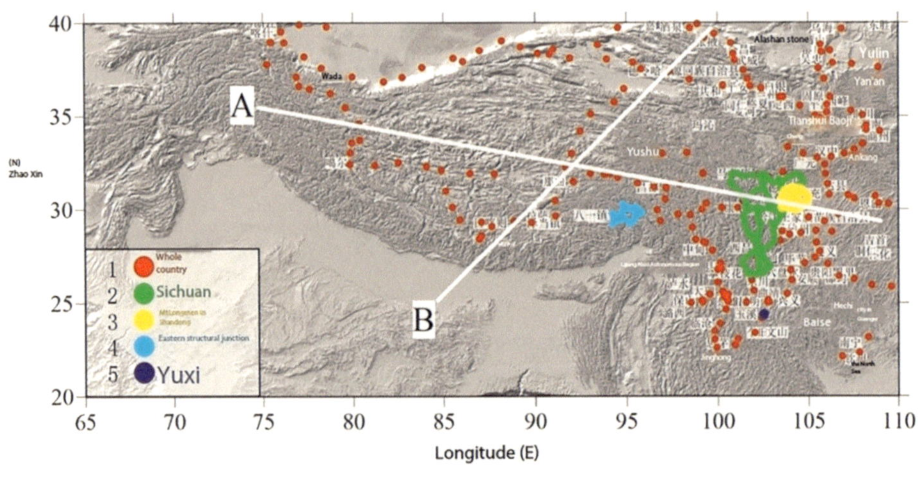

Figure 28.
Gravity data distribution in the Tibetan Plateau and adjacent areas.

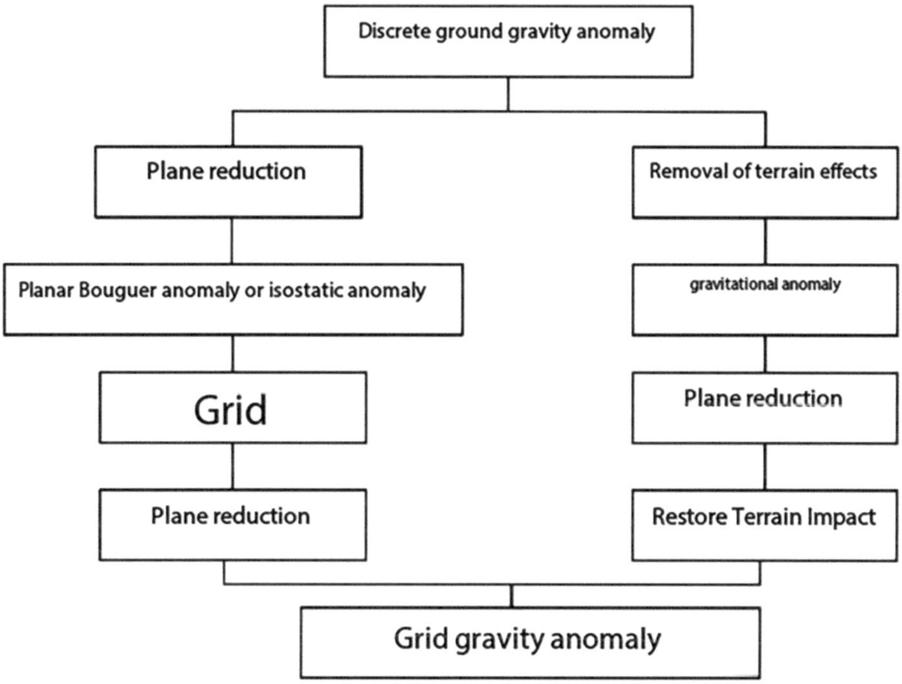

Figure 29.
Schematic diagram of the plane gravity reduction scheme.

anomaly data, with surface gravity measurements combined with leveling elevation data. The surface gravity data is distributed across a rectangular area (65 to 110°E, 20 to 40°N) with a resolution of approximately 1 arc-minute by 1 arc-minute. A 15 arc-minute by 15 arc-minute digital terrain model of the Tibetan Plateau and its adjacent areas is also available.

5.3.2 Determination of the regional gravity field

To determine a high-precision, high-resolution regional gravity field, we integrate discrete ground gravity data with a gravity field model. We employ the plane gravity reduction method [72–74], which uses the "remove-restore" approach to determine the regional gravity field (**Figure 29**). The processing flow for discrete ground gravity data in this paper is as follows:

1. Calculation of the average ground spatial anomaly

    - Calculate the discrete ground spatial anomaly from discrete ground gravity data.

    - Using a digital terrain model, compute the plane Bouguer correction and plane topographic correction for discrete points to obtain the Bouguer anomaly for these points.

    - Grid the data to obtain a $1' \times 1'$ regional average Bouguer anomaly.

- Subtract the plane topographic correction and plane Bouguer correction to obtain the average ground spatial anomaly.

2. Calculation of the residual/zero-order height anomaly

- Use the digital terrain model and reference gravity field coefficients to calculate the $1' \times 1'$ model spatial anomaly and $1' \times 1'$ model height anomaly within the region.
- Subtract the model spatial anomaly from the average ground spatial anomaly to obtain the residual spatial anomaly.
- Using spherical Stokes' integration, calculate the residual height anomaly from the residual spatial anomaly and digital terrain model.
- Add the model height anomaly to obtain the zero-order height anomaly.

3. Calculation of the topographic correction for height anomaly

- Morozovskiy Integration: Obtain the Morozovskiy first-order term by combining the zero-order height anomaly with the average ground spatial anomaly.
- Stokes' Integration: Add the Morozovskiy first-order term and the digital terrain model to calculate the topographic correction for the height anomaly.

4. Integration of GPS leveling and gravity data to determine the regional gravity field

- Calculate the model height anomaly at GPS leveling points using the reference gravity field.
- Use the average ground spatial anomaly and the Morozovskiy first-order term to calculate the residual height anomaly at GPS leveling points.
- Through the least squares collocation method, integrate the observed field elements and collocation field elements to obtain the height anomaly correction factors.
- Determine the regional gravity field using the model height anomaly, residual height anomaly, topographic correction for height anomaly, and height anomaly correction factors.

Gravity Field Forward and Inversion The gravity anomaly values obtained from the previous section's data processing are a comprehensive reflection of the Earth's internal density anomalies (local gravity field) and the undulations of density interfaces (regional gravity field). The upward continuation technique for potential fields can highlight the anomalous characteristics of regional or deep-seated large-scale geological bodies. To extract gravity information between the target study layers, this paper performs forward modeling on the gravity field of the Tibetan Plateau and adjacent areas, which involves multiple continuation processes of the Bouguer gravity anomaly data to separate the gravity anomalies caused by density structure changes at different depths. During this process, the gravity anomalies caused by the average density of each layer are also eliminated. The density contrast across the interfaces is set to 0.04 g/cm^3, and the average depth is taken as 100 km [75]. Referring to the crustal models obtained from existing gravity profiles and seismic interpretations [76–80], gravity forward modeling is conducted on the three-dimensional density structure of the Tibetan Plateau. This method allows for the acquisition of gravity anomalies caused by density differences in each layer, providing a more intuitive and effective analysis of the tectonic change characteristics of the region [81–83]. Moreover, this layered gravity processing method optimizes the large-scale gravity inversion and model construction for the Tibetan Plateau and adjacent areas in the following text, improving the inversion's temporal and spatial efficiency.

In the gravity inversion, the density converted from the velocity model obtained by seismology is used as the initial value for the inversion model [84–87]. The inversion model is set as follows: 65 to 110°E, 20 to 40°N, with a grid resolution of 3 arc-minutes. The density anomaly distributions at depths of 5, 15, 25, 35, 45, 65, 85 km, and below 85 km in the Tibetan Plateau and adjacent areas obtained from the inversion are shown in the plane map of density anomaly changes in the Tibetan Plateau and adjacent areas.

6. Regional gravity field and density structure

6.1 Gravity anomalies in the Tibetan plateau and adjacent areas

In this study, a density value of 2.67 g/cm^3 and a filter radius of 90 arc-minutes (166.7 km) were used for the Bouguer correction and topographic correction, and the normal gravity values from were used for the latitude correction. The spatial anomalies in the Tibetan Plateau and adjacent areas (**Figure 30**) and the Bouguer anomalies (**Figure 31**) were obtained. Compared with the gravity field model, the resolution and accuracy were effectively improved in local areas (**Figure 32**).

The Moho depth in the Tibetan Plateau and adjacent areas was calculated using 1 arc-minute by 1 arc-minute average Bouguer gravity anomalies (**Figure 33**). The Moho beneath the plateau is smooth and steeply inclined, with the greatest burial depth located in the northern Tibetan Plateau, reaching up to 72 km. Considering the topography of the area, the crustal thickness could potentially reach 78 km. Along the margins of the Tibetan Plateau, where the elevations are higher, the depth of the Moho interface is not significant.

6.2 Separation of the gravity potential field in the Tibetan plateau and adjacent areas

This study applied the method of upward continuation of gravity to obtain low-frequency gravity anomaly information at different frequencies. By subtracting the

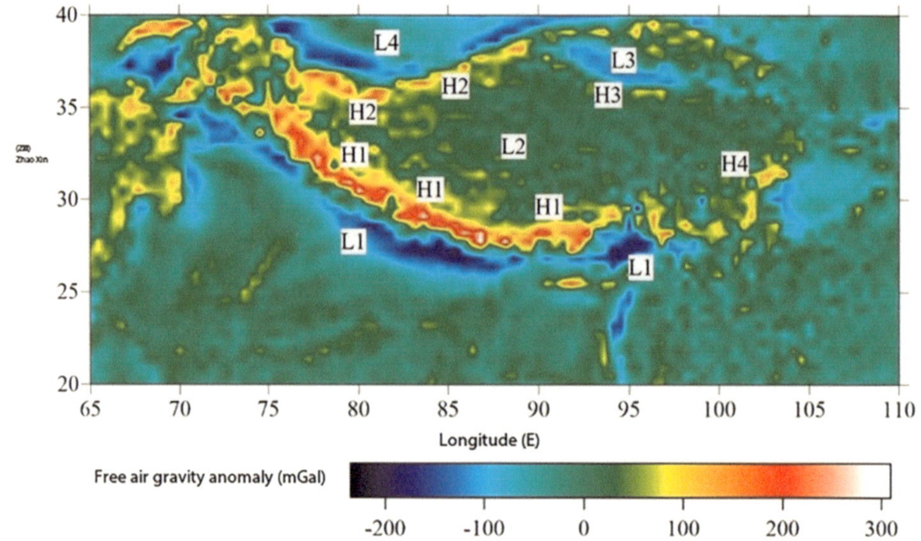

Figure 30.
Free-air gravity anomaly characteristics in the Tibetan Plateau and adjacent areas (H1 ~ H4 and L1-L4 represent high and low gravity value areas, respectively.)

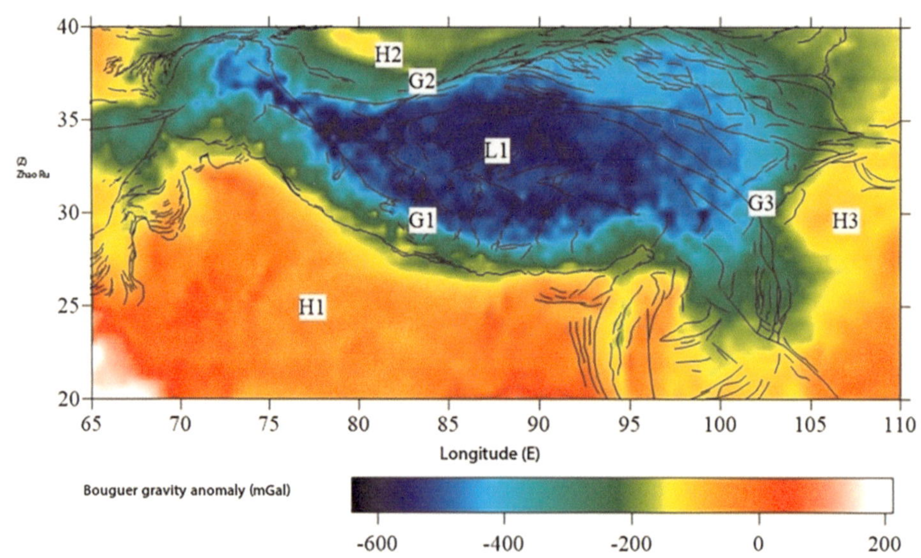

Figure 31.
Bouguer gravity anomaly characteristics in the Tibetan Plateau and adjacent areas (H1 ~ H3 and L1 represent high and low gravity value areas, respectively. G1 ~ G3 represent high gravity gradient zones.)

original Bouguer gravity anomaly field from the upward continued Bouguer gravity anomaly, the changes in gravity anomalies caused by density structure variations in different layers were obtained. As shown in **Figure 34**, the gravity anomaly changes caused by a total of 8 density layers are presented: 0 ~ 5, 5 ~ 15, 15 ~ 25, 25 ~ 35, 35 ~ 45, 45 ~ 65, 65 ~ 85 km, and below 85 km.

DOI: http://dx.doi.org/10.5772/intechopen.1008859

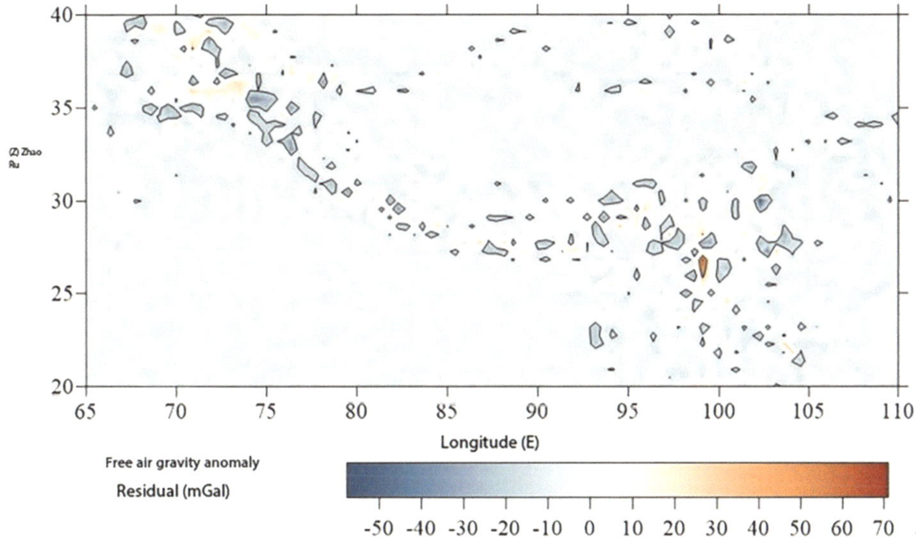

Figure 32.
Comparison of the Tibetan Plateau regional gravity field with the EGM2008 model (free-air gravity anomaly).

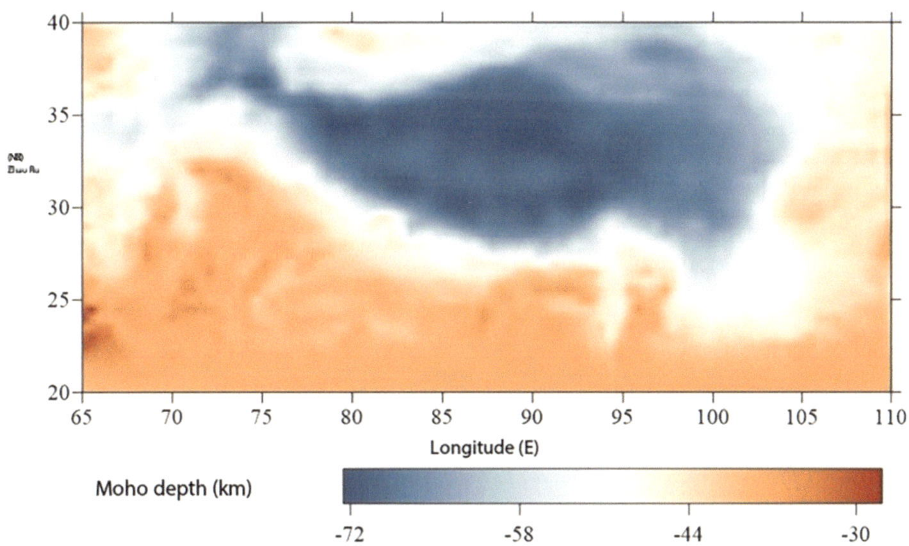

Figure 33.
The Moho depth in the Tibetan Plateau and adjacent areas.

Gravity anomalies below 85 km are regional gravity anomalies caused by the overall density distribution differences in the subsurface space (Moho interface) (**Figure 34(h)**), while other layers are due to gravity anomalies caused by density contrasts. By subtracting the gravity anomalies generated by the undulations of the interface below 85 km depth, the residual Bouguer gravity anomaly is obtained. This is considered to be the gravity anomaly produced by the uneven density of materials

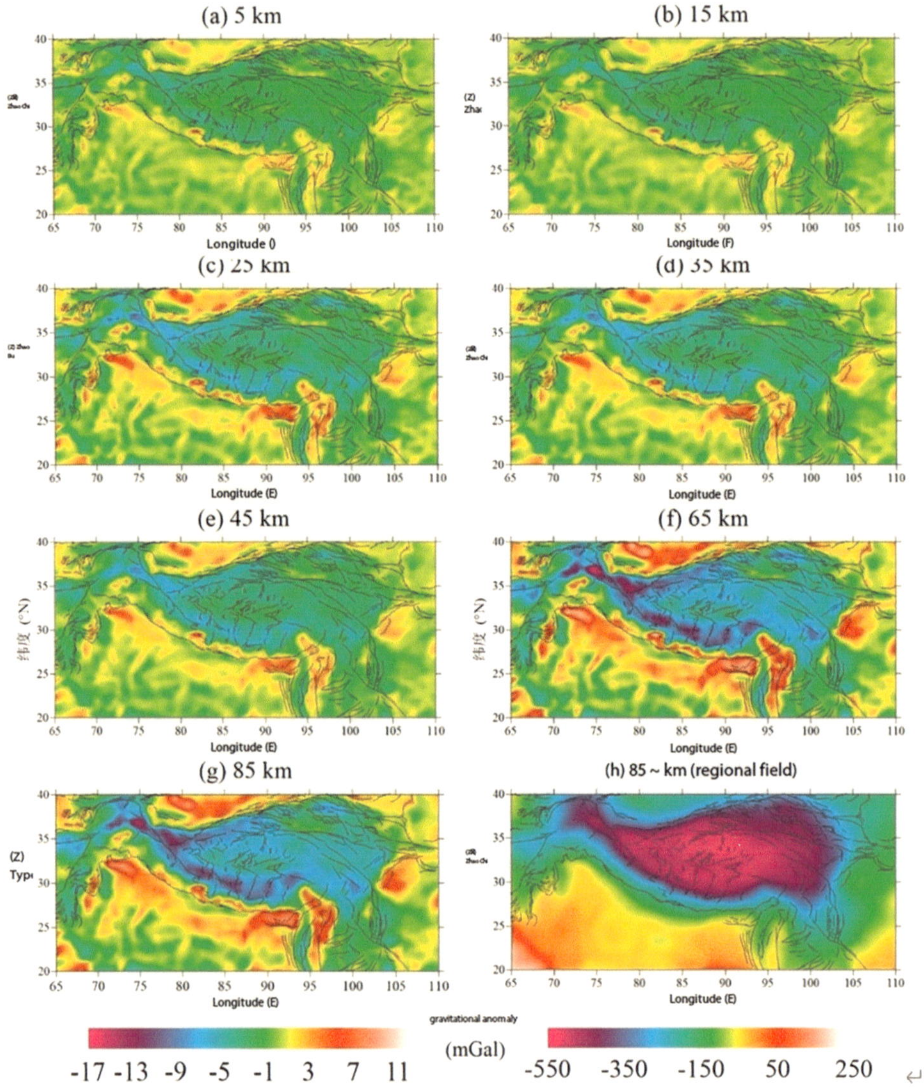

Figure 34.
Gravity potential field separation in the Tibetan Plateau and adjacent areas.

within the crust and is used in the inversion of the three-dimensional crustal density structure in the Tibetan Plateau and adjacent areas.

6.3 Density structure in the Tibetan plateau and adjacent areas

The three-dimensional density structure of the Tibetan Plateau and adjacent areas obtained from gravity inversion. The depths correspond to the density anomaly distributions at 5, 10, 15, 35, 45, 65, 85 km, and below 85 km, as shown in **Figures 35** and **36**.

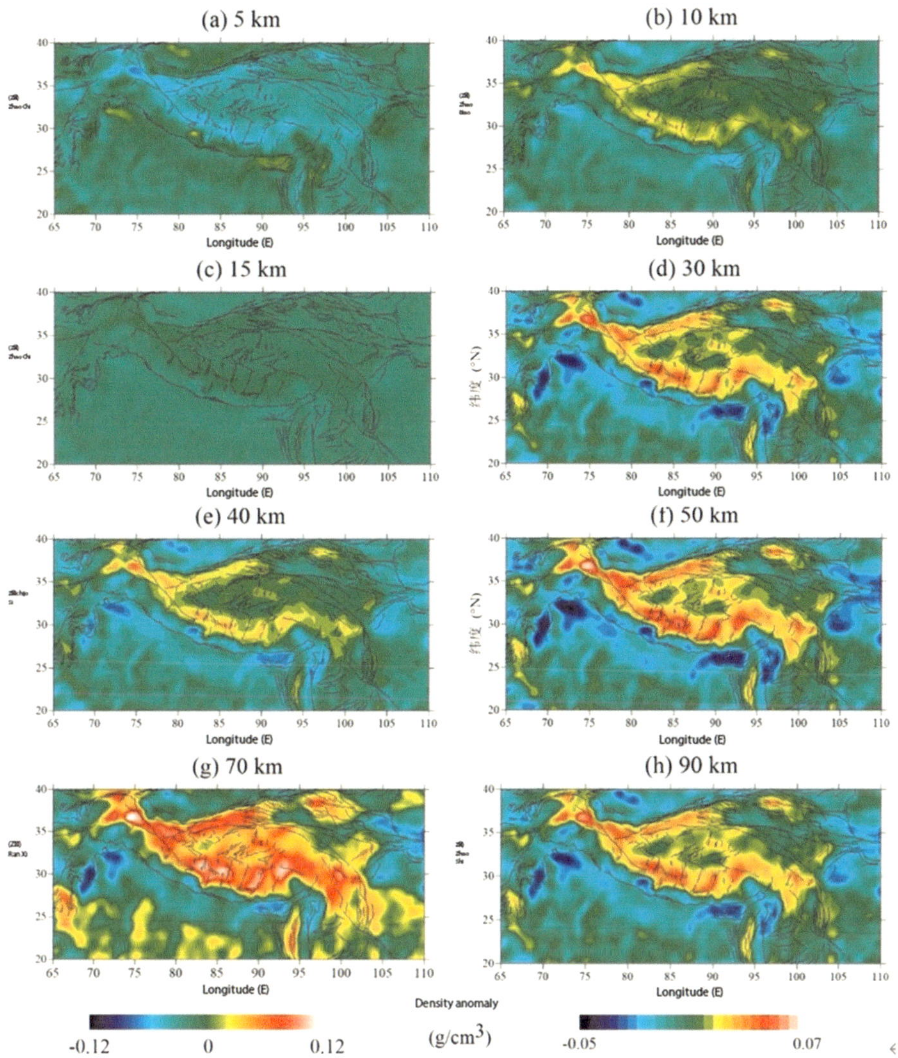

Figure 35.
Plan view of density anomaly variations in the Tibetan Plateau and adjacent areas (0–5, 5–10, 10–15, 15–30, 30–40, 40–50, 50–70, and 70–90 km depth ranges).

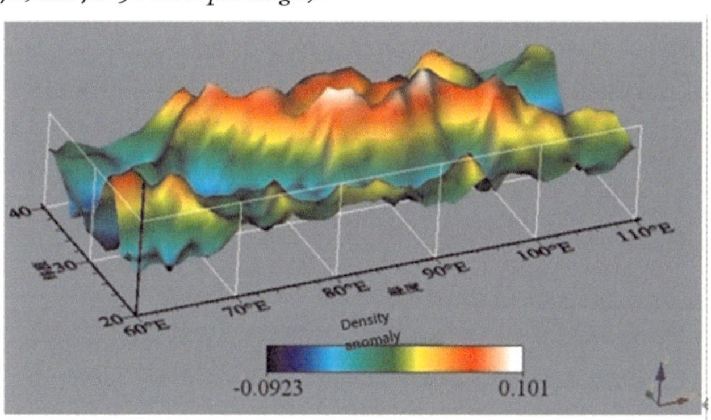

Figure 36.
3D surface plot of density anomaly variations in the Tibetan Plateau and adjacent areas.

7. Analysis and discussion

7.1 Characteristics of the gravity field

7.1.1 Regional free-air gravity anomalies

Combining regional geotectonic features and geological background, we analyze the regional structural characteristics using the representation of gravity anomaly values. From the regional free-air gravity anomalies shown in **Figure 30**, the following observations can be made:

The high-gradient gravity zones are related to the fold and thrust belts of the Eurasian and Indian plates. The main boundaries of the Tibetan Plateau and adjacent areas, such as the Himalayan piedmont, the Yarlung Zangbo suture zone, the Altyn Tagh fault, the Tianshan suture zone, and the Qilian-Qinling ranges, show a good consistency with the distribution of regional gravity anomalies. The free-air gravity anomalies over the Himalayan fold-thrust belt, the Indian-Yarlung Zangbo suture zone (H1), the Altyn Tagh-Kunlun suture zone (H2, H3), the Qilian-Qinling suture zone (H2), and the Longmenshan tectonic complex zone (H4) all exhibit significant changes. Geological studies indicate that these are caused by high-density rocks along their growth paths. Particularly, the gravity anomaly changes dramatically in the region from the Ganges Basin to the Himalayan piedmont. This phenomenon suggests that the gravity field of the Tibetan Plateau and adjacent areas is similar to that of island arc regions or mountainous convergent zones.

In contrast, the basins within the region generally reflect low gravity values, such as the Ganges Basin (L1), Qaidam Basin (L3), and Tarim Basin (L4), mainly due to the presence of low-density sediments.

The Tibetan Plateau is characterized by small gravity values (L2), varying within the range of −20 to −30 mGal, indicating that the regional crust is almost in a state of isostatic equilibrium, despite the negative free-air gravity anomalies suggesting the presence of isostatic compensation in the region. However, the high gravity values on the Himalayas indicate insufficient root compensation in the crust. This is consistent with most orogenic belts, where the backside plateau is almost isostatically compensated, but there is insufficient compensation at the edges, leading to stress and strain transformation zones and seismically active areas.

The low gravity value (L1) obtained in the Namcha Barwa mountain range at the eastern structural node of the Himalayas is the only free-air anomaly low-value area outside the basins and fold-thrust belts. Geological research indicates that there are no high-density mafic rocks in this region. Seismic activity shows that deep-source earthquakes are concentrated in the subduction zones of the Tibetan Plateau lithosphere, similar to oceanic subduction zones, which can be attributed to the rapid and low-angle subduction of the Indian Plate [88, 89]. Meanwhile, the characteristic high gravity value of the Himalayan thrust belt (H1) does not exist in the eastern structural node of the Himalayas.

7.1.2 Regional Bouguer gravity anomalies

Analyzing the Bouguer gravity anomalies in the Tibetan Plateau and adjacent areas (**Figure 31**), we find that the plateau exhibits a large-scale low gravity value (L1), which is due to the isostatic thickening of the crust, while its sides are associated with the G1 and G2 gravity gradient zones related to thrust and suture belts.

The gradient changes at the intersections of different blocks, such as the Himalayan orogenic belt, the East Kunlun fault zone, and the Longmenshan tectonic complex zone, vary. The Himalayan orogenic belt has a gradient of approximately 1.85 mGal/km. Even within the same gradient zone, the changes differ in different block regions; for example, the gradient value is about 2.5 mGal/km at the northeastern end of the Longmenshan zone and about 0.8 mGal/km at the southwestern end.

Considering the geological background of the Tibetan Plateau and its adjacent areas, there are two main parts within its interior: subduction complex zones and active continental margin accretionary belts. These two have significant density differences and can form gravity anomalies of different natures. Therefore, gravity anomaly gradient zones often exist at collision zones. Within the Tibetan Plateau itself, there are multiple nearly parallel gravity gradient zones that correspond to surface fault traces, all running east-west across the entire plateau.

Under the influence of free-air gravity anomalies, these faults and suture zones also exhibit high gravity values. The relative high gravity value H1 is associated with the Tarim Basin, as there are various intrusive bodies in this area [90]. It is noteworthy that the distribution locations of free-air anomalies related to high-density rocks in thrust faults and suture zones (H1, H2, H3, and H4, **Figure 30**) are not reflected in the Bouguer gravity anomalies. This can be attributed to the influence of isostatic compensation. There is still an approximately 120 mGal positive isostatic anomaly near the Himalayan structural belt, especially around Mount Everest, indicating an excess of underground mass in this area. The analysis suggests that this is due to the northward compression of the Indian Plate.

Overall, the Bouguer gravity anomalies reflect the main geological and structural trends of the Tibetan Plateau and its adjacent areas.

7.2 Gravity field separation

The gravity field of the Tibetan Plateau presents a pattern of higher values around the edges and lower values in the center, with the lowest anomaly reaching −590 mGal. The anomaly pattern is complex but exhibits clear regularities. To facilitate interpretation, the gravity field is divided into four regions-east, west, south, and north-for analysis. The plateau is divided into distinctly different northern and southern regions by the Bangong-Nujiang suture zone. To better understand the internal gravity field details of the Tibetan Plateau and its adjacent areas, we separated the gravity field through upward continuation, deducting the deep influence to obtain the shallow crustal residual gravity anomaly, and analyzed the overall characteristics of the lateral structure of the plateau.

By visually analyzing the gravity anomaly information from different layers (**Figure 34**), it can be seen that the interfaces of the crustal layers have significant undulations, indicating strong crustal deformation in the plateau. The interface of the southern Himalayan block is relatively shallow, in the steep interface zone, and becomes deeper towards the north into the Lhasa block. In the forward stratification images, large interface undulations are related to major fault structures. The gravity anomalies at the upper crust depth (**Figure 34(a)**, **(b)**) clearly reflect the northeast-trending faults in the region adjacent to the Ganges Basin. At the middle crust depth, an obvious low gravity anomaly is observed in the West Kunlun fault area and is well-imaged between 25 $\sim$ 45 km depths (**Figure 34(d)**, **(e)**). At the lower crust depth (**Figure 34(f)**, **(g)**), the distribution pattern of the residual gravity anomaly is two nearly NWW-SEE trending sausage-shaped bands, which have been proven to be the result of multiple factors such as the subduction complex zone formed by the collision of the Eurasian and Indian plates, the crust-mantle transition fault zone, the thicker lower crust, or the high-speed crust-mantle mixture [91, 92].

Indirectly analyzing the impact of different layers on surface gravity anomaly changes from a structural perspective: for the sedimentary layers of the Tibetan Plateau and its adjacent areas (**Figure 34(a)**, **(b)**), the material changes little at 5 km, causing local gravity anomalies within the range of 4 $\sim$ 5 mGal, with a spatial extent of several tens of kilometers. However, for the crystalline basement and Moho undulations (**Figure 33**), the Moho undulations in the study area are significant [93], and the density difference across this boundary is about 0.3 g/cm^3 [94, 95], which may cause the gravity anomaly changes in the Tibetan Plateau and its adjacent areas to reach 12 $\sim$ 16 mGal, with a spatial extent of several hundred kilometers, showing a very obvious regional anomaly. Inside the upper mantle (**Figure 34(g)**, **(h)**), evidence from seismic and electromagnetic studies of velocity and resistivity variations [85–87] reveals that the gravity anomalies in the upper mantle change dramatically laterally. The subduction of lithospheric plates results in the subducted plate having a higher density than the surface density at the same depth, and when the residual density reaches 0.05 g/cm3, it can produce $\pm$2 mGal gravity anomaly changes, with a spatial extent of over a hundred kilometers.

The gravity stratification (forward modeling) results (**Figure 34**) show that the interfaces of the crustal layers have significant undulations, indicating strong crustal deformation in the plateau with a pronounced vertical stratification characteristic, which is consistent with the typical "sandwich" structure [96–98]. Similarly, seismic results also indicate that the upper crust and upper mantle within the Tibetan Plateau lithosphere are hard and brittle layers, while the lower crust is a weak and ductile layer [99, 100]. From the density structure profile line B (**Figure 37**), it can be seen that the interface of the southern Himalayan block of the Tibetan Plateau is relatively shallow, located in the steep interface zone, and becomes deeper towards the north into the Lhasa terrane, until the Songpan-Ganzi block, where the regional interface shows another steep zone. In the forward modeling profiles, large interface undulations are associated with major fault structures, such as the India-Yarlung Zangbo suture zone and the Bangong-Nujiang suture zone areas. The steep changes in the Moho also suggest that there may be dislocations in the Moho. Statistics reveal that gravity anomalies generated by the crust below 30 km account for more than 80% of the total anomalies, and the gravity anomalies produced by low-density layers and high-density bodies can account for about 10% of the total anomalies (**Figures 34** and **35**).

7.3 Density structure characteristics

Analyzing the calculated gravity field and density anomaly structure from a longitudinal perspective, the NWW-SEE trending Profile A (**Figure 38**) and the SW-NE trending Profile B (**Figure 37**) are located at the positions of the two white solid lines in **Figure 28**, which shows the distribution of gravity data in the Tibetan Plateau and adjacent areas. It can be observed that with the dramatic change in altitude (**Figures 37(a)** and **38(a)**), the free-air anomalies also change (**Figures 37(b)** and **38(b)**), and after the Bouguer correction, the Bouguer gravity anomalies show a distinct mirror image change (**Figures 37(c)** and **38(c)**).

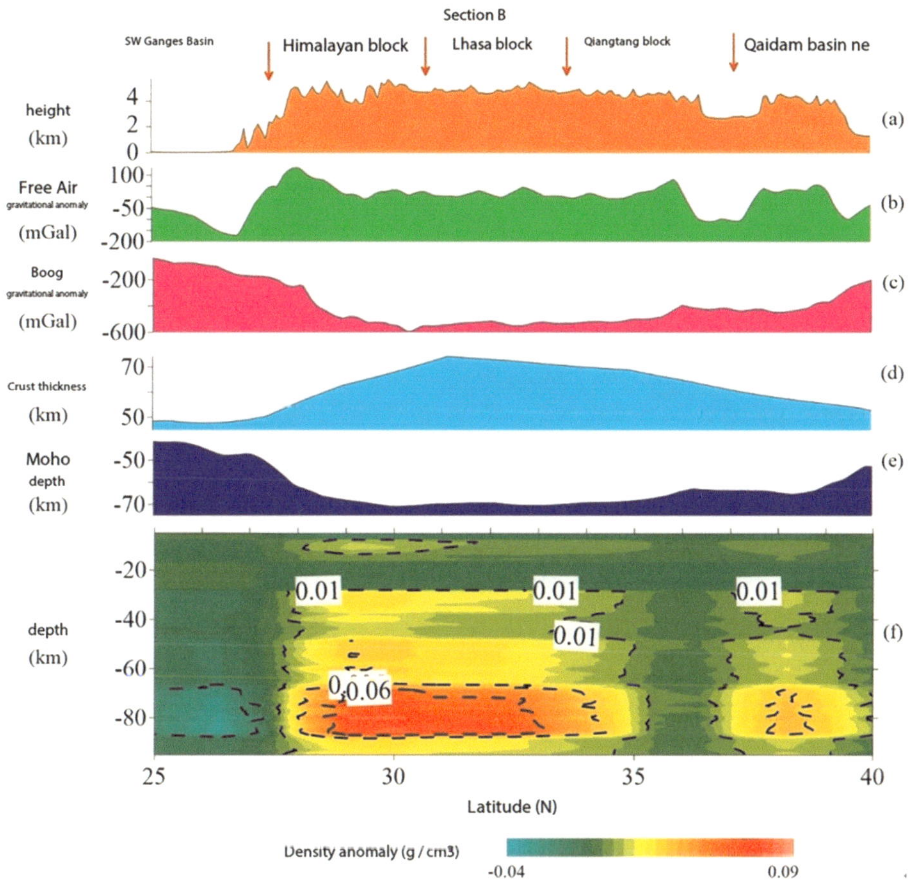

Figure 37.
Density structure of the Tibetan Plateau and adjacent areas – Profile B.

Figure 37 is a comparison chart of the Bouguer anomaly, topography, free-air gravity anomaly, crustal thickness, and Moho depth crossing the Tibetan Plateau from southwest to northeast. From the figure, it can be seen that the crustal thickness of the Tibetan Plateau increases from about 35 km in the south to around 70 km in the north, with the Himalayas located in the gradient zone of the gravity anomaly. It is also evident that the Bouguer gravity anomaly (**Figure 37(c)**) shows a negative correlation with the regional crustal thickness (**Figure 37(d)**) and a good consistency with the regional Moho depth (**Figure 37(e)**). It should be noted that both the east-west and north-south structural profiles show that the Moho of the Tibetan Plateau has a symmetrical basin-like structural characteristic (**Figures 37(e)** and **38(e)**), and the steep changes in the Moho suggest the presence of dislocations in the Moho. In the longitudinal density imaging (**Figures 37(f)** and **38(f)**), it can also be seen that due to the existence of the crust or Moho, there are significant differences in the deep density anomalies. Analyzing the changes in crustal thickness and the heterogeneity of density in the upper mantle, it can be determined that the main deep geological factors causing the gravity anomaly are the changes in crustal thickness. Additionally, the variation in the density of upper mantle materials also affects the distribution of

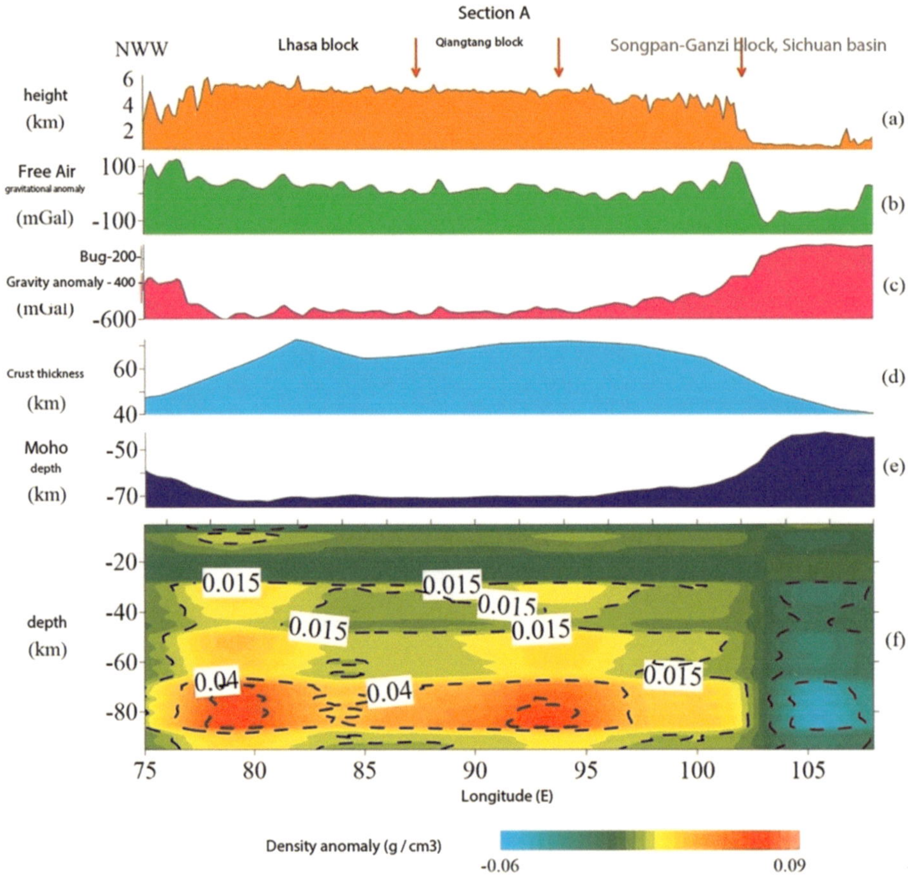

Figure 38.
Density Structure of the Tibetan Plateau and adjacent areas – Profile A.

gravity anomalies to some extent. Measurements indicate that the average density of the upper crust is 2.6 ~ 2.7 g/cm^3, the lower crust is 2.9 g/cm^3, and the upper mantle is 3.31 g/cm^3. It is evident that both the Conrad surface and the Moho are distinct density interfaces, and their undulations play a decisive role in the basic background of the gravity field. Thicker crust shows lower gravity and higher density anomalies; conversely, thinner crust shows higher gravity and lower density anomalies. The crustal thickness can vary from 50 km to about 70 km in the Tibetan Plateau region (**Figure 39(c)**), and the corresponding Bouguer anomaly also changes from +400 mGal to around −500 mGal (**Figure 39(d)**).

From the cross-sections shown in **Figures 37** and **38**, it can also be observed that at the depth of the middle and lower crust, different blocks have distinct density anomaly characteristics. The Himalayas, Lhasa, and Qiangtang blocks are areas of high positive density anomalies, while the eastern part of the Songpan-Garze block has areas of lower negative density anomalies, and the Tarim Basin has areas of higher positive density anomalies. From the center to the east, the materials in the middle and upper crust of the study area differ, with complex distributions of high and low density anomalies, while the lower crust materials show a certain degree of continuity, characterized by intense low density anomalies (**Figure 35(h)**).

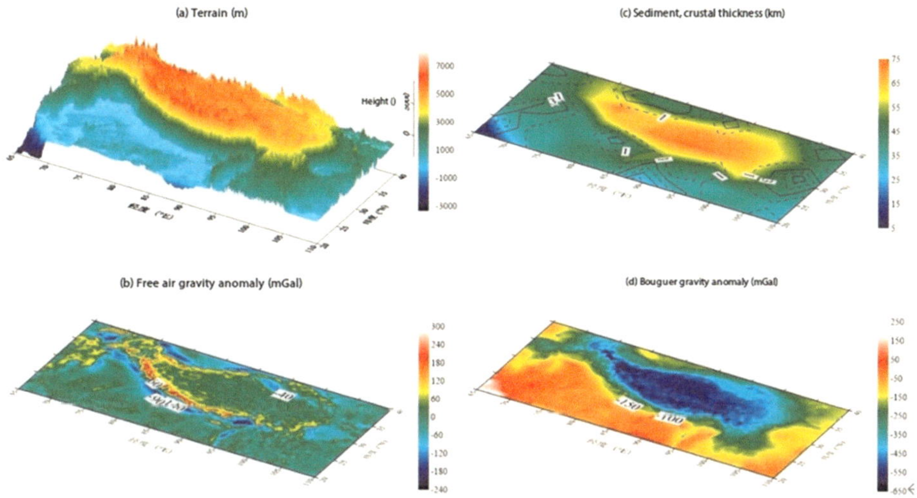

Figure 39.
Geophysical field distribution map of the Tibetan Plateau and adjacent areas.

A careful analysis of the eastern edge of the Tibetan Plateau reveals that both the Bouguer gravity anomaly and the three-dimensional density structure show (**Figures 34** and **35**, **38**) that the Sichuan Basin has lower density anomalies in the middle and upper crust, while the Songpan-Garze block has relatively higher density anomalies. The Longmenshan tectonic complex zone is located on a density anomaly gradient zone, which is the result of the Songpan-Garze block thrusting southeastward towards the Sichuan Basin. At the depth of the lower crust, the Sichuan Basin is an area of low density anomalies, indicating that the crustal material is hard, while the Songpan-Garze block is an area of high density anomalies, indicating that the material is relatively soft. The presence of low density anomaly blocks may hinder the flow of lower crustal material in the eastern part of the Tibetan Plateau.

For the large density differences ($-0.12 \sim 0.12$ g/cm^3) existing in the deep part of the Tibetan Plateau and adjacent areas, based on other geophysical and geological data, it is inferred that there may be mantle intrusive bodies. The large-scale negative Bouguer gravity anomalies that coexist could be due to the thinning of the rigid lithosphere in the southern margin of the Tibetan Plateau, in the Himalayan orogenic belt region, leading to the upwelling of hot, high-density mantle material. Geological evidence has revealed the presence of abnormally high pore fluid pressures in the sediments beneath the large strike-slip faults at the eastern edge of the Tibetan Plateau [101, 102], which necessarily leads to a reduction in interlayer frictional forces, while also having a strong buoyancy effect. This could ultimately result in the shear stress of the rock layers tending toward zero, making them prone to sliding [103]. Due to the gravitational imbalance, the lower density layers intrude into the higher density layers above them, leading to gravitational instability.

Despite compensating for weathering erosion and material flow, the Tibetan Plateau still maintains a trend of rapid uplift, indicating that the "mountain root" of the plateau is continuously replenished by deep materials. Therefore, after the plateau reaches a certain height, the role of the crustal structural force source also gradually strengthens [104]. Current seismological research also shows that the bottom of the plateau crust is molten near the mantle [105–108], with higher crustal temperatures [109, 110], accompanied by a large-scale negative gravity equilibrium. These results all indicate that the tectonic activities in the deep part of the Tibetan Plateau are ongoing, with many surplus materials distributed at the bottom of the plateau crust, and the uplift of the plateau caused by equilibrium adjustment continues. Therefore, the formation of the Tibetan Plateau may be related to the differentiation process of mantle materials. After the formation of the plateau, the thick highland crust produces horizontal extension movement under the action of gravity, and the entire process exhibits the characteristic of the conversion of Earth materials from vertical to horizontal. Considering the southeastern margin of the Tibetan Plateau, the most prominent feature of its geological structure is the NNW-NNE trending structures, and referring to the seismic wave velocity results, there are local low-velocity zones within the crust of the southeastern margin of the Tibetan Plateau [55, 111]. This suggests that, in the southern margin of the Tibetan Plateau, the movement state of the crust and mantle in the central part of the Himalayan orogenic belt may have a decoupling phenomenon, due to the differences in the strength of different layers in the region and their varying responses to loading. Meanwhile, the Bouguer gravity anomaly shows (**Figure 31**) that there is an obvious additional stress field formed by the self-weight (vertical gravity) within the Tibetan Plateau [112, 113], which could lead to stress field changes in its deep part, and subsequently drive the material at the depth of the lower crust and upper mantle to migrate [107, 114]. Considering the current deformation field, stress field, and geological background of the Tibetan Plateau and adjacent areas, under the action of gravity, there may always be a huge residual gravitational potential energy and a powerful horizontal stress system within the crust of the Tibetan Plateau, controlling internal tectonic movements and seismic activities [115, 116].

"Density heterogeneity and gravitational instability" have been considered as the controlling forces of tectonics [117–121]. In the study of the dynamic mechanism of the Tibetan Plateau, factors such as lithospheric delamination, asthenospheric material action, and partial melting have been involved. The delamination process is a gravitational collapse caused by density differences within the lithosphere. During delamination, the gradually thickening lower crust slowly transforms into other materials with higher density and delaminates into the asthenospheric mantle along with the lithospheric mantle, while the upwelling asthenospheric materials and partially molten materials have lower density anomalies. Under the control of the gravity field, the internal vertical density of the Tibetan Plateau and adjacent areas shows stratification or density inversion, and the horizontal heterogeneity may serve as a regional tectonic driving force.

The regional gravity field and density structure both indicate the differences in the tectonic blocks of the Tibetan Plateau and its adjacent areas. Gravity anomaly gradient zones at different depths can effectively delineate the regional tectonic units of the Tibetan Plateau and its neighboring regions. The density distribution shows stratification vertically and significant heterogeneity horizontally. Within the depth range of 10 to 85 km, there is a large area of low-density regions, and the positive density anomaly areas are concentrated between the 70 to 90 km layers, forming a density difference of -0.12 to 0.12 g/cm^3 with the surrounding areas. There may be a decoupling phenomenon between the crust and mantle in the central part of the Himalayan orogenic belt. Within the crust, the trend of density anomaly contour lines is basically consistent with the trend of surface fractures. After entering the mantle, the trend of density anomaly contour lines has undergone a clockwise rotation.

The three-dimensional density structure of the Tibetan Plateau lithosphere is conducive to deepening the understanding of regional tectonic movements and analyzing dynamic mechanisms.

8. Future directions and challenges in earthquake disaster research based on computational gravimetry

8.1 Gravity inversion of density and seismogenic environment in the Tibetan plateau and its adjacent areas

In this study, based on the theory of geodetic geophysics, we investigated the tectonic movements in the Tibetan Plateau and its adjacent areas using the method of studying the regional gravity field and the three-dimensional density structure of the lithosphere. By integrating the measured gravity values into existing gravity field models, we effectively filled the gravity gap areas in the study region and established a regional gravity field model. Corrections were made to the gravity field using regional geological and geophysical data, obtaining high-resolution Bouguer gravity anomaly information for the study area. Using the interface inversion method, we analyzed the contributions of materials at different depths to the regional gravity field. Furthermore, through gravity inversion methods, we calculated the three-dimensional density structure of the Tibetan Plateau and its adjacent areas, comparing the three dimensional density differences between different blocks and deep, large faults. We studied the tectonic deformation characteristics of the Tibetan Plateau as a whole and within different blocks within the region. From the perspective of tectonics, we compared the density variation characteristics with the tectonic framework of the region, thereby gaining a deeper understanding of the tectonic deformation characteristics and kinematic models of the Tibetan Plateau and its adjacent areas.

Using the method of regional gravity field modeling, our obtained regional gravity field model shows that within the tectonic units of the Tibetan Plateau and its adjacent areas, the deep, large faults are basically consistent with the distribution of gravity anomaly gradient zones. There are three particularly special regions:

1. The northeastern margin of the Tibetan Plateau, where the gravity field indicates that the Kunlun-Altun-Qilian large strike-slip fault has penetrated the Moho and entered the upper mantle, dividing the Tibetan Plateau block. At this time, the northward extrusion of the Tibetan Plateau will be blocked by the fault, causing the material migration to turn from north to northeast, and the extruded material from the Tibetan Plateau will simultaneously affect the crust and upper mantle structures at the northeastern margin of the plateau.

2. The southeastern margin of the Tibetan Plateau (Sanjiangyuan region), where the Sichuan Basin has a thick sedimentary layer composed of hard bedrock and is a stable block, playing an obstructive tectonic role at the eastern edge of the Tibetan Plateau, that is, causing a certain obstacle to the eastward migration of underground materials in the Tibetan Plateau and its adjacent areas, resulting in the direction of material migration changing from east to northeast and southeast. Both the extrusion of rigid blocks and gravitational collapse coexist in this region, and the driving effect of “uneven density and unstable gravity” in the lower crust of this block is the main reason for the rigid block extrusion and

gravitational collapse in this area. Moreover, the stress accumulation rate in this region is rapid, causing frequent seismic activities. By studying the tectonic landforms in the research area, it is found that the current active tectonics in the Sanjiang area are mainly erosive, while the tectonic uplift caused by crustal thickening is relatively weak.

3. The southern margin of the Tibetan Plateau, which belongs to the tectonic deformation after the orogeny of the Himalayan tectonic belt. The compressive deformation of the Himalayan tectonic belt in this region is the result of the collision between the Eurasian continent and the Indian Plate. The regional gravity field shows that the transverse structures of this tectonic belt are the most prominent, differing from the classic low-angle thrust compressive tectonic deformation produced by the low Himalayas in the frontal zone, but are dominated by tensile and transverse structures characterized by differential uplift and normal faulting, especially with obvious deformation characteristics of northeast-trending fault blocks. Geologically and morphologically, this transverse fault not only controls the development of water systems crossing the Himalayas but is also significantly related to the distribution of moderate-to-strong earthquakes. We believe that the transverse structures of this orogenic belt are also important active structures that control and modify the southern margin of the Tibetan Plateau.

At the same time, our calculations also indicate that the gravity field and density structure of the Tibetan Plateau and its adjacent areas exhibit layered characteristics vertically and inhomogeneity horizontally, showing a linear variation from the southwest (southern margin of the Tibetan Plateau/Lhasa block) to the northeast (northern margin of the Tibetan Plateau/Alxa block). The study area is basically divided into two major parts, north and south, with the east-west-trending Bangong-Nujiang suture zone as the boundary. The differences in the gravity field and density between the Lhasa block and the Alxa block also indicate differences in characteristics such as lithospheric structure, fault development, and intensity of tectonic activity between the blocks. The Qinghai-Tibet block is a rigid and stable landmass with very intense tectonic movements, frequent seismic activities, and developed faults. Its northern margin, the Kunlun-Altun-Qilian fault zone, is the main deformation zone with a relatively small crustal thickness. At the northeastern margin of the Tibetan Plateau, the block is also a rigid and stable Precambrian ancient landmass, with tectonic movements and fault development far less intense than those of the Qinghai-Tibet block, but it blocks the northward movement of the Tibetan Plateau and redirects its movement towards the northeast.

8.2 Potential improvements and innovations in model construction

In the field of earthquake disaster research based on computational gravimetry, there are several potential improvements and innovations that can be made in the construction of models:

1. Integration of multi-scale and multi-parameter data: Models could be enhanced by integrating data from various sources and scales, including satellite gravity data, ground-based measurements, and geological and geophysical parameters. This multi-disciplinary approach can provide a more comprehensive understanding of the subsurface structures relevant to earthquake hazard assessment.

2. Advancement in inversion techniques: Developing more sophisticated inversion algorithms that can handle complex geological structures and nonlinear relationships in the data. This could improve the accuracy and resolution of density models derived from gravity data.

3. Incorporation of time-dependent processes: Models could be extended to account for time-dependent changes in the Earth's gravity field, such as those caused by post-seismic relaxation, fluid migration, or ground deformation, which are critical for understanding the evolution of earthquake processes.

4. Utilization of machine learning and artificial intelligence: Applying machine learning algorithms and artificial intelligence to process and interpret large datasets, identify patterns, and predict earthquake-related phenomena could lead to more efficient and accurate model construction.

5. High-resolution gravity field modeling: Advancements in the resolution of gravity field models could provide more detailed insights into the shallow subsurface structures that are often associated with earthquake damage.

6. Integration of 3D and 4D modeling: Transitioning from 2D models to fully 3D and even 4D (including temporal changes) models can capture the complexity of the Earth's crust and the dynamics of earthquake processes more effectively.

7. Improved physical parameterization: Enhancing the physical basis of the models by incorporating a wider range of physical parameters that influence gravity signals, such as temperature, pressure, and phase changes in the Earth's interior.

8. Validation and uncertainty quantification: Developing robust methods for validating models against independent data and quantifying the uncertainties associated with model predictions is crucial for the reliability of earthquake disaster research.

8.3 Challenges in computational complexity, convergence, and algorithm stability

In the realm of earthquake disaster research utilizing computational gravimetry, several challenges related to computational complexity, convergence, and algorithm stability must be addressed to enhance the effectiveness and reliability of the models and analyses.

Here are some of the key challenges:

1. Computational complexity:

    - High-dimensional inversion problems: Gravity inversion problems often involve high-dimensional parameter spaces, which can be computationally intensive and require significant computational resources.

    - Large datasets: The integration of multi-source and multi-scale data increases the volume of data that needs to be processed, which can overwhelm traditional computational capabilities.

2. Convergence:

    - Non-linearity: Gravity inversion is inherently non-linear, and finding a global minimum for the inversion problem can be challenging, leading to convergence on local minima instead of the global solution.
    - Multiple solutions: The non-uniqueness of gravity inversion solutions can lead to multiple plausible models, making it difficult to determine the most accurate representation of the subsurface.

3. Algorithm stability:

    - Numerical instability: Inversions can be sensitive to initial model assumptions and parameterizations, leading to numerical instabilities that can affect the convergence and reliability of the results.
    - Regularization: Choosing the appropriate level of regularization to balance data fitting and model smoothness is critical. Too much regularization can oversmooth the model, while too little can lead to overfitting and noise amplification.

Addressing these challenges will require the following strategies:

- Development of efficient algorithms: The creation of more efficient inversion algorithms that can handle large datasets and high-dimensional problems without compromising on accuracy.
- Parallel computing: Utilizing parallel processing and high-performance computing to manage the computational load of complex inversion problems.
- Advanced regularization techniques: Implementing advanced regularization methods that can help stabilize the inversion process and provide more reliable solutions.
- Uncertainty analysis: Incorporating uncertainty analysis into the inversion process to assess the confidence of the model predictions and to guide the choice of inversion parameters.
- Model validation: Rigorous testing and validation of inversion models against seismic, geological, and geophysical data to ensure the stability and accuracy of the results.

Acknowledgements

The author acknowledges (or authors acknowledge) the use of ChatGLM (AI tool) for language polishing of the manuscript.

Thanks to the support of the following funds: The National Natural Science Foundation of China (No. 41930101); China Postdoctoral Science Foundation (No. 2019M660091XB); The Key Research and Development Project of Ecological Civilization Construction in Gansu Province (No. 24YFFA054); The Natural Science

Foundation of Gansu Province (Nos. 23JRRA857; 21JR7RA317); The Gansu Province Higher Education Institutions Young Doctor (2024QB-046); Wuhan Gravitational Field and Solid Tides National Field Observation and Research Station Open Fund (WHYWZ202403); The National Cryosphere Desert Data Center (No. E01Z790201/2021kf07); Lanzhou Talent Innovation and Entrepreneurship (No. 2022-RC-73); "Young Scientific and Technological Talents Supporting Project" Project of Gansu Province (Li Wei). Thanks to the editors and reviewers for their comments and suggestions.

Author details

Wei Li
Faculty of Geomatics, Lanzhou Jiaotong University, Lanzhou, China

*Address all correspondence to: geosci.wli@lzjtu.edu.cn

References

[1] Arzi A. Microgravimetry for engineering applications. Geophysical Prospecting. 1975;**23**(3):408-425

[2] Shichi R. Bouguer gravity anomaly map of the industrial belt zone from Kanto to Kyushu, southwest Japan. Earthq. J. 2019;**67**:54-57 (in Japanese)

[3] Okubo S. Advances in gravity analyses for studying volcanoes and earthquakes. Proceedings of the Japan Academy, Series B. 2020;**96**(2):50-69

[4] Devyatisil'ny AS. On the problem of computational vector gravimetry. Technical Physics. 2006;**51**(7):938-940

[5] Eigenmann R, Irfanoglu A. Computational earthquake and tsunami research. Computing in Science & Engineering. 2011;**13**(4):11-13

[6] Okada Y. Surface deformation due to shear and tensile faults in a half-space. Bulletin of the Seismological Society of America. 1985;**75**(4):1135-1154

[7] Okubo S. Potential and gravity changes raised by point dislocations. Geophysical Journal International. 1991; **105**:573-586

[8] Okubo S. Gravity and potential changes due to shear and tensile faults in a half-space. Journal of Geophysical Research. 1992;**97**(B5):7137-7144

[9] Chinnery M. The deformation of the ground around surface faults. Bulletin of the Seismological Society of America. 1961;**51**(3):355-372

[10] Sun W, Okubo S. Effects of earth's spherical curvature and radial heterogeneity in dislocation studies—For a point dislocation. Geophysical Research Letters. 2002;**29**(12):46-41-46-44

[11] Kanamori H, Anderson DL. Theoretical basis of some empirical relations in seismology. Bulletin of the Seismological Society of America. 1975; **65**(5):1073-1095

[12] Lay T, Kanamori H. Insights from the great 2011 Japan earthquake. Physics Today. 2011;**64**(12):33-39

[13] Ozawa S, Nishimura T, Suito H, Kobayashi T, Tobita M, Imakiire T. Coseismic and postseismic slip of the 2011 magnitude-9 Tohoku-Oki earthquake. Nature. 2011;**475**(7356):373-376

[14] Sun W, Okubo S. Surface potential and gravity changes due to internal dislocations in a spherical earth—I. Theory for a point dislocation. Geophysical Journal International. 1993; **1**(114):569-592

[15] Sun W, Okubo S. Surface potential and gravity changes due to internal dislocations in a spherical earth—II. Application to a finite fault. Geophysical Journal International. 1998;**132**(1):79-88

[16] Peng Z, Lei X. Physical mechanisms of earthquake nucleation and foreshocks: Cascade triggering, aseismic slip, or fluid flows? Earthquake Research Advances. 2024:100349

[17] Stein RS, Bird P. Why do great continental transform earthquakes nucleate on branch faults? Seismological Research Letters. 2024;**95**(6):3406-3415

[18] Bouchon M, Karabulut H, Aktar M, Özalaybey S, Schmittbuhl J, Bouin M-P. Extended nucleation of the 1999 Mw 7.6 Izmit earthquake. Science. 2011;**331** (6019):877-880

[19] Barcheck G, Brodsky EE, Fulton PM, King MA, Siegfried MR, Tulaczyk S.

Migratory earthquake precursors are dominant on an ice stream fault. Science Advances. 2021;**7**(6):eabd0105

[20] Li W, Huang Y, Wang X, Jiang X, Li X, Xie X, et al. Source model and Seismogenic environment of the Ms 6.4 Yangbi earthquake in Yunnan, China—Based on InSAR observation. Applied Sciences. 2022;**12**(12):5908

[21] Wang K, Peng Z, Liang S, Luo J, Zhang K, He C. Migrating foreshocks driven by a slow slip event before the 2021 MW 6.1 Yangbi, China earthquake. Journal of Geophysical Research: Solid Earth. 2024;**129**(1):e2023JB027209

[22] Ito R, Kaneko Y. Physical mechanism for a temporal decrease of the Gutenberg-Richter b-value prior to a large earthquake. Journal of Geophysical Research: Solid Earth. 2023;**128**(12): e2023JB027413

[23] Antonioli A, Belardinelli ME, Bizzarri A, Vogfjord KS. Evidence of instantaneous dynamic triggering during the seismic sequence of year 2000 in South Iceland. Journal of Geophysical Research: Solid Earth. 2006;**111**:B3

[24] Ding X, Xu S, Xie Y, van den Ende M, Premus J, Ampuero J-P. The sharp turn: Backward rupture branching during the 2023 Mw 7.8 Kahramanmaraş (Türkiye) earthquake. Seismica. 2023;**2**(3). DOI: 10.26443/seismica.v2i3.1083

[25] Dong P, Chen R, Xia K, Yao W, Peng Z, Elsworth D. Earthquake delay and rupture velocity in near-field dynamic triggering dictated by stress-controlled nucleation. Seismological Research Letters. 2024;**94**(2A):913-924

[26] Walter JI, Meng X, Peng Z, Schwartz SY, Newman AV, Protti M. Far-field triggering of foreshocks near the nucleation zone of the 5 September 2012 (MW 7.6) Nicoya peninsula, Costa Rica earthquake. Earth and Planetary Science Letters. 2015;**431**:75-86

[27] Martínez-Garzón P, Poli P. Cascade and pre-slip models oversimplify the complexity of earthquake preparation in nature. Communications Earth & Environment. 2024;**5**(1):120

[28] Peng Z, Zhao P. Migration of early aftershocks following the 2004 Parkfield earthquake. Nature Geoscience. 2009;**2**(12):877-881

[29] Ross ZE, Rollins C, Cochran ES, Hauksson E, Avouac J-P, Ben-Zion Y. Aftershocks driven by afterslip and fluid pressure sweeping through a fault-fracture mesh. Geophysical Research Letters. 2017;**44**(16):8260-8267

[30] Perfettini H, Frank WB, Marsan D, Bouchon M. A model of aftershock migration driven by afterslip. Geophysical Research Letters. 2018;**45**(5):2283-2293

[31] Itoh Y, Socquet A, Radiguet M. Largest aftershock nucleation driven by afterslip during the 2014 Iquique sequence. Geophysical Research Letters. 2023;**50**(24):e2023GL104852

[32] Liu Y, McGuire JJ, Behn MD. Aseismic transient slip on the Gofar transform fault, East Pacific rise. Proceedings of the National Academy of Sciences. 2020;**117**(19):10188-10194

[33] Lei X. Fluid-driven fault nucleation, rupture processes, and permeability evolution in oshima granite - Preliminary results and acoustic emission datasets. Geohazard Mechanics. 2024;**2**(3):164-180

[34] Sibson RH. An episode of fault-valve behaviour during compressional inversion? - The 2004 MJ6.8 mid-Niigata

prefecture, Japan, earthquake sequence. Earth and Planetary Science Letters. 2007;**257**(1):188-199

[35] Kato A. Implications of fault-valve behavior from immediate aftershocks following the 2023 M6.5 earthquake beneath the Noto peninsula, Central Japan. Geophysical Research Letters. 2024;**51**(1):e2023GL106444

[36] Sirorattanakul K, Ross ZE, Khoshmanesh M, Cochran ES, Acosta M, Avouac J-P. The 2020 Westmorland, California earthquake swarm as aftershocks of a slow slip event sustained by fluid flow. Journal of Geophysical Research: Solid Earth. 2022;**127**(11): e2022JB024693

[37] Peng Z, Lei X, Wang Q-Y, Wang D, Mach P, Yao D, et al. The evolution process between the earthquake swarm beneath the Noto peninsula, Central Japan and the 2024 M 7.6 Noto Hanto earthquake sequence. Earthquake Research Advances. 2025;**5**(1):100332

[38] Ma K-F, von Specht S, Kuo L-W, Huang H-H, Lin C-R, Lin C-J, et al. Broad-band strain amplification in an asymmetric fault zone observed from borehole optical fiber and core. Communications Earth & Environment. 2024;**5**(1):402

[39] Ozacar AA, Beck SL. The 2002 Denali fault and 2001 Kunlun fault earthquakes: Complex rupture processes of two large strike-slip events. Bulletin of the Seismological Society of America. 2004;**94**(6B):S278-S292

[40] Hussain E, Kalaycıoğlu S, Milliner CWD, Çakir Z. Preconditioning the 2023 Kahramanmaraş (Türkiye) earthquake disaster. Nature Reviews Earth & Environment. 2023;**4**(5):287-289

[41] Pan G, Li X, Wang L, et al. Preliminary division of tectonic units of the Qinghai-Tibet Plateau and its adjacent regions. Geological Bulletin of China. 2022;**21**(11):701-707 (in Chinese)

[42] Wang P, Scherler D, Liu-Zeng J, Mey J, Avouac JP, Zhang Y, et al. Tectonic control of Yarlung Tsangpo Gorge revealed by a buried canyon in Southern Tibet. Science. 2014;**346** (6212):978-981

[43] Chung SL, Chu MF, Zhang Y, et al. Tibetan tectonic evolution inferred from spatial and temporal variations in post-collisional magmatism[J]. Earth-Science Reviews. 2005;**68**(3-4):173-196

[44] Rowley DB. Age of initiation of collision between India and Asia: A review of stratigraphic data. Earth and Planetary Science Letters. 1996;**145**(1-4):1-13

[45] Rowley DB. Minimum age of initiation of collision between India and Asia north of Everest based on the subsidence history of the Zhepure Mountain section. The Journal of Geology. 1998;**106**(2):220-235

[46] DeCelles PG, Robinson DM, Zandt G. Implications of shortening in the Himalayan fold-thrust belt for uplift of the Tibetan Plateau. Tectonics. 2002;**21** (6):12-11

[47] Bai L, Li G, Khan NG, Zhao J, Ding L. Focal depths and mechanisms of shallow earthquakes in the Himalayan–Tibetan region. Gondwana Research. 2017;**41**:390-399

[48] Hodges KV, Parrish RR, Housh TB, et al. Simultaneous Miocene extension and shortening in the Himalayan orogen. Science. 1992;**258**:1446-1470

[49] Hodges KV, Hurtado JM, Whipple KX. Southward extrusion of Tibetan crust and its effect on Himalayan tectonics. Tectonics. 2001;**20**:799-809

[50] Shi F, Wang Y, Yu T, et al. Lower-crustal earthquakes in southern Tibet are linked to eclogitization of dry metastable granulite[J]. Nature Communications. 2018;**9**(1):3483

[51] Yin A. Cenozoic tectonic evolution of the Himalayan orogen as constrained by along-strike variation of structural geometry, exhumation history, and foreland sedimentation[J]. Earth-Science Reviews. 2006;**76**(1-2):1-131

[52] Gupta H, Gahalaut VK. Seismotectonics and large earthquake generation in the Himalayan region[J]. Gondwana Research. 2014;**25**(1):204-213

[53] Hatzfeld D, Molnar P. Comparisons of the kinematics and deep structures of the Zagros and Himalaya and of the Iranian and Tibetan plateaus and geodynamic implications[J]. Reviews of Geophysics. 2010;**48**(2)

[54] Zheng G, Wang H, Wright TJ, et al. Crustal deformation in the India-Eurasia collision zone from 25 years of GPS measurements[J]. Journal of Geophysical Research: Solid Earth. 2017;**122**(11):9290-9312

[55] Wang S, Xu X, Hu J. Review on the study of crustal structure and geodynamic models for the southeast margin of the Tibetan Plateau. Chinese Journal of Geophysics. 2015;**58**(11):4235-4253 (in Chinese)

[56] Li W, Xu C, Yi L, et al. Source parameters and seismogenic structure of the 2017 Mw 6.5 Mainling earthquake in the Eastern Himalayan Syntaxis (Tibet, China)[J]. Journal of Asian Earth Sciences. 2019;**169**:130-138

[57] Xu C. Crustal Movement Model and Tectonic Stress Field of Qinghai-Tibet Plateau [M]. Surveying and Mapping Press; 2002

[58] Zhu S, Cai Y, Shi Y. Computation of the present-day strain rate field of the Qinghai Tibetan Plateau and its geodynamic implications. Chinese Journal of Geophysics. 2005;**48**(5):1053-1061 (in Chinese)

[59] Tanaka Y, Okubo S, Machida M, Kimura I, Kosuge T. First detection of absolute gravity change caused by earthquake. Geophysical Research Letters. 2001;**28**(15):2979-2981

[60] Schultz KW, Sachs MK, Yoder MR, Rundle JB, Turcotte DL, Heien EM, et al. Virtual quake: Statistics, co-seismic deformations and gravity changes for driven earthquake fault systems. In: International Symposium on Geodesy for Earthquake and Natural Hazards (GENAH) Proceedings of the International Symposium on Geodesy for Earthquake and Natural Hazards (GENAH), Matsushima, Japan, 22-26 July, 2014; 2017. Matsushima, Japan: Springer; 2017. pp. 29-37

[61] Schultz KW et al. Virtual quake: Statistics, co-seismic deformations and gravity changes for driven earthquake fault systems. In: Hashimoto M, editor. International Symposium on Geodesy for Earthquake and Natural Hazards (GENAH). International Association of Geodesy Symposia. Vol. 145. Cham: Springer; 2015. DOI: 10.1007/1345_2015_134

[62] Imanishi Y, Sato T, Higashi T, Sun W, Okubo S. A network of superconducting gravimeters detects submicrogal coseismic gravity changes. Science. 2004;**306**(5695):476-478

[63] Sun W, Okubo S. Coseismic deformations detectable by satellite gravity missions: A case study of Alaska (1964, 2002) and Hokkaido (2003) earthquakes in the spectral domain. Journal of Geophysical Research: Solid

Earth. 2004;**109**:B04405. DOI: 10.1029/2003JB002554

[64] Sun W, Okubo S, Sugano T. Determining dislocation love numbers using satellite gravity mission observations. Earth, Planets and Space. 2006;**58**(5):497-503

[65] Sun W, Okubo S, Fu G, Araya A. General formulations of global co-seismic deformations caused by an arbitrary dislocation in a spherically symmetric earth model-applicable to deformed earth surface and space-fixed point. Geophysical Journal International. 2009;**177**(3):817-833

[66] Tapley BD, Bettadpur S, Watkins M, Reigber C. The gravity recovery and climate experiment: Mission overview and early results. Geophysical Research Letters. 2004;**31**(9):L09607. DOI: 10.1029/2004GL019920

[67] Han SC, Shum C, Bevis M, Ji C, Kuo CY. Crustal dilatation observed by GRACE after the 2004 Sumatra-Andaman earthquake. Science. 2006;**313**: 658-662

[68] Heki K, Matsuo K. Coseismic gravity changes of the 2010 earthquake in Central Chile from satellite gravimetry. Geophysical Research Letters. 2010;**37**: L24306. DOI: 10.1029/2010GL045335

[69] Matsuo K, Heki K. Coseismic gravity changes of the 2011 Tohoku-Oki earthquake from satellite gravimetry. Geophysical Research Letters. 2011;**38**:L00G12. DOI: 10.1029/2011GL049018

[70] Fu G, Gao S, Freymueller JT, et al. Bouguer gravity anomaly and isostasy at western Sichuan Basin revealed by new gravity surveys[J]. Journal of Geophysical Research: Solid Earth. 2014; **119**(4):3925-3938

[71] Fu G, She Y. Gravity anomalies and isostasy deduced from new dense gravimetry around the Tsangpo Gorge, Tibet[J]. Geophysical Research Letters. 2017;**44**(20):10,233-10,239

[72] Heiskanen WA, Moritz H. Physical geodesy. Bulletin Géodésique. 1967;**86** (1):491-492

[73] Li J, Chen J, Ning J, Chao D. Approximation Theory of Earth Gravity Field and Determination of China 2000 Quasi-Geoid. Wuhan University Press; 2003

[74] Zhang C, Chao D, Ding J, et al. Arithmetic and characters analysis of terrain effects for CM-order precision height anomaly. Acta Geodaetica et Cartographica Sinica. 2006;**35**(4):308-314 (in Chinese)

[75] Braitenberg C, Zadro M, Fang J, et al. The gravity and isostatic Moho undulations in Qinghai–Tibet plateau[J]. Journal of Geodynamics. 2000;**30**(5): 489-505

[76] Meng L, Gao R, Zhou F, et al. Study on Yadong-Golmud crustal structure by gravity anomaly [J]. Acta Geologica Sinica. 1990;**2**:149-161

[77] Peng C, Gao R. Study on the Lateral Variations of the Lithosphere/Asthenosphere Structure in Mainland China and Adjacent Seas [J]. Beijing: Seismological Press; 2000

[78] Wang Q, An Y. Gravity field and deep structure of madoi-shama region in eastern Qinghai-xizang (Tibetan) Plateau. Progress in Geophysics. 2001;**16** (4):4-10 (in Chinese)

[79] Liu H, Kong X, Ma X, et al. Physical structure of the crust in the southeast of the Qinghai-Tibet Plateau. Science in

China (Series D). 2001;**31**(B12):61-65 (in Chinese)

[80] Yang L. Numerical simulation of crust-mantle deformation and stratospheric coupling over the Tibetan Plateau [doctoral dissertation]. Institute of Geology and Geophysics, Chinese Academy of Sciences; 2002

[81] Parker RL. Inverse theory with grossly inadequate data[J]. Geophysical Journal International. 1972;**29**(2):123-138

[82] Fang J, Xu H. Three-dimensional density structure of lithospheric layers in the Tibetan Plateau and its adjacent areas. Chinese Journal of Geophysics. 1997;**40**(5):660-666 (in Chinese)

[83] Fang J. Isostatic gravity anomalies and their geodynamic characteristics in China and adjacent regions [Ph. D. thesis]. China Earthquake Administration: Institute of Geology; 2006

[84] Nafe JE, Drake CL. Physical Properties of Marine Sediments (No. TR-2). Lamont Geological Observatory Palisadesny; 1961

[85] Bao X, Song X, Xu M, Wang L, Sun X, Mi N, et al. Crust and upper mantle structure of the North China Craton and the NE Tibetan Plateau and its tectonic implications. Earth and Planetary Science Letters. 2013;**369**:129-137

[86] Bao X, Sun X, Xu M, Eaton DW, Song X, Wang L, et al. Two crustal low-velocity channels beneath SE Tibet revealed by joint inversion of Rayleigh wave dispersion and receiver functions. Earth and Planetary Science Letters. 2015;**415**:16-24

[87] Li J, Song X. Tearing of Indian mantle lithosphere from high-resolution seismic images and its implications for lithosphere coupling in southern Tibet. Proceedings of the National Academy of Sciences. 2018;**115**(33):8296-8300

[88] Jin Y, McNutt MK, Zhu YS. Mapping the descent of Indian and Eurasian plates beneath the Tibetan Plateau from gravity anomalies. Journal of Geophysical Research: Solid Earth. 1996;**101**(B5): 11275-11290

[89] Mishra DC, Ravi Kumar M. Geodynamics of Indian Plate and Tibet: Buoyant lithosphere, rapid drift and channel flow from gravity studies. Memoirs of the Geological Survey of India. 2008;**68**:151-172

[90] Yang SF, Li Z, Chen H, Santosh M, Dong CW, Yu X. Permian bimodal dyke of Tarim Basin, NW China: Geochemical characteristics and tectonic implications. Gondwana Research. 2007;**12**(1):113-120

[91] Kind R, Yuan X, Saul J, Nelson D, Sobolev SV, Mechie J, et al. Seismic images of crust and upper mantle beneath Tibet: Evidence for Eurasian plate subduction. Science. 2002;**298** (5596).1219-1221

[92] Li Y, Tian X, Wu Q, et al. The Poison ratio and crustal structure of the central Qinghai-Xizang inferred from INDEPTH-III teleseismic waveforms: Geological and geophysical implications. Chinese Journal of Geophysics. 2006;**49** (4):1037-1044 (in Chinese)

[93] Chen W, Tenzer R. Moho modeling using FFT technique. Pure and Applied Geophysics. 2017;**174**(4):1743-1757

[94] Laske G, Masters G, Ma Z, Pasyanos ME. CRUST1. 0: An updated global model of Earth's crust. Geophysical Research Abstracts. 2021;**14**:3743

[95] Laske G, Masters G, Ma Z, Pasyanos M. Update on CRUST1.0—A 1-degree

global model of Earth's crust. In: Geophysical Research Abstracts. Vol. 15. Vienna, Austria: EGU General Assembly; 2013. p. 2658

[96] Goetze C, Evans B. Stress and temperature in the bending lithosphere as constrained by experimental rock mechanics. Geophysical Journal International. 1979;**59**(3):463-478

[97] Brace W, Kohlstedt DL. Limits on lithospheric stress imposed by laboratory experiments. Journal of Geophysical Research: Solid Earth. 1980;**85**(B11): 6248-6252

[98] Ranalli G, Murphy DC. Rheological stratification of the lithosphere. Tectonophysics. 1987;**132**(4): 281-295

[99] Chen W, Molnar P. Focal depths of intracontinental and intraplate earthquakes and their implications for the thermal and mechanical properties of the lithosphere. Journal of Geophysical Research: Solid Earth. 1983;**88**(B5): 4183-4214

[100] Chen W, Yang Z. Earthquakes beneath the Himalayas and Tibet: Evidence for strong lithospheric mantle. Science. 2004;**304**(5679):1949-1952

[101] Xu Z, Wu Z, Li H, et al. The most rapid respond to a large earthquake—— The Wenchuan earthquake fault scientific drilling project. Chinese Journal of Geophysics. 2018;**61**(5):1666-1679 (in Chinese)

[102] Li H, Xu Z, Ma S, et al. Research progress on faulting and dynamic processes of the 2008 Wenchuan and 2017 Jiuzhaigou earthquakes: Lo commemorate the 10th anniversary of the Wenchuan earthquake. Chinese Journal of Geophysics. 2018;**61**(5):1653-1665 (in Chinese)

[103] Rey P, Vanderhaeghe O, Teyssier C. Gravitational collapse of the continental crust: Definition, regimes and modes. Tectonophysics. 2001;**342**(3-4):435-449

[104] Liu Q, Van Der Hilst R, Li Y, et al. Eastward expansion of the Tibetan Plateau by crustal flow and strain partitioning across faults. Nature Geoscience. 2014;**7**(5):361

[105] Sun X, Bao X, Xu M, Eaton DW, Song X, Wang L, et al. Crustal structure beneath SE Tibet from joint analysis of receiver functions and Rayleigh wave dispersion. Geophysical Research Letters. 2014;**41**(5):1479-1484

[106] Bao X, Shen Y. Foundering of the lithospheric mantle under the Eastern Tibetan Plateau revealed by full-wave Pn tomography. In: AGU Fall Meeting Abstracts. 2017

[107] Chen M, Niu F, Tromp J, et al. Lithospheric foundering and underthrusting imaged beneath Tibet. Nature Communications. 2017;**8**:15659

[108] Chen M, Niu F, Tromp J, et al. Publisher correction: Lithospheric foundering and underthrusting imaged beneath Tibet. Nature Communications. 2018;**9**(1):3443

[109] Clark MK, Royden LH. Topographic ooze: Building the eastern margin of Tibet by lower crustal flow. Geology. 2000;**28**(8):703-706

[110] Schoenbohm LM, Burchfiel BC, Liangzhong C. Propagation of surface uplift, lower crustal flow, and Cenozoic tectonics of the southeast margin of the Tibetan Plateau. Geology. 2006;**34**(10): 813-816

[111] Wang C, Lou H, Silver PG, et al. Crustal structure variation along 30 N in the eastern Tibetan Plateau and its

tectonic implications. Earth and Planetary Science Letters. 2010; **289**(3-4):367-376

[112] England P, Molnar P. Active deformation of Asia: From kinematics to dynamics. Science. 1997a;**278**(5338): 647-650

[113] England P, Molnar P. The field of crustal velocity in Asia calculated from quaternary rates of slip on faults. Geophysical Journal International. 1997b;**130**(3):551-582

[114] Chen M, Niu F, Liu Q, et al. Multiparameter adjoint tomography of the crust and upper mantle beneath East Asia: 1. Model construction and comparisons. Journal of Geophysical Research: Solid Earth. 2015;**120**(3): 1762-1786

[115] Poirier JP. Introduction to the Physics of the Earth's Interior. Cambridge University Press; 2000

[116] Teng J. The research of deep physics of Earth's interior and dynamics in China: The sexten major thesis evidences and scientifc quide. Progress in Geophysics. 2009;**3**:002 (in Chinese)

[117] Li S. Crustal structure and movement (in Chinese). Science in China (Series A). 1973a;**16**(4):400

[118] Li S. Introduction to Geomechanics. China: Science Press; 1973b

[119] Zoback MD. State of stress in the lithosphere. Reviews of Geophysics. 1983;**21**(6):1503-1511

[120] Artemjev ME, Kaban MK, Kucherinenko VA, et al. Subcrustal density inhomogeneities of Northern Eurasia as derived from the gravity data and isostatic models of the lithosphere. Tectonophysics. 1994;**240**(1-4):249-280

[121] Kaban MK, Schwintzer P, Reigber C. A new isostatic model of the lithosphere and gravity field. Journal of Geodesy. 2004;**78**(6):368-385

Chapter 4

Concealed Active Fault Related to Surface Deformation Patterns Detected through Satellite SAR Differential Interferometry

Yasuto Itoh and Takumi Onuma

Abstract

Satellite synthetic aperture radar (SAR) differential interferometry has successfully depicted surface deformation areas around the termination of an active transcurrent fault in southwest Japan, which is under strong tectonic stresses linked to recent oblique subduction of the Philippine Sea Plate. Sentinel-1 CSAR data were used to generate and stack images to create interferograms for the period from May 2015 to November 2021. Among the ongoing complicated elevation increases and decreases distributed over alluvial plains surrounding Osaka Bay, an enormous neotectonic basin, a rhomboidal area of upheaval was clearly recognized around an urban district. Geological information demonstrates that the conspicuous superficial displacement is in perfect agreement with the Riedel shear structures within the Quaternary unconsolidated sediment piles provoked by right-lateral intermittent slips on the deep-rooted Koyo Fault, which has recently had a high level of activity. Analyses for gravity anomalies showed that the Koyo Fault is accompanied by a noteworthy horizontal gravity gradient, suggesting a sharp density contrast across the fault plane. Additional satellite SAR data from the 1990s, obtained by ERS1/2 satellites, have also showed the growth of a rhomboidal structure postdating the 1995 Great Hanshin Earthquake and imply a cumulative buildup of the active landform under vigorous tectonism.

Keywords: differential interferometry SAR, surface displacement, active fault, neotectonics, Southwest Japan

1. Introduction

Because of their cloud-penetrating ability, frequent observation opportunities, and superb resolution, satellites equipped with synthetic aperture radar (SAR) are regarded as a centerpiece of remote sensing of Earth's vast surface. For example, instantaneous land motions have been successfully detected in connection with the migration of formation fluid [1], volcanic activity [2], and pre-/postseismic changes in deformation trends [3, 4]. Itoh and Onuma [5] analyzed the interferometric SAR

(InSAR) data around Osaka Bay, southwest Japan, in exacting detail. These motions indicate that subtle surface deformations with multi-year accumulation possess a tight linkage with the longstanding deep-rooted structural buildup. This paper focuses on a conspicuous but unsolved topographic anomaly within the southwest Japan study area and aims to elucidate its origin based on an integrated geological and geophysical study. Our approach paves a pathway to quantitative tectonic research using satellite-borne SAR data.

2. Background

2.1 Regional geology

Figure 1 depicts the area of differential InSAR (DInSAR) analysis executed by Itoh and Onuma [5]. It is about 100 km square and covers the Quaternary sedimentary basins surrounding Osaka Bay. Among the significant active faults shown in the figure, the Median Tectonic Line (MTL) has an exceptionally long active period since ≈100 Ma and stands as an arc-bisecting fault. The northern side of the MTL is called the Inner Zone, and it consists of a Paleozoic/Mesozoic accretionary complex intruded by igneous rocks and younger sedimentary rocks, whereas the southern side (Outer Zone) is a Mesozoic/Cenozoic gigantic accretionary prism partly subjected to high-pressure metamorphism.

As for the neotectonic context of southwest Japan, including the present study area, pioneering works were presented by Huzita [6, 7]. He ascribed the Quaternary

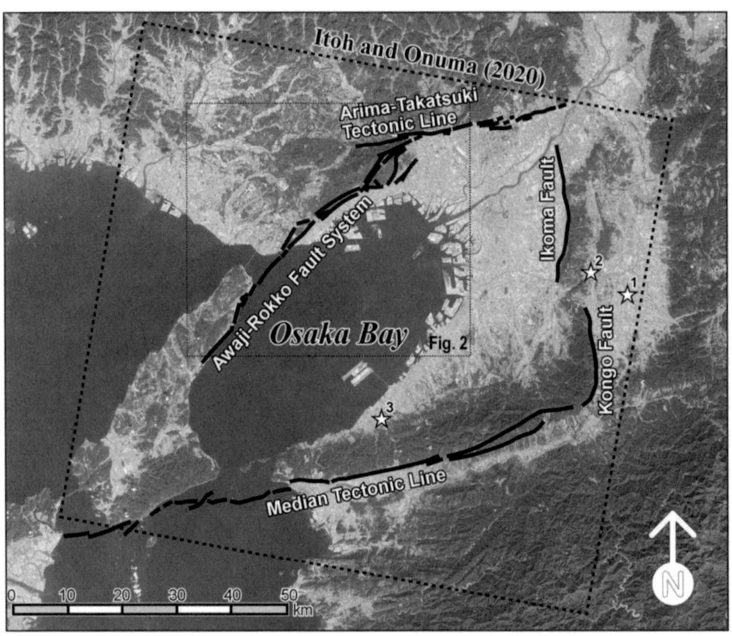

Figure 1.
Area of DInSAR analysis in southwest Japan by Itoh and Onuma [5]. Their observation points of remarkable deformation are denoted as numbered stars. Thick lines are major active fault systems. Background image: Landsat 8 OLI Panchromatic Band acquired on 2020/10/27. Landsat 8 OLI image data courtesy of the U.S. Geological Survey.

tectonism of this island arc to E-W compressive stress provoked by subduction of the Pacific Plate and defined a middle portion of the arc with intensive faulting as the Kinki Triangle, the southwestern part of which coincides with the study area. On the other hand, Itoh et al. [8] pointed out that the late Pleistocene change in the converging direction of the Philippine Sea Plate activated strike-slips on the arc-parallel active faults. Motion sensor data on the notable faults in **Figure 1** (Awaji-Rokko Fault System, Arima-Takatsuki Tectonic Line, Ikoma Fault, Kongo Fault, and MTL) are concordant with these tectonic schemes.

Huzita and Kasama [9] presented a reliable estimate on the activity level of the Awaji-Rokko Fault System, the main target of our DInSAR analysis, based on displacements of key beds within the Quaternary Osaka Group. They found that slip rates for the major NE-SW components (Gosukebashi, Ashiya, and Koyo Faults) of the regional fault system are approximately 0.1 mm/yr. It was also confirmed that the sum of individual movements exceeds 1 mm/yr., which is regarded as the highest level of active movement in the Japanese Islands.

2.2 Seismic surveys

Itoh and Takemura [10] performed three-dimensional structural analysis on a diverse dataset of reflection seismic lines deployed within Osaka Bay. They clarified the development processes of an asymmetric basin deepening toward the Osaka Bay Fault off the eastern coast of Awaji Island. In addition, geological surveys of seismic profiles obtained from the surrounding land areas were sporadically attempted for the Osaka Plain [11] and vicinity of the MTL [12], but a comprehensive interpretation has not yet been established. As for the south flank of the Rokko Mountains, which is evaluated in this study, Yokota et al. [13] reported the outline of a series of seismic shootings, which were planned after the disaster of the 1995 Great Hanshin Earthquake (Hyogo-ken Nanbu Earthquake).

3. Satellite SAR differential interferometry

3.1 Sentinel-1 CSAR dataset

Generation and stacking of interferogram were executed using 11 scenes of Sentinel-1 CSAR data (wavelength of 5.6 cm), acquired with descending orbits, from November 2014 to October 2019 (see **Table 1**). Due to microwave scattering of C-band radar by branches and leaves, coherence levels in vegetated areas remain low, whereas urban areas covered by solid structures are characterized by high coherence and provide precise estimates of surface movements. To reduce the phase noise and improve the coherence of an interferogram, a multi-look process was applied for all the interferometry pairs with eight looks in the range direction and two looks in the azimuth direction, resulting in a ground pixel size of 30 m square. A 10-m mesh digital elevation model (DEM) was used for the reference datum.

The Sentinel-1 satellite has a revisit cycle of 12 days. In total, Itoh and Onuma [5] prepared 28 interferometry pairs with an observation interval longer than 600 days (50 cycles, **Table 2**). A refined baseline was estimated based on the relationship between the interferometric phase and the range distance for ground control points selected from the survey area having a coherence higher than 0.7. After this analytical sequence, they found several points with time-progressive surface deformations,

Orbit	Year	Month	Day	Bp(m)	Days since 2014/11/23
3414	2014	11	23	88	0
5864	2015	5	10	−39	168
8664	2015	11	18	−47	360
11,464	2016	5	28	−57	552
13,914	2016	11	12	−1	720
16,714	2017	5	23	0	912
19,339	2017	11	19	32	1092
21,964	2018	5	18	−45	1272
24,064	2018	10	9	4	1416
27,389	2019	5	25	−36	1644
29,314	2019	10	4	10	1776

Bp: Perpendicular baseline of the interferogram with the scene of 2017/5/23 as the master data.

Table 1.
Sentinel-1 CSAR data used.

which are depicted by stars in **Figure 1**. Land motions around stars 1, 2, and 3 are related to reactivation of the Miocene backarc rift system under strong neotectonic stress, confined deformation at the termination of a strike-slip active fault, and an active undulation within a Quaternary basin, respectively. In contrast, a peculiar rhomboidal area of upheaval around Takarazuka (**Figure 2**) was out of their research

No.	First	Second	Bp(m)	Days
1	23-11-2014	12-11-2016	−89	720
2	23-11-2014	23-05-2017	−88	912
3	23-11-2014	19-11-2017	−56	1092
4	23-11-2014	18-05-2018	−133	1272
5	23-11-2014	09-10-2018	−84	1416
6	23-11-2014	25-05-2019	−123	1644
7	23-11-2014	04-10-2019	−78	1776
8	10-05-2015	23-05-2017	39	744
9	10-05-2015	19-11-2017	71	924
10	10-05-2015	18-05-2018	−7	1104
11	10-05-2015	09-10-2018	43	1248
12	10-05-2015	25-05-2019	3	1476
13	10-05-2015	04-10-2019	49	1608
14	18-11-2015	19-11-2017	79	732
15	18-11-2015	18-05-2018	2	912
16	18-11-2015	09-10-2018	51	1056
17	18-11-2015	25-05-2019	12	1284
18	18-11-2015	04-10-2019	57	1416
19	28-05-2016	18-05-2018	12	720

No.	First	Second	Bp(m)	Days
20	28-05-2016	09-10-2018	61	864
21	28-05-2016	25-05-2019	21	1092
22	28-05-2016	04-10-2019	67	1224
23	12-11-2016	09-10-2018	5	696
24	12-11-2016	25-05-2019	−34	924
25	12-11-2016	04-10-2019	11	1056
26	23-05-2017	25-05-2019	−36	732
27	23-05-2017	04-10-2019	10	864
28	19-11-2017	04-10-2019	−22	684

Table 2.
Interferometry pairs used for stacking operation. Pairs with the interval longer than 600 days were chosen.

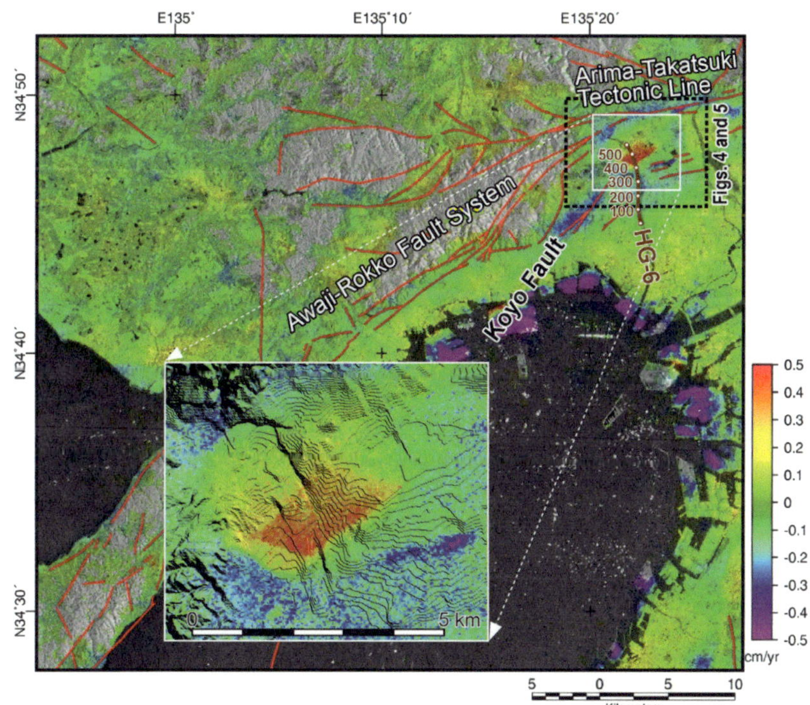

Figure 2.
Vertical displacement rate during the monitoring period of Itoh and Onuma [5] around Takarazuka City. See **Figure 1** *for the mapped area. Red lines are active faults documented by the Research Group for Active Faults [14]. The thick brown line represents a reflection seismic survey track (HG-6) across the rhomboidal upheaval. The inset map shows DEM data as gray contours.*

scope because linkages between deep-rooted structure and surface distribution of active faults (e.g., [14]) were not established.

In the present study, we added Sentinel-1 CSAR data with ascending orbits from May 2015 to November 2021. Data for 14 scenes from the ascending orbit and 15 scenes from the descending orbit were collected at about a half-year interval. As a result, 80 pairs of the descending and ascending orbits, with observation intervals spanning from 132 days to 2370 days (**Tables 3** and **4**), were processed through the

#	First	Second	Bp(m)	ΔT(days)	#	First	Second	Bp(m)	ΔT(days)	#	First	Second	Bp(m)	ΔT(days)
1	20150510	20151025	20.072	168	31	20160528	20170523	57.0943	360	61	20171119	20211110	−82.9625	1452
2	20150510	20160528	−18.1966	384	32	20160528	20181009	60.9751	864	62	20180518	20181009	49.3586	144
3	20150510	20170523	38.8977	744	33	20160528	20190525	21.4526	1092	63	20180518	20190525	9.8361	372
4	20150510	20171119	70.6766	924	34	20160528	20191004	67.1838	1224	64	20181009	20190525	−39.5225	228
5	20150510	20181009	42.7785	1248	35	20160528	20200519	47.2486	1452	65	20181009	20191004	6.2087	360
6	20150510	20190525	3.256	1476	36	20160528	20201115	92.4714	1632	66	20181009	20201115	31.4963	768
7	20150510	20191004	48.9872	1608	37	20160528	20210514	35.3122	1812	67	20181009	20210514	−25.6629	948
8	20150510	20200519	29.052	1836	38	20160528	20211110	5.9107	1992	68	20181009	20211110	−55.0644	1128
9	20150510	20201115	74.2748	2016	39	20161112	20170523	1.3571	192	69	20190525	20191004	45.7312	132
10	20150510	20210514	17.1156	2196	40	20161112	20171119	33.136	372	70	20190525	20200519	25.796	360
11	20150510	20211110	−12.2859	2376	41	20161112	20181009	5.2379	696	71	20190525	20210514	13.8596	720
12	20151025	20160528	−38.2686	216	42	20161112	20190525	−34.2846	924	72	20190525	20211110	−15.5419	900
13	20151025	20161112	17.4686	384	43	20161112	20191004	11.4466	1056	73	20191004	20200519	−19.9352	228
14	20151025	20171119	50.6046	756	44	20161112	20200519	−8.4886	1284	74	20191004	20201115	25.2876	408
15	20151025	20181009	22.7065	1080	45	20161112	20201115	36.7342	1464	75	20191004	20211110	−61.2731	768
16	20151025	20190525	−16.816	1308	46	20161112	20210514	−20.425	1644	76	20200519	20201115	45.2228	180
17	20151025	20191004	28.9152	1440	47	20161112	20211110	−49.8265	1824	77	20200519	20210514	−11.9364	360
18	20151025	20200519	8.98	1668	48	20170523	20171119	31.7789	180	78	20201115	20210514	−57.1592	180
19	20151025	20201115	54.2028	1848	49	20170523	20180518	−45.4778	360	79	20201115	20211110	−86.5607	360
20	20151025	20210514	−2.9564	2028	50	20170523	20190525	−35.6417	732	80	20210514	20211110	−29.4015	180
21	20151025	20211110	−32.3579	2208	51	20170523	20191004	10.0895	864					
22	20151118	20171119	78.9766	732	52	20170523	20200519	−9.8457	1092					

#	First	Second	Bp(m)	ΔT(days)	#	First	Second	Bp(m)	ΔT(days)	#	First	Second	Bp(m)	ΔT(days)
23	20151118	20181009	51.0785	1056	53	20170523	20201115	35.3771	1272					
24	20151118	20190525	11.556	1284	54	20170523	20210514	−21.7821	1452					
25	20151118	20191004	57.2872	1416	55	20170523	20211110	−51.1836	1632					
26	20151118	20200519	37.352	1644	56	20171119	20180518	−77.2567	180					
27	20151118	20201115	82.5748	1824	57	20171119	20181009	−27.8981	324					
28	20151118	20210514	25.4156	2004	58	20171119	20200519	−41.6246	912					
29	20151118	20211110	−3.9859	2184	59	20171119	20201115	3.5982	1092					
30	20160528	20161112	55.7372	168	60	20171119	20210514	−53.561	1272					

Table 3.
Interferometry pairs used for stacking operation (Sentinel-1 descending orbit).

#	First	Second	Bp(m)	ΔT(days)	#	First	Second	Bp(m)	ΔT(days)	#	First	Second	Bp(m)	ΔT(days)
1	20150522	20151013	45.7912	144	31	20160516	20201121	−58.5818	1650	61	20180512	20201121	−59.0186	924
2	20150522	20160516	−14.9318	360	32	20160516	20210520	−60.085	1830	62	20180512	20210520	−60.5218	1104
3	20150522	20170517	−62.3211	726	33	20160516	20211116	41.5297	2010	63	20180512	20211116	41.0929	1284
4	20150522	20171113	30.4114	906	34	20161130	20170517	−146.3742	168	64	20181120	20190624	−38.2359	216
5	20150522	20180512	−14.495	1086	35	20161130	20171113	−53.6417	348	65	20181120	20191115	−35.8056	360
6	20150522	20181120	29.0232	1278	36	20161130	20181120	−55.0299	720	66	20181120	20201121	−102.5368	732
7	20150522	20190624	−9.2127	1494	37	20161130	20190624	−93.2658	936	67	20181120	20210520	−104.04	912
8	20150522	20191115	−6.7824	1638	38	20161130	20191115	−90.8355	1080	68	20181120	20211116	−2.4253	1092
9	20150522	20200525	16.7906	1830	39	20161130	20200525	−67.2625	1272	69	20190624	20191115	2.4303	144
10	20150522	20201121	−73.5136	2010	40	20161130	20201121	−157.5667	1452	70	20190624	20200525	26.0033	336
11	20150522	20210520	−75.0168	2190	41	20161130	20210520	−159.0699	1632	71	20190624	20210520	−65.8041	696
12	20150522	20211116	26.5979	2370	42	20161130	20211116	−57.4552	1812	72	20190624	20211116	35.8106	876
13	20151013	20160516	−60.723	216	43	20170517	20171113	92.7325	180	73	20191115	20200525	23.573	192
14	20151013	20161130	38.2619	414	44	20170517	20180512	47.8261	360	74	20191115	20201121	−66.7312	372
15	20151013	20171113	−15.3798	762	45	20170517	20190624	53.1084	768	75	20191115	20211116	33.3803	732
16	20151013	20180512	−60.2862	942	46	20170517	20191115	55.5387	912	76	20200525	20201121	−90.3042	180
17	20151013	20181120	−16.768	1134	47	20170517	20200525	79.1117	1104	77	20200525	20210520	−91.8074	360
18	20151013	20190624	−55.0039	1350	48	20170517	20201121	−11.1925	1284	78	20201121	20210520	−1.5032	180
19	20151013	20191115	−52.5736	1494	49	20170517	20210520	−12.6957	1464	79	20201121	20211116	100.1115	360
20	20151013	20200525	−29.0006	1686	50	20170517	20211116	88.919	1644	80	20210520	20211116	101.6147	180
21	20151013	20201121	−119.3048	1866	51	20171113	20180512	−44.9064	180					
22	20151013	20210520	−120.808	2046	52	20171113	20181120	−1.3882	372					

#	First	Second	Bp(m)	ΔT(days)	#	First	Second	Bp(m)	ΔT(days)	#	First	Second	Bp(m)	ΔT(days)
23	20151013	20211116	−19.1933	2226	53	20171113	20191115	−37.1938	732					
24	20160516	20161130	98.9849	198	54	20171113	20200525	−13.6208	924					
25	20160516	20170517	−47.3893	366	55	20171113	20201121	−103.925	1104					
26	20160516	20180512	0.4368	726	56	20171113	20210520	−105.4282	1284					
27	20160516	20181120	43.955	918	57	20171113	20211116	−3.8135	1464					
28	20160516	20190624	5.7191	1134	58	20180512	20181120	43.5182	192					
29	20160516	20191115	8.1494	1278	59	20180512	20190624	5.2823	408					
30	20160516	20200525	31.7224	1470	60	20180512	20200525	31.2856	744					

Table 4.
Interferometry pairs used for stacking operation (Sentinel-1 ascending orbit).

DInSAR analytical flow. After stacking, the line-of-sight (LOS) displacement rate was calculated for every orbit, and a 2.5-dimensional evaluation was executed based on the geometric relation of the orbits for the isolation of quasi-vertical/quasi-lateral (E-W) movements.

3.2 ERS1/2 AMI dataset

To evaluate land movements for longer periods, we used data from another satellite-borne SAR. The ERS1 and 2 satellites were equipped with a C-band radar (wavelength of ≈6 cm). ERS1 was operated from 1991 to 2000, and the ERS2 mission began in 1995 and ended in 2011. Fifteen scenes of ERS1/2 Active Microwave Instrument (AMI) data from May 1992 to March 2000 (**Table 5**) were analyzed for 22 interferometry pairs having an observation interval longer than ≈300 days (**Table 6**). The data coverage is equivalent to the northern part of the survey area of Itoh and Onuma [5]. A multi-look process was applied for all the interferometry pairs, with one look in the range direction and five looks in the azimuth direction, resulting in a ground pixel size of 20 m square. After the DInSAR analytical procedures were applied, LOS displacements around the rhomboidal landform anomaly around Takarazuka showed a clear tendency approaching the satellite postdating the 1995 Great Hanshin Earthquake (Hyogo-ken Nanbu Earthquake). The tectonic context of these results obtained from our DInSAR study is discussed in the following sections.

Satellite	Year	Month	Day
ERS1	1992	5	10
ERS1	1992	9	27
ERS1	1992	11	1
ERS1	1993	7	4
ERS1	1993	8	8
ERS1	1993	9	12
ERS1	1995	5	23
ERS1	1995	6	27
ERS1	1995	8	1
ERS1	1996	2	27
ERS1	1996	4	2
ERS2	1996	4	3
ERS1	1996	5	7
ERS2	1997	9	10
ERS2	2000	3	8

Table 5.
ERS1/2 AMI Data used.

First	Second	Bp(m)	Days
Periods before the Hyogo-ken Nambu Earthquake			
10-05-1992	08-08-1993	22	455
10-05-1992	12-09-1993	409.5	490
27-09-1992	04-07-1993	−566.3	280
27-09-1992	08-08-1993	367.9	315
01-11-1992	08-08-1993	−529.4	280
01-11-1992	12-09-1993	−141.9	315
Periods after the Hyogo-ken Nambu Earthquake			
23-05-1995	02-04-1996	−127.7	315
23-05-1995	03-04-1996	−239.8	316
23-05-1995	08-03-2000	131.4	1751
01-08-1995	10-09-1997	−67.3	771
02-04-1996	08-03-2000	259	1436
10-09-1997	08-03-2000	−296.6	910
Periods spanning over the Hyogo-ken Nambu Earthquake			
10-05-1992	01-08-1995	57.4	1178
10-05-1992	10-09-1997	−9.9	1949
27-09-1992	23-05-1995	−92.6	968
27-09-1992	08-03-2000	38.7	2719
01-11-1992	27-02-1996	−101.7	1213
01-11-1992	07-05-1996	−39.3	1283
08-08-1993	01-08-1995	35.1	723
08-08-1993	10-09-1997	−32.2	1494
12-09-1993	27-02-1996	40.6	898
12-09-1993	07-05-1996	102.9	968

Table 6.
Interferometry pairs used for stacking operation of ERS1/2 AMI.

4. Surface deformation pattern and geological structure around Takarazuka

The area of uplift in Takarazuka is divided from the surrounding areas by a sharp quadrangular border. This topographic anomaly is not simply interpreted as a landform related to the fluvial terrace because it spans the Mukogawa River. The anomaly is also not attributed to other superficial undulations because it has no correlation with the DEM data (see the inset of **Figure 2**). As for the information on the local subsurface structure, a line of the reflection seismic survey by Yokota et al. [13] (HG-6, **Figure 2**) ran over the noted upheaval. Shooting on the seismic line, which is located along the river reservation apart from the residential quarter, was conducted at 70–90% of the maximum power of the vibrator (IVI Y-2400) and achieved relatively high resolution. **Figure 3** shows that the top of the acoustic basement in the profile deepens southward

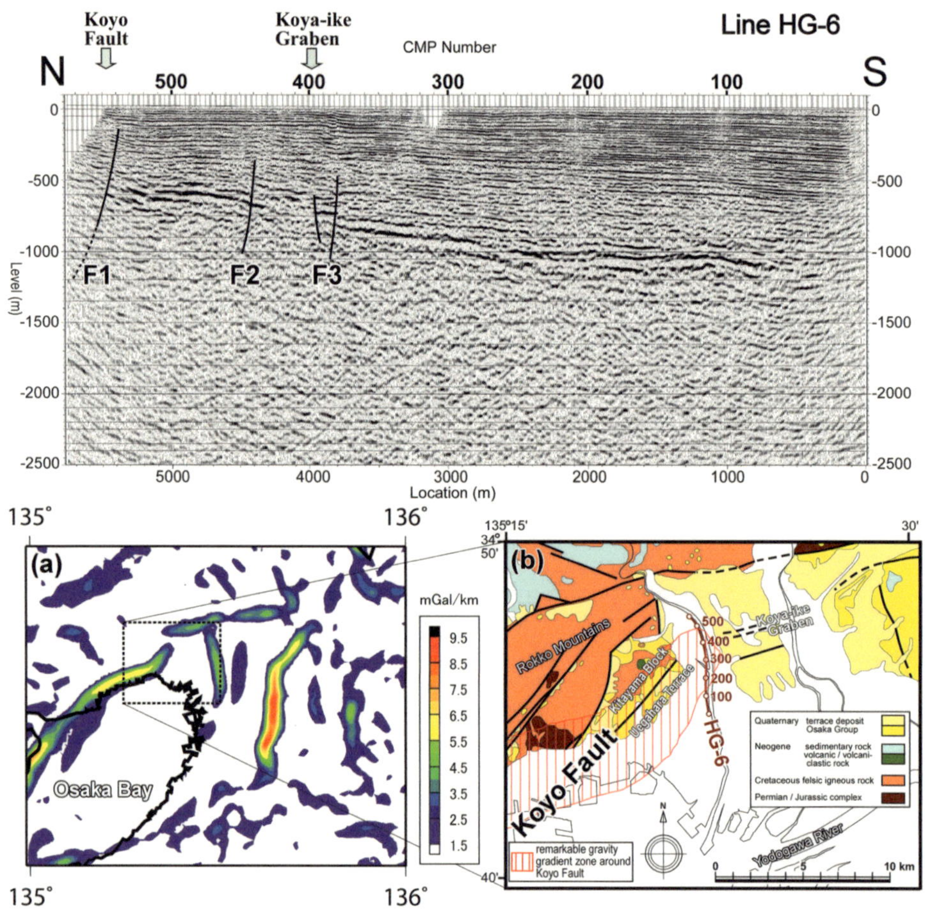

Figure 3.
Seismic section (HG-6) across the rhomboidal upheaval without vertical exaggeration. See **Figure 2** *for the line location. The geologic interpretation (Koyo Fault and Koya-ike Graben) attached to the section is after Yokota et al. [13]. The structural interpretation of the present study (fractures F1, F2, and F3) is depicted in the section. (a) Horizontal gravity gradient map around Osaka Bay [15] for values over 1.5 mGal/km. (b) Index for gravity gradient, surface geology, and the discussed seismic line location. The geologic map is compiled after Kawada et al. [16].*

up to 1000 m. Although the stratigraphic ages and structural buildup processes of the basin fill are hardly elucidated on this isolated section, the overlying strata are probably part of the Plio-Pleistocene Osaka Group that has a lack of cumulative deformation around the observed rhomboidal bulge (CMP No. 410–520).

Huzita and Kasama [9] presented a geological description of the Koyo Fault and Koya-ike Graben depicted in **Figure 3**. The Koyo Fault defines the front line of the Awaji-Rokko Fault System running along the boundary between the Kitayama Block and Uegahara Terrace, whereas its northeast extension under the alluvial plain is obscured. The Osaka Group adjacent to this fault suffers complicated deformation and slips recorded on basement granitic rocks in accord with reverse faulting. In contrast with a subtle structural disturbance on the seismic profile, analyses based on gravity anomalies provide us with insightful features originating from deep-rooted fault architecture. **Figure 3a** shows the horizontal gravity gradient map around Osaka Bay [15]. The map clearly shows that the Awaji-Rokko Fault System, including the Koyo Fault, corresponds to an abrupt change in the gravitational trend, the maximum of which reaches 7.0 mGal/km, and its far northeastern tip reaches the alluvium (see the geologic map [16] in **Figure 3b**). As suggested by Kusumoto [17], the horizontal first derivation of the Bouguer anomaly is effective for detection of faults and/or tectonic lines dividing regional geologic provinces. Hence, the high gravity gradient around the marked rhomboidal upheaval implies density contrast within basement rocks across a strike-slip boundary concealed under unconsolidated sediment.

Koya-ike Graben is an E-W valley having many small reservoirs and is interpreted as a depression linked to subsurface transcurrent motions. In an active fault database [18], the Koyo Fault and the Koya-ike Graben are regarded as parts of the Nishinomiya (reverse faulting) and Itami (dextral faulting) segments, respectively, and are unified into the regional Rokko seismogenic fault system.

5. Fault movements related to surface deformation

As mentioned in the previous section, the growing bulge around Takarazuka detected through our DInSAR analysis possesses a quite sharp rhomboidal outline, which implies structural development driven by tectonic stress. However, a causal relationship between the long-term prevailing stress field and real-time ground motion has not been established. From a regional point of view, the NE-SW-trending Awaji-Rokko Fault System is a basically compressive crustal rupture that is right-laterally activated under an E-W stress provoked by the Quaternary plate motion [7]. Generally speaking, a strike-slip master fault cutting through basement rocks is accompanied by an initial set of subsidiary conjugate faults in cover beds, which are called Riedel shears (e.g., [19]). **Figure 4** delineates a hypothetical master fault under

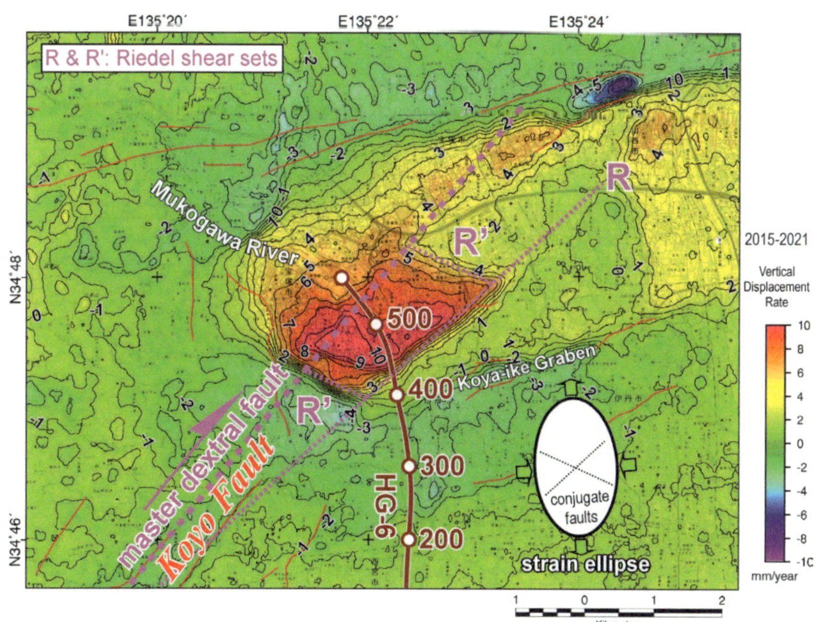

Figure 4.
Probable structural interpretation of the conspicuous vertical displacement pattern around Takarazuka City. See **Figure 2** *for the mapped area. Magenta broken lines represent a Riedel shear set (R and R') and master dextral fault. The base map contours show the quasi-vertical displacement rate. Red lines are active faults after the Active Fault Map in the Urban Area "Osaka-Seihokubu" [20].*

the alluvial plain as an extension of the Koyo Fault together with its Riedel shears (R and R'). Azimuth angles of the fault set and the outline of the distinct upheaval exhibit a striking coincidence, which proves their causal linkage. Although deep-seated transcurrent faults are associated with upward diverging flower structures in the sediment cover (e.g., [21]), cumulative separation on the master fault in this study area is still under the level that compels development of such features discernible on seismic profiles because the recent tectonic stress in southwest Japan has persisted since ≈1 Ma. Yokota et al. [13] exclusively recognized the Koyo Fault and Koya-ike Graben on the seismic record shown in **Figure 3**, though these structures in **Figure 4** [20] do cross over the survey line. We identified three fractures (F1-F3, see **Figure 3**), among which F2 is located at the southeast border of the topographic anomaly parallel to the Riedel shear (R). Thus, interpretation of subtle stratal deformation should be reexamined based on high-resolution seismic data.

Our DInSAR study demonstrated that accurate sensing of land surface is effective for detection of significant tectonic movements with cumulative crustal deformation. During our analytical period, the Northern Osaka Earthquake (M6.1) occurred on June 18, 2018, the epicenter of which is located in the northeastern Osaka Plain. Fujiwara et al. [22] stated that surface displacement along the Arima-Takatsuki Tectonic Line was detected by means of InSAR analysis of ALOS-2 satellite data. Their region of interest is, however, outside our study area. However, Hashimoto [23] analyzed Sentinel-1 SAR data for the same period and stated that no LOS displacement was found related to the damaging earthquake. Morishita [24], who performed nationwide prolonged ground monitoring based on Sentinel-1 data, also found no direct linkage between the earthquake event and surface motions around the distant Takarazuka City.

Hashimoto [23] pointed out that the Northern Osaka Earthquake (M6.1) was too small and deeply focused to detect significant coseismic deformation based

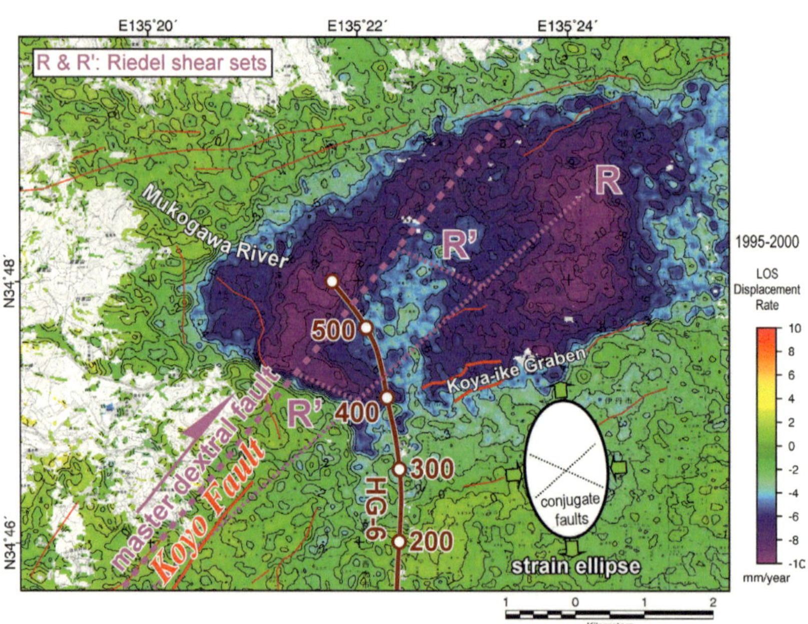

Figure 5.
LOS displacement pattern around Takarazuka City after the 1995 Hyogo-ken Nanbu Earthquake (M7.3). The mapped area is the same as that in **Figure 4**.

on the C-band DInSAR analysis. Thus, we focused on the 1995 Hyogo-ken Nanbu Earthquake (M7.3) by means of the ERS1/2 AMI data. As for the seismogenic fault of this devastating event, Huzita and Sano [25] assumed a number of concealed ruptures around the area of this study. Their existence, however, has not been verified through seismic surveys (e.g., [13]). **Figure 5** presents the postseismic (1995–2000) land motions around the rhomboidal area of interest. Although the pattern of upheavals and subsidences around Takarazuka is slightly distinct from that in later periods (**Figure 4**), it seems that the differentially moving portions are bounded by Riedel flexures above the master fault. It is plausible that short-term surface displacements can serve as an indicator of subsurface structural buildup over geological timescales.

6. Conclusion

In the present study, the authors demonstrated a close relationship between long-standing land motion along a regional subsurface fault cutting through the basement rocks and instantaneous deformations of the land surface. Although the approach described in this work is generally applicable to the quantitative assessment of countless active faults in mobile belts, the tectonic context of subtle displacements detected through satellite DInSAR data should be carefully verified by means of independent geophysical and/or geological information. We have shown that the horizontal gravity gradient provides a snapshot of deep-seated crustal structures controlling surface movements. Our area of interest lies within the seriously damaged zone of the 1995 Great Hanshin Earthquake (Hyogo-ken Nanbu Earthquake). To elucidate the linkages between complicated and transient upheavals and subsidences delineated by DInSAR analysis and the heterogeneous nature of filled Quaternary basins, an in-depth examination of stratigraphic databases of civil engineering boreholes has great importance and creates a pathway to the evaluation of future seismic events.

Conflict of interest

The authors declare no conflict of interest.

Author details

Yasuto Itoh[1*] and Takumi Onuma[2]

1 Osaka Metropolitan University, Osaka, Japan

2 JGI, Inc., Tokyo, Japan

*Address all correspondence to: yasutokov@yahoo.co.jp

References

[1] Onuma T, Ohkawa S. Detection of surface deformation related with CO_2 injection by DInSAR at In Salah, Algeria. Energy Procedia. 2009;**1**:2177-2184

[2] Tobita M, Murakami M, Nakagawa H, Yarai H, Fujiwara S, Rosen PA. 3-D surface deformation of the 2000 Usu eruption measured by matching of SAR images. Geophysical Research Letters. 2001;**28**:4291-4294

[3] Ozawa S, Murakami M, Fujiwara S, Tobita M. Synthetic aperture radar interferogram of the 1995 Kobe earthquake and its geodetic inversion. Geophysical Research Letters. 1997;**24**:2327-2330

[4] Onuma T. Surface deformation patterns associated with the 2007 Noto Hanto earthquake detected by satellite SAR differential interferometry. Bulletin of the Earthquake Research Institute, University of Tokyo. 2007;**82**:361-368

[5] Itoh Y, Onuma T. Accurate Sensing of Active Movements of the Land Surface: Basic Theory of Satellite SAR Interferometry and its Application to a Mobile Belt. New York: Nova Science Publishers, Inc.; 2020. p. 75

[6] Huzita K. Role of the Median Tectonic Line in the Quaternary Tectonics of the Japanese islands. Memoirs of the Geological Society of Japan. 1980;**18**:129-153

[7] Huzita K. The Quaternary tectonic stress states of Southwest Japan. Journal of Geoscience, Osaka City University. 1976;**20**:93-103

[8] Itoh Y, Kusumoto S, Takemura K. Evolutionary process of Beppu Bay in Central Kyushu, Japan: A quantitative study of the basin-forming process controlled by plate convergence modes. Earth, Planets and Space. 2014;**66**:74. DOI: 10.1186/1880-5981-66-74

[9] Huzita K, Kasama T. Geology of the Osaka-Seihokubu District, with Geological Sheet Map at 1:50,000. Tsukuba: Geological Survey of Japan, AIST; 1982. p. 112

[10] Itoh Y, Takemura K. Three-Dimensional Architecture and Paleoenvironments of Osaka Bay: An Integrated Seismic Study on the Evolutionary Processes of a Tectonic Basin. Singapore: Springer Nature Singapore Pte Ltd.; 2019. p. 119

[11] Itoh Y, Takemura K. Subsurface structure of the Osaka plain - its perspective based on geophysical data and future problems. News of Osaka Micropaleontologists. 2016;**17**:1-74

[12] Itoh Y, Iwata T, Takemura K. Three-dimensional architecture of the Median Tectonic Line in Southwest Japan based on detailed reflection seismic and drilling surveys. In: Itoh Y, editor. Evolutionary Models of Convergent Margins: Origin of their Diversity. Rijeka: InTech; 2017. DOI: 10.5772/67434

[13] Yokota H, Ikawa T, Sano M, Takemura K. Seismic reflection profiles across the Northern part of Osaka Bay and Rokko Mountain foothill. In: Museum of Nature and Human Activities, Hyogo, editor. The Great Hanshin-Awaji Earthquake Disaster and Rokko Movements: Report on the Research of Active Tectonics around the Damaged Area by the 1995 Hyogoken-Nanbu Earthquake. Kobe: Kobe Civil Engineering Office; 1997. pp. 57-89

[14] Research Group for Active Faults. The Active Faults in Japan: Sheet Maps and Inventories. Revised ed. Tokyo: University of Tokyo Press; 1991. p. 437

[15] Itoh Y, Kusumoto S, Takemura K. Tectonically controlled asymmetric basin formation and evolution: An example from an active plate margin. In: Veress B, Szigethy J, editors. Horizons in Earth Science Research. Vol. 14. New York: Nova Science Publishers, Inc.; 2015. pp. 123-141

[16] Kawada K, Miyamura M, Yoshida F. Geological Map of Japan 1:200,000, Kyoto-Osaka. Tsukuba: Geological Survey of Japan, AIST; 1986

[17] Kusumoto S. Estimation of dip angle of fault or structural boundary by eigenvectors of gravity gradient tensors. Butsuri-Tansa. 2015;**68**:277-287

[18] National Institute of Advanced Industrial Science and Technology (AIST), editor. Active Fault Database of Japan (April 26, 2024 Version). Tsukuba: AIST; 2024. Available from: https://gbank.gsj.jp/activefault/index_e_gmap.html

[19] Katz Y, Weinberger R, Aydin A. Geometry and kinematic evolution of Riedel shear structures, Capitol Reef National Park, Utah. Journal of Structural Geology. 2004;**26**:491-501

[20] Nakata T, Okada A, Suzuki Y, Watanabe M, Ikeda Y. Active Fault Map in Urban Area 1:25,000, Osaka-Seihokubu. Tsukuba: Geospatial Information Authority of Japan; 1996

[21] Harding TP. Identification of wrench faults using subsurface structural data: Criteria and pitfalls. AAPG Bulletin. 1990;**74**:1590-1609

[22] Fujiwara S, Hayashi K, Morishita Y, Yarai H, Nakano T, Une H. Linear surface ruptures along Arima-Takatsuki fault associated with Osaka-fu-hokubu earthquake detected by ALOS-2 InSAR. In: Programme and Abstracts, Seismological Society of Japan, 2018 Fall Meeting. Saitama, Japan: Seismological Society of Japan; 2018. pp. S24-06

[23] Hashimoto M. What SAR could see and could not see: Northern Osaka and eastern Iburi, Hokkaido earthquakes. DPRI Annuals. 2019;**62B**:296-305

[24] Morishita Y. Nationwide urban ground deformation monitoring in Japan using Sentinel-1 LiCSAR products and LiCSBAS. Progress in Earth and Planetary Science. 2021;**8**:6. DOI: 10.1186/s40645-020-00402-7

[25] Huzita K, Sano M. The great Hanshin-Awaji earthquake disaster and Rokko movements. In: Museum of Nature and Human Activities, Hyogo, editor. The Great Hanshin-Awaji Earthquake Disaster and Rokko Movements: Report on the Research of Active Tectonics around the Damaged Area by the 1995 Hyogoken-Nanbu Earthquake. Kobe: Kobe Civil Engineering Office; 1997. pp. 97-106

Chapter 5

A Critical Review of Geomagnetic and Ionospheric Anomalies as Earthquake Precursors

Yiyang Luo

Abstract

This chapter presents a critical review of research on geomagnetic and ionospheric anomalies as potential precursors to earthquakes. Despite significant interest and numerous studies, these phenomena have not been conclusively proven as definitive indicators of seismic activity. The chapter explores historical doubts, instances of misjudgment, and conflicting perspectives within the scientific community. It delves into advancements in understanding these anomalies, highlighting specific cases and theoretical mechanisms. Additionally, the chapter examines the potential connection between solar activity, geomagnetic storms, and seismic events. By addressing these complexities, the chapter aims to provide a comprehensive overview of current knowledge and identify areas requiring further research to improve the reliability of earthquake precursor studies.

Keywords: ionospheric anomalies, infrasonic waves, geomagnetic anomalies, global scale disturbances, atmospheric gravity waves, potential earthquake precursors, space weather conditions, solar activity, seismic activity prediction

1. Introduction

Recent advancements in earthquake (EQ) research and engineering have significantly enhanced our understanding of seismic hazards, fault mechanics, and early warning systems [1–11]. Among the most notable developments are the innovations in remote sensing technologies, such as Interferometric Synthetic Aperture Radar (InSAR) [5] and global seismic networks [4, 8, 12–15], which have greatly improved our capability to detect and monitor EQ precursors. These technologies enable precise measurements of ground deformations and subsurface changes, thus contributing to more accurate seismic hazard assessments and the potential for improved EQ forecasting [16–35]. **Table 1** provides a comparative summary of the various EQ prediction methodologies, including their respective methodologies, advantages, and limitations. This table offers a structured overview of the different techniques and highlights their relative effectiveness and applicability.

As illustrated in **Table 1**, the integration of machine learning (ML) and big data analytics enhances the detection of subtle patterns in seismic and geomagnetic

	Methodology	Description	Advantages	Limitations
Seismic Network	Seismic Network	Records seismic waves via network of seismometers	Wide coverage, real-time	Limited spatial resolution
	Seismic Tomography	Uses wave velocities to image subsurface structures	Detailed imaging	Requires dense network
	Moment Tensor Inversion	Estimates EQ source characteristics	Quantitative source data	Requires well-recorded events
	Ambient Noise Tomography	Uses background noise for subsurface imaging	Continuous, passive	Limited frequency range
Geodetic Methods	GPS/GNSS	Measures ground displacement	High precision, wide coverage	Low vertical resolution
	InSAR	Radar for surface deformation	High resolution, large coverage	Limited to radar availability
	GPS-based Early Warning	Uses GPS for EQ alerts	Real-time, early warning	Limited to nearby events
	Geodetic Strain Analysis	Monitors long-term crustal strain	Regional, long-term monitoring	Requires extended observation
Geophysical Modeling	Finite Element Method	Models EQ dynamics	Captures complex waveforms	Computationally heavy
	Boundary Element Method	Models waves near fault boundaries	Efficient near-fault modeling	Needs detailed fault data
	Dynamic Rupture Modeling	Simulates rupture and ground motion	Detailed rupture dynamics	Requires detailed fault parameters
	Wave Propagation Models	Simulate wave propagation	Capture complex wave behavior	Need accurate velocity models
EQ Prediction	Statistical Analysis	Analyze historical patterns	Identify trends	Limited short-term accuracy
	Seismic Precursors	Study pre-EQ phenomena	Potential early warning	Uncertain EQ correlation
	Machine Learning	Applies algorithms for prediction	Pattern recognition	Limited by training data

Table 1.
A comparison of existing methods, models, theory, and techniques for studying EQs.

datasets, thereby improving EQ prediction capabilities [5, 28–31, 35]. ML algorithms applied to large-scale datasets uncover previously undetectable correlations and anomalies, providing valuable insights for identifying trends and potential precursors that traditional approaches might overlook. This advancement offers a promising avenue for refining EQ forecasting methodologies [31].

In addition, analyzing ionospheric and magnetospheric anomalies has emerged as a significant research area in EQ prediction. These disturbances, observed before major EQs, are attributed to the massive energy release during seismic events. The hypothesis that magnetic anomalies could serve as EQ precursors was first proposed by Moore [36] following the 1964 Alaska EQ, which revealed disturbances in the geomagnetic field. Subsequent studies by Antselevich [37] further supported this notion by documenting changes in ionospheric parameters before the 1966 Tashkent EQ.

Systematic research on ionospheric disturbances began in the 1980s, with Pulinets et al. [35] reporting ionospheric anomalies recorded by satellites, highlighting the potential for satellite-based EQ forecasting.

Currently, over 30 countries are engaged in ionospheric precursor research, with substantial evidence indicating a statistically significant correlation between ionospheric disturbances and EQs, particularly those with a magnitude exceeding 5. Magnetospheric anomalies are also investigated as potential indicators of large EQs, arising from interactions between seismic waves and the Earth's magnetic field [38–56]. Despite growing interest in these disturbances as potential EQ precursors, the complexity of interpreting ionospheric and magnetospheric data necessitates further investigation to confirm their reliability for accurate prediction.

This chapter provides a comprehensive review of the current state of knowledge regarding ionospheric and geomagnetic anomalies as potential EQ precursors. It is important to note that the study of the ionosphere, magnetosphere, near-Earth space, and upper atmosphere may be less familiar to traditional geoscientists. Researchers in fields such as radio physics, radio astronomy, and space physics are typically more acquainted with these topics, which may affect the development of this research area. Consequently, this chapter will offer an overview of current EQ research advancements, a brief introduction to the magnetosphere and ionosphere, potential triggers of geomagnetic and ionospheric anomalies, and analytical cases of geomagnetic anomalies related to EQ. It will critically discuss these aspects and provide step-by-step explanations to enhance the reader's understanding. Additionally, the chapter identifies gaps in current research and proposes future directions to advance our comprehension and application of these phenomena in EQ prediction.

2. A brief introduction to the magnetosphere and ionosphere

The Earth's magnetosphere is a vast region of space dominated by the Earth's geomagnetic field, extending from the upper atmosphere into the vacuum of space. It functions as a protective shield against the solar wind, which is a continuous stream of charged particles emitted by the Sun. The interaction between the solar wind and the geomagnetic field (GMF) shapes a complex structure comprising the bow shock, magnetosheath, magnetopause, and magnetotail, as illustrated in **Figure 1**. The GMF is generated by the movement of molten iron in the Earth's outer core and resembles a tilted dipole, extending thousands of kilometers into space. This magnetic field effectively deflects the majority of the solar wind, preventing it from stripping away the Earth's atmosphere and mitigating the effects of space weather phenomena.

It is important to note that the magnetosphere is a dynamic region containing Earth's plasma that interacts with the solar wind. As illustrated in **Figure 1**, on the Sun-facing side of Earth, the solar wind slows and diverts around the magnetosphere, creating a bow shock approximately 13.5 Earth radii from the planet's center. This region, known as the magnetosheath, impacts and weakens Earth's magnetic field. The magnetotail, extending away from the Sun, displays asymmetry due to the solar wind's influence and is subject to magnetic reconnection and plasma instabilities. These processes facilitate the transfer of significant energy from the solar wind to the magnetosphere and subsequently to the ionosphere and neutral atmosphere.

The ionosphere is a region of Earth's atmosphere extending approximately 30–1000 kilometers above the surface, as illustrated in **Figure 2**, characterized by ionization caused by solar radiation. It is divided into distinct layers: the D layer (48–90 km),

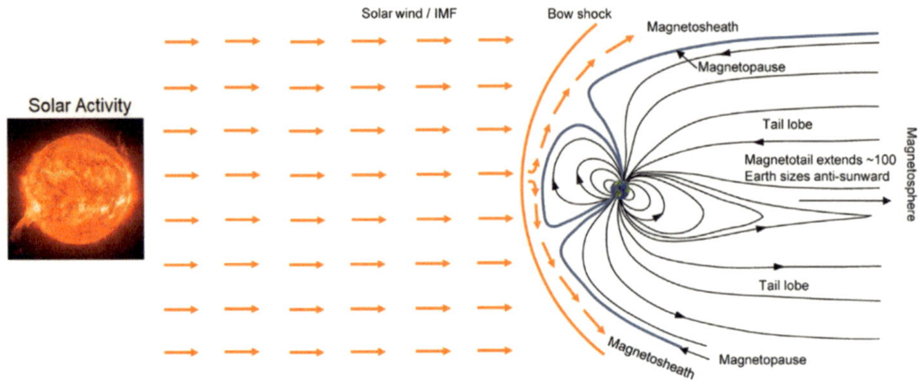

Figure 1.
Schematic diagram of the relationship between the components of the Earth's magnetosphere and solar activity.

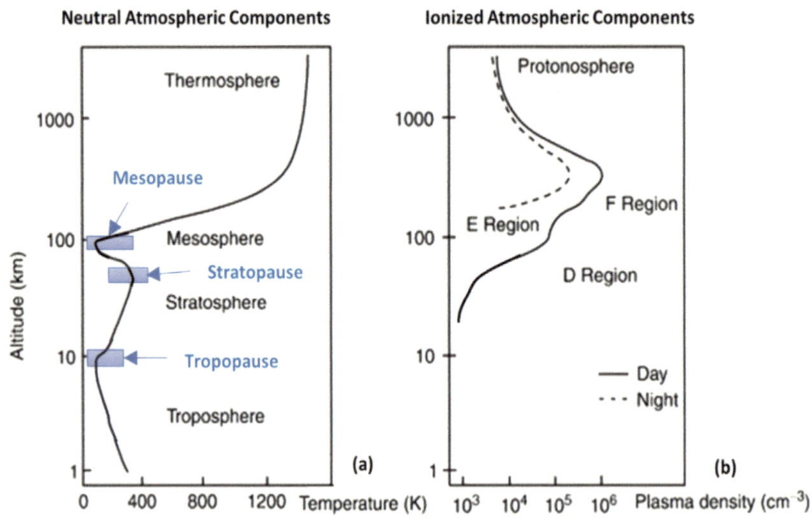

Figure 2.
Schematic representation of the typical vertical distribution of neutral atmospheric temperature (a) and ionospheric plasma density (b).

which experiences high recombination rates and significantly affects radio wave propagation; the E layer (90–150 km), which reflects medium- and high-frequency radio waves; and the F layer (150–1000 km), which supports long-distance communication. The electron density and ionization levels in the ionosphere are influenced by solar activity and exhibit regular diurnal, seasonal, and latitudinal variations, collectively referred to as space weather or ionospheric climate [3].

Given the critical role of the ionosphere in communications, navigation and satellite imaging calibration, understanding its interaction with the magnetosphere is vital for evaluating space weather impacts on communication systems. Magnetic reconnection occurs when the interplanetary magnetic field (IMF) and the GMF have opposing directions, facilitating the transfer of energy and momentum between the solar wind and the magnetosphere. This process induces ionospheric disturbances, acting as a top-down forcing factor. Additionally, natural events such as EQs or typhoons release substantial energy, causing further disturbances in the ionosphere and atmosphere [1]. These disturbances stem from nonlinear processes as energy propagates from the source through the surface atmosphere to the ionosphere and magnetosphere, resulting in changes to ionospheric parameters and geomagnetic anomalies, functioning as bottom-up forcing factors. The mechanisms behind ionospheric and GMF anomalies will be explored in detail in the following section.

3. Potential triggers for ionospheric and magnetospheric anomalies

The Earth and its surrounding environment form the Earth-atmosphere-ionosphere-magnetosphere (EAIM) system, comprising subsystems such as the lithosphere, atmosphere, ionosphere, magnetosphere, and oceans. These layers interact through complex mechanisms, creating a system with memory and feedback where natural or human-induced events trigger both positive and negative couplings. Based on the system paradigm developed by Chernogor [1], we explore further interconnections among the subsystems, emphasizing the crucial role of geophysical fields and energy release (i.e., high-energy processes). The system structure is schematically illustrated in **Figure 3**. The main features include openness to external inputs, dynamic behavior governed by nonlinear processes, and self-organizing behavior driven by energy inflows. Understanding these interactions requires a holistic approach, as analyzing isolated layers fails to capture the system's complex dynamics.

Coupling within the EAIM system occurs through processes such as wave interactions, electromagnetic fields, and particle fluxes. Studies, including those by Burmaka et al. [57, 58], demonstrate that wave-like activity (WLA) is critical for energy and momentum transfer within the system, particularly during significant natural events. These interactions become more pronounced when involving high-energy sources, highlighting the importance of studying high-energy natural hazard events such as severe space weather events, meteoroids explosion, EQs, and typhoons.

Natural events such as strong EQs or typhoons release substantial amounts of energy, leading to disturbances in the ionosphere and atmosphere that cause changes in ionospheric parameters. These changes arise from nonlinear processes as energy propagates from the source through the atmosphere to the ionosphere. For typhoons, atmospheric gravity waves (AtGWs) generated by nonlinear processes [1, 58] propagate upward into the ionosphere, where some of their energy is dissipated, heating the neutral air and modulating electron density, potentially causing plasma irregularities [2]. Similarly, during continental EQs, ground shaking induces acoustic and atmospheric gravity waves that reach the ionosphere. Seafloor disturbances during seismic events generate underwater seismic waves and subsequent tsunamis, which also induce atmospheric disturbances that propagate to ionospheric altitudes.

The EAIM paradigm provides a comprehensive framework for understanding these interrelated phenomena, facilitating interdisciplinary collaboration among seismology, meteorology, and space physics. Within the EAIM system, subsystems exhibit strong coupling through various mechanisms [1], including the following:

1. Waves or wave activities from different sources that impact medium parameters and undergo transformations

2. Quasi-steady-state electromagnetic fields

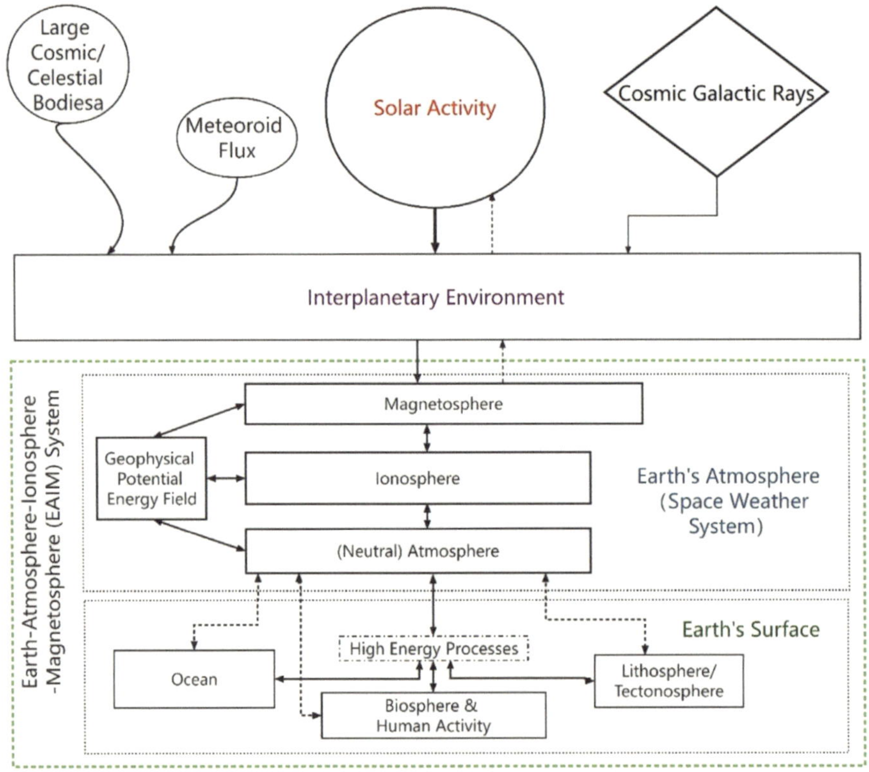

Figure 3.
Schematic diagram of the EAIM system.

3. Particle fluxes that propagate between the ionosphere and the plasmasphere

4. High-energy particle precipitation from the magnetosphere.

Systematic experiments conducted by researchers such as Burmaka et al. [57, 58] highlight the significant role of WLA in the EAIM system. WLA not only transfers energy and momentum but also reflects the state and changes in the space weather system. Sudden energy sources in the ionosphere notably alter WLA characteristics, leading to significant variations in spectral content and amplitude. When large amounts of energy are involved, such as during high-energy natural disaster events (e.g., EQs, typhoons, extreme space weather events, and meteoroid explosions), the coupling between EAIM subsystems becomes more pronounced, necessitating system-level studies that emphasize subsystem interactions.

The EAIM paradigm is ideal for examining these interactions, offering rigorous and objective analysis of various event types. It enables simultaneous study of diverse high-energy disturbance sources and provides a comprehensive understanding of the interconnections between these events and systems. These events impact the Earth's atmosphere and geospace through mechanisms such as wave perturbations, electromagnetic interactions, acoustic-gravity waves (AGWs), atmospheric heating and ionization, and modulation of electric fields and currents.

Previous studies [1, 10, 20, 23–26, 33, 34, 57, 58] have shown that applying the EAIM paradigm underscores the importance of investigating events such as EQs under relatively stable space weather conditions (SWCs), characterized by minimal geomagnetic activity indices (e.g., V_{sw}, IMF B_y, IMF B_z, D_{st}, and K_p). This approach enhances scientific understanding for readers, engineers, and researchers by providing detailed insights into the research subjects and instruments used. Additionally, the paradigm promotes collaboration among experts in seismology, geography, meteorology, and space physics. By studying ionospheric and magnetospheric effects within this framework, we can better understand these phenomena and develop strategies to mitigate their impacts on communication systems, power grids, navigation, and space technology.

4. Analysis of physical mechanisms behind potential earthquake-induced ionospheric and magnetospheric effects

EQs have significantly enhanced our understanding of energy flows within the Earth, with the most powerful EQs releasing up to 19 × 10^{10} J of energy and 17 × 10^{10} W of power [1]. Besides triggering secondary effects such as landslides and tsunamis, EQs reveal interactions between the EAIM and tectonics-atmosphere-ionosphere-magnetosphere (TAIM) systems [20]. These interactions are influenced by energetic particle precipitation from radiation belts, which is crucial for understanding the lithosphere-atmosphere-ionosphere-magnetosphere coupling (LAIMC) system [8, 35, 59–61]. Despite advances, many aspects of EQ effects remain poorly understood.

Historically, ionospheric disturbances linked to seismic activity have been considered potential EQ precursors. Moore [36] and Bolt [62] first identified a connection between ionospheric disturbances and EQs during the 1964 Alaska EQ. This finding was further explored by Antselevich [37] and Hirshberg et al. [63]. Systematic research on these disturbances began in the 1980s [1, 38, 56], with the identification of short-term precursors detectable by satellites [56]. Recent studies suggest that ionospheric anomalies can be observed and predicted from days to hours before earthquakes of magnitude 5 or greater [7, 20, 33]. However, many studies have overlooked the influence of space weather, solar activity, and other factors that affect ionospheric dynamics.

Predicting strong EQs remains a challenge, and current methods are speculative [7, 10, 17, 53]. Numerous precursors—ranging from mechanical deformation and gas emissions to electromagnetic signals and ionospheric changes—have been examined, but no consistent prediction technique has been confirmed. Early reports of electromagnetic signals associated with earthquakes by Park et al. [13], Smirnova et al. [19], and Guglielmi [14] generated significant interest, though observations were often isolated [11].

This section emphasizes the importance of both preseismic and postseismic phenomena in understanding LAIMC. Major precursors include seismic activity (foreshocks and aftershocks), mechanical deformation [5], and variations in groundwater levels and gas emissions, particularly radon [35]. Temperature fluctuations related to groundwater or steam, as well as AGWs, are additional potential indicators [8]. Significant changes have also been observed in electromagnetic signals and ionospheric parameters, such as total electron content (TEC) [60], increased electromagnetic emissions [64] and variations in very low frequency (VLF) reflection height [65]. Pre-EQ effects—such as stress release, crustal deformation, and radon emissions—can induce anomalies in the ionosphere and magnetosphere. Stress release may result in gas emissions that alter ionospheric electron density [66]. Ionospheric fluctuations might also be linked to meteorological changes caused by seismic waves [1], while magnetospheric disturbances can affect the Earth's magnetic field, propagating to the ionosphere and causing further anomalies.

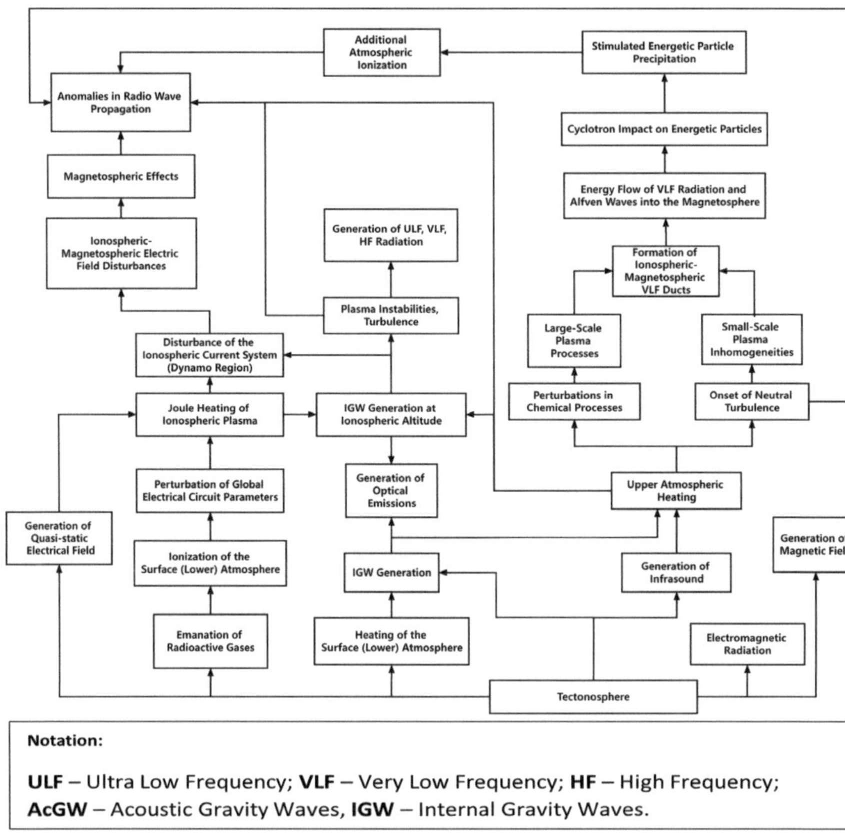

Figure 4.
Block diagram of geosphere interactions and key physical processes during an earthquake event. Further modified from the diagram proposed by Chernogor [1].

Post-EQ effects influence the space environment through several channels, as illustrated in **Figure 4**, which depicts the interaction of Earth's geospheres and the primary physical processes during an EQ event. These include AGWs, quasi-steady electric and magnetic fields, electromagnetic waves resulting from tectonic strain changes, and magnetohydrodynamic (MHD) waves induced by AGWs [10]. It is important to note that **Figure 4** will be further refined as our understanding of these interactions improves.

As highlighted by Odintsov et al., the relationship between solar activity, particularly geomagnetic storms, and seismic activity has been extensively examined in solar-terrestrial science literature for many years (e.g., Sytinskii [67]) [68] in the field of solar-terrestrial science. Despite the extensive history of seismological research, many conclusions are primarily based on statistical results and lack a thorough physical explanation. Specifically, there is a notable lack of substantial evidence supporting the ionospheric and magnetic effects associated with preseismic activity.

5. Analytical methods and case studies for investigating potential EQ-related ionospheric and magnetospheric anomalies

EQs, as high-energy natural events, can potentially induce ionospheric and magnetospheric effects. Our research uses traditional case/event analysis methods to study these effects. We will outline fundamental research processes from a radio engineering perspective, present case studies using fluxgate magnetometer data, and emphasize the need for space weather analysis. We will also briefly review cases lacking space weather consideration.

5.1 Research process and analytical methods

The specific research process is outlined as follows:

5.1.1 Set up a continuous real-time monitoring system

Build equipment, debug equipment, test equipment, set up preprocessing procedures (signal processing technology, band-pass filtering, etc.), parameter calculation, and passive monitoring system.

5.1.2 Identify the specific events and anomalies of observations

On one hand, we continue to focus on our real-time monitoring system and closely observe any detected anomalies. Concurrently, we remain vigilant in obtaining advisories and warnings from various online platforms, including NOAA and NASA alert systems, the International Seismological Center News, daily newspapers, and browser notifications. Despite significant advancements in ionospheric research, challenges and controversies persist in accurately identifying ionospheric disturbances caused by different types of high-energy processes.

Given the current state of research on ionospheric and magnetospheric anomalies, our study aims to examine the ionospheric and geomagnetic effects specifically caused by single high-energy sources, avoiding the uncertainties and complexities associated with mixed high-energy sources. Essential information on high-energy natural disaster events can be accessed through multiple open-source platforms, such as the Goddard Space Flight Center, the Space Weather Prediction Center, the Geomagnetic and Space Magnetic Data Analysis Center, the Ionosphere Working Group, the Global Ionospheric Radio Astronomy Observatory, the Center for Near-Earth Object Studies, the GEOSCOPE Observatory, the United States Geological Survey, and the Digital Typhoon Platform. It is also important to note that non–high-energy natural disasters, including noncatastrophic high-energy processes such as the solar terminator, can induce ionospheric and magnetospheric anomalies. Additionally, inherent factors of the ionosphere, such as diurnal variations (e.g., the presence of the D layer during local daytime), also impact the ionosphere's dynamic properties.

5.1.3 Further process the experimental data, seek, and verify a scientific interpretation of the identified events and anomalies

Different instruments have varying detection capabilities and focus on different aspects of the ionosphere, depending on their design and the technology they employ. Specialized signal preprocessing techniques significantly enhance our ability to detect meaningful changes. Detailed examples will be provided later in this section. Additionally, it is crucial to account for the cumulative impact of natural variations and the selection of reference days used in the study. Key parameters of monitored quasi-periodic variations (QPV) or WLAs include period (frequency), wavelength, amplitude, time delay, propagation speed, and propagation direction. Furthermore, supplementary information from other monitoring tools, such as ionospheric sounding instruments, noncoherent radars, GNSS total electron content maps, medium–/high–/ultra-high–frequency Doppler radars, and sounding rockets, and verification from related studies by other researchers provide valuable support for our scientific interpretation.

5.1.4 Documentation of the event, discussion, and publication

It's important to note that the observation results encompass a wealth of content. During the analysis, researchers prioritize the description and analysis of the most significant features. Due to limitations in knowledge and experimental conditions, there may be insufficient discussion and further analysis of certain characteristics reflected in the observation results.

Below, we illustrate the analysis methods and findings using data from the Kharkiv National University (KhNU) fluxgate magnetometer (FGM) (@49.64°N, 36.93°E). We use observations from the 2003 Chuya EQ and compare them with data from the 2019 Albania EQ and the 2020 Elazığ EQ. This example illustrates the complexity and current challenges in analyzing geomagnetic and ionospheric anomalies, demonstrates the analysis technique, and encourages further discussion.

5.2 Case study: 2003 Chuya EQ

The 2003 Chuya/Altai EQ, which occurred in Altai, Russia, had its epicenter at approximately 50.17°N, 87.80°E, and a depth of about 12.3 km. The mainshock, with a magnitude of $Mw \approx 7.3$, occurred at 18:33 local time (11:33 UTC) on September 27, 2003. This was followed by approximately 140 aftershocks, with the three most significant recorded as follows: $Mw = 6.7$ at 18:52 UTC on September 27, $Mw = 7.0$ at 01:03 UTC on October 1, and $Mw = 4.5$ at 03:55 UTC on October 1.

While evidence for geomagnetic anomalies preceding EQs is limited, the mechanisms behind geomagnetic effects induced by EQs are well-established [10–17]. Seismic waves (SW) and AGWs propagate through the Earth's surface and atmosphere, respectively, affecting atmospheric pressure and modulating ionospheric electron concentrations and currents, leading to QPVs in the GMF.

Comprehensive space weather analysis is crucial for understanding ionospheric effects of EQs. For the Chuya EQ, additional data from NOAA's Space Physics Data Facility [Source: https://omniweb.gsfc.nasa.gov/form/dx1.htm] were used, with the observation period spanning from September 25 to October 3, 2003, during the descending phase of Solar Cycle 23. Key space weather indices, including Bz, Kp, Ap, Dst, and AE, are used to assess SWCs during the Chuya/Altai EQ (refer to **Figure 5**).

This period saw decreased solar flare and coronal mass ejection activity, with solar activity ranging from low to moderate. Region 464 produced a C6/Sf flare on September 28 and an M-class flare on October 1. Geomagnetic field conditions varied from quiet to unsettled between September 27 and 28, reaching G1 storm levels. Geomagnetic storms (GMS) progress through three stages based on Dst index: (1) sudden storm commencement, indicated by a positive Dst change from increased solar wind pressure; (2) the main phase, defined by the lowest Dst value as the ring current intensifies; and (3) the recovery phase, where Dst returns to normal.

Previous studies on minor GMS events like the 2017 GMS and 2019 GMS highlighted minor perturbations in GMF horizontal components during the later recovery

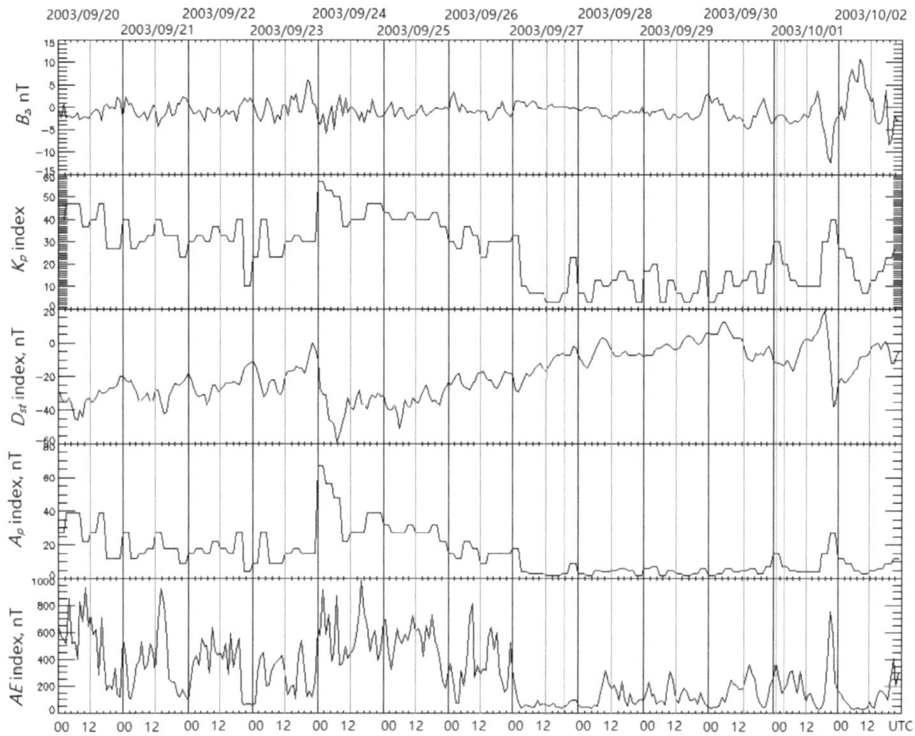

Figure 5.
Space weather conditions from December 20 to October 2, 2003. (Note: The yellow and orange vertical lines indicate the mainshock and aftershock times, respectively).

phase. As a result, September 27, 2003, is categorized as a magnetically quiet day, while October 1 showed slight geomagnetic disturbances.

FGM aids in studying QPVs by analyzing geomagnetic effects (GMEs) in the "dynamo region" of the ionosphere (refer to the block diagram in **Figure 4**). **Figure 6** illustrates FGM measurement interpretations, where system spectra analysis (SSA) was applied to time variations in the D component (West–East) of the GMF, focusing on periods of 100–1000 seconds, following Chernogor's methods [10].

These results (**Figure 6**) show the output of band-pass filtering within the 100–1000 second (1–10 mHz) frequency range from FGM data, highlighting QPVs. The panels display the original signal, windowed Fourier transform, adaptive Fourier transform, and Morlet wavelet transform, alongside relative oscillation energy. The dotted line marks sunset at 300 km altitude, and the solid line denotes the EQ aftershock at ~18:53 UTC ($M_w \approx 6.7$). Two significant QPVs in the GMF D-component fluctuations followed this EQ aftershock, likely linked to seismic activity. To estimate the propagation velocities of the associated ionospheric disturbances (IDs), delay times were analyzed as follows:

1. For the first QPV (QPV-1), the time difference is Δt_1 = 19:13 UT – 18:53 UT = 20 minutes.

2. For the second QPV (QPV-2), Δt_2 = 20:25 UT – 18:53 UT = 92 minutes.

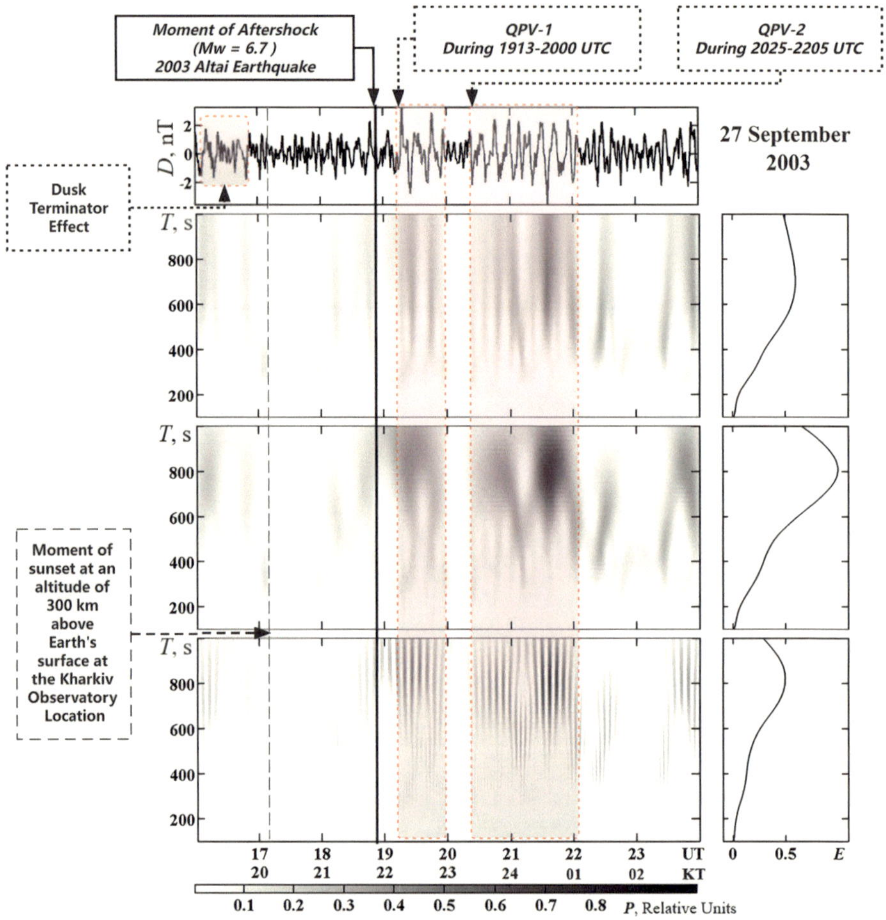

Figure 6.
SSA results of the GMF D-component fluctuation over Kharkiv measured by FGM during the Chuya/Altai EQ on 27 September 2003. Further Modified from previous analysis results [24].

3. Using the IRI 2016 model, the propagation time for AGWs from the Earth's surface to the E-region height is $\Delta t_0 \approx 5$ minutes. Given the epicenter's distance to the Kharkiv Observatory ($R_D \approx 3500$ km), the apparent propagation velocity for QPV-1 (with a period $Ta_1 \approx 6.6$–15 min) is $v_1 = R_D/(\Delta t_1 - \Delta t_0) \approx 3.9$ km/s, consistent with Rayleigh wave speeds (2–5 km/s). This suggests the first QPV could be linked to seismic wave–induced disturbances. The upward propagation mechanism of AGWs remains complex, and linear theory may not fully capture the dynamics [60]. The QPV-1 observed may have resulted from interactions between these waves and ionospheric plasma density modulations, affecting ionospheric currents in the dynamo region, inducing GMEs.

4. For QPV-2 (with a period $Ta_2 \approx 6.6$–15 min), the estimated apparent velocity is $v_2 = R_D/(\Delta t_2 - \Delta t_0) \approx 670$ m/s, which aligns with atmospheric gravity waves (AtGWs). This QPV could be attributed to disturbances generated by AGWs at ionospheric heights above the epicenter.

5. The velocities (v_1, v_2) and periods (Ta_1, Ta_2) provide theoretical support, suggesting a connection between these QPVs and IDs triggered by the EQ. Further details on the Chuya EQ can be found in Ref. [24]. However, it is important to note that the underlying physical processes are highly complex.

6. Regarding perturbations in electron concentration, gas movement within the AGW field captures charged plasma components, leading to periodic modulation of electron concentration (N). The relative amplitude δN of these variations can be estimated using Maxwell's rotary eq. [1].

$$\Delta B = \mu_0 \Delta j \Delta z = \mu_0 j_0 \delta_N \Delta z, \qquad (1)$$

where $\mu_0 = 4\pi \cdot 10^{-7}$ H/m represents the magnetic constant, Δj corresponds to the perturbation of the ionospheric current density in the dynamo region of the atmosphere, j_0 represents the undisturbed value of the ionospheric current density, and Δz represents the thickness of the dynamo region. Assuming $\Delta B \approx 0.4$ nT, $j_0 \approx 10–7$ A/m2, and $\Delta z \approx 30$ km, we get $\delta N \approx 10.6\%$.

5.3 Comparison and discussion

Except for the 2003 Chuya EQ [24], previous studies analyzed two other EQ events (2019 Albania EQ [25] and 2020 Turkey EQ [26]) using the FGM, as shown in **Figure 7**. Observation results, compared under calm SWCs, are summarized in **Table 2**. Note: QPV—Quasi-Periodic Variations; M_w—Magnitude of EQ mainshock and aftershocks; Z—Depth of the EQ epicenter; R_D—Distance between the FGM instrument and the epicenter; $v_{1,2,3}$—Apparent characteristic propagation speeds; ΔT—Duration of QPVs; T_a—Period of QPVs; δN_a—Relative amplitude of electron concentration disturbances in the ionosphere associated with QPVs.

The primary findings align with previous research results [24–26]. Observations from these three EQs generally show agreement. However, the GMEs and IDs associated with the 2019 Albania EQ were weaker compared to those of the Turkey EQ. This variation can be attributed to several factors:

- The magnitude of the Albania EQ (M_w = 6.5) was slightly lower than that of the Turkey EQ (M_w = 6.7).

- The epicenter of the Albania EQ was somewhat deeper (24 km) compared to the Turkey EQ. (23 km).

- The distance from the epicenter to FGM was greater for the Albania EQ. (1620 km) than for the Turkey EQ. (1350 km).

- GMEs and IDs from the Albania EQ occurred during relatively calmer SWCs.

Furthermore, the observed characteristic speeds are influenced by various factors. Comparing seismic data with FGM measurements is essential for understanding the relationship between EQ activity and GMEs and IDs induced by EQs. For instance, after the 2020 Turkey EQ, damped oscillations with a period of approximately 2 minutes were detected 104 minutes postevent at Kharkiv, located 1350 km from the epicenter. These oscillations, occurring after the initial GME, are linked to the EQ and

Figure 7.
The distribution of measurement equipment and the mainshocks' epicenter of 2003 Chuya EQ, 2019 Albania EQ, and 2020 Turkey EQ.

Earthquake	Chuya EQ	Albanian EQ	Turkey EQ
Time	2003-2009-27 11:33:25 UTC	2019-2111-26 02:54:10 UTC	2020-2101-24 17:55:14 UTC
Location	Altai Republic, Russia	Albania	Elazığ Province, Turkey
Type	Oblique-slip	Thrust (Reverse)	Strike-slip
Fault	—	Shijak Fault	East Anatolian Fault
Epicenter	50.17°N, 87.80°E	41.42°N, 19.48°E	38.33°N, 39.08°E
Z, km	12.3	24	23
Aftershocks	~140	~1300	~390
R_D, km	~3500	~1620	~1350
Mw	7.3 // 6.7 // 7.0	6.5	6.7
v_1, km/s	3.2 // 3.9–5.3 // 1.9	—	—
v_2, m/s	634–711// 450–670// -	—	—
v_3, m/s	320–350// - // -	267–293	230–320
δN_a, %	12.5–25	5.3	8
T_a, min	2.5–4.2; 6.6–15	7–14	1.5–2.5; 11.6–15
ΔT, min	44–49; 75–105; // - //	130–140	75
References	[24]	[25]	[26]

Table 2.
A comparison of the EQ basic information and EQ-related GMEs and IDs.

AGW. Similar waveform patterns were observed at the RER station (Riviere de l'Est—Sainte Rose—La Reunion Island, source: http://geoscope.ipgp.fr/index.php/en) and appeared as trailing waves in the seismic data, as illustrated in **Figure 8**. These findings warrant further study and interpretation, highlighting lack of current research.

Additionally, the dynamics of the ionosphere should not be underestimated. Often, the observed effects in the ionosphere and magnetosphere result from a combination of factors occurring before and after an EQ, rather than being solely

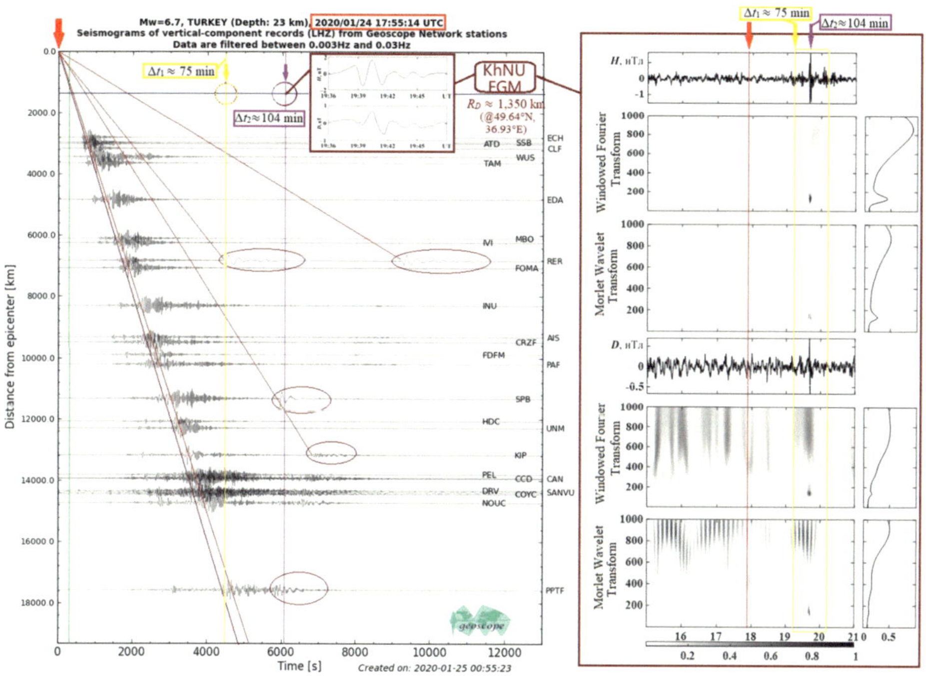

Figure 8.
FGM observations and seismograms of the 2020 Turkey EQ. Note: Modified from [26].

attributable to the EQ itself. As shown in **Figure 6**, dusk terminators may also contribute to the potential GMEs and QPVs.

Furthermore, measurements from a single observatory are insufficient for a comprehensive understanding of potential EQ-induced IDs and GMEs. A network of magnetometers, spaced at appropriate distances, is necessary for accurate detection. The delay time between the GME/ID and the wave source increases with the distance between the EQ epicenter and the observatory. Networks such as the International Network of Real-Time Magnetic Observations can filter geomagnetic variations effectively. However, even with a detection network, interpreting results can be contentious, as illustrated by the 2011 Japanese tsunami EQ [15, 17, 69, 70].

The author acknowledges the limitations of current knowledge and recognizes that not all observed phenomena can be fully explained. Open discussion and collaboration with readers are encouraged to deepen our understanding of the complex interactions between high-energy natural hazards, such as EQs, and GMEs and IDs. An interdisciplinary approach is advocated to enhance understanding among readers, engineers, and experimentalists and to foster collaboration among experts in seismology, meteorology, and space physics. The next subsection will critically reflect on the current state of research and underscore the need for a more comprehensive understanding and interdisciplinary collaboration.

6. Discussion and critical thinking

First and foremost, it is essential to emphasize that pre-EQ geomagnetic anomalies (GMAs) and ionospheric anomalies (IAs) have not been definitively proven as reliable precursors of seismic activity. Historically, doubts, misjudgments, and conflicting views have arisen due to insufficient evidence and lack of thorough analysis of SWCs, as highlighted by Romanova et al. [16]. Specific instances of these issues include the following:

Doubt: In 1964, Moore [36] observed a short-term increase in GMF prior to the Great Alaska EQ. However, subsequent research did not corroborate this finding, leading to skepticism about its validity.

Misjudgment: Fraser-Smith [48, 49] detected a GMA before the Loma Prieta EQ, which was later attributed to a magnetic storm by Campbell [18]. Shestopalov et al. [39] observed a 690 nT GMF perturbation an hour before the 2010 Chilean EQ, which Romanova et al. [16] suggested was more likely related to substorms rather than the EQ itself.

Lack of SWC analysis: Sobisevich et al. [40–43] reported pre-EQ GMAs with amplitudes ranging from 1 to 5 nT but did not perform comprehensive SWC analyses. Astafyeva [28] discussed future applications of ionospheric measurements for detecting seismic activity but did not address SWCs.

Conflicting perspectives: Fraser-Smith et al. [48, 50, 51] and Karakelian et al. [12], studied ultra-low-frequency (ULF) GMAs with periods ranging from 0.1 to 100 s, attributing these oscillations to upper atmospheric processes rather than EQ preparation.

Theoretical mechanisms lacking examples: Kunitsyn et al. [71] proposed that AGWs might play a significant role in generating ULF variations in the GMF but did not provide specific examples.

Advancements and caveats in research on GMAs and ionospheric effects preceding EQs include the following:

Pulse-type GMAs: Molchanov et al. [49] observed distinct ULF electromagnetic emissions hours before the 1988 Spitak EQ (Mw $\approx$ 6.9, depth 6 km). These anomalies were unrelated to geomagnetic pulsations from magnetospheric sources.

Quasi-periodic GMAs: Kopytenko et al. [47] reported GMAs with an amplitude of $\Delta B \approx 0.03$–0.2 nT and a duration of $T \approx 1$–10 s at 129 km from the 1988 Spitak EQ epicenter. Other studies [10, 17, 44, 45] identified quasi-periodic GMAs ($T \approx 20$–50 s, $\Delta B \approx 0.1$ nT) occurring a few days before EQs.

Caveats: Periods of 40–250 seconds are generally associated with geomagnetic pulsations of cosmic origin, observed by ground-based and space-based magnetometers. Sobisevich et al. [40–43] emphasized the uncertainty surrounding the mechanisms responsible for these GMAs.

Air heating mechanism: Pulinets et al. [2, 35, 60] suggested that radon gas release ionizes air molecules, forming ion clusters that serve as nuclei for water vapor condensation. Chernogor [10] expanded on this mechanism, with pre-EQ GMAs ranging from ~0.4 to 4 nT and lasting minutes to hours.

E × B drift: Heki [70] found that pre-EQ total electron content changes resulted in positive electron density anomalies above fault zones, correlating with EQ magnitudes. These anomalies likely result from E × B drift caused by surface electric fields penetrating the ionosphere.

ULF-ELF-VLF-duct: Hegai et al. [72] analyzed geophysical data from two major Alaska EQs (Mw = 8.2, Mw = 9.2) and proposed that favorable pre-EQ conditions allowed seismogenic electric fields to penetrate the ionosphere, forming ULF-ELF-VLF ducts that amplified waves in these ranges. Measurements from the China-Seismo-Electromagnetic Satellite supported these findings for the 2021 Alaska EQ.

Additionally, numerous researchers [6, 13, 17, 19, 52] have explored ULF seismomagnetic phenomena. However, the mechanisms behind ULF precursors to EQs remain elusive. Given these uncertainties, further investigation is essential for a thorough understanding of these mechanisms.

Recent study [4] has identified potential lithosphere-atmosphere-ionosphere coupling effects before the Jamaica Mw = 7.7 EQ, including sulfur dioxide emissions, increased ionospheric electron density, and electromagnetic signals detected by CSES and Swarm. However, like most researchers [28, 73–80], the authors did not account for the potential impact of space weather on ionospheric disturbances, leaving their conclusions incomplete and undermining the reliability of their analysis of ionospheric effects as precursors to EQs.

Future Research Directions:

- Emphasize the need to analyze space weather conditions prior to studying ionospheric and magnetospheric anomalies.

- Further explore the physical mechanisms underlying EQ-related ionospheric and magnetospheric anomalies, including the impact of EQ magnitude, depth, fault type, location, season, etc.

- Develop new seismic observation methods and employ multi-instrument observations along with artificial intelligence [81] to enhance the understanding of ionospheric responses.

- Integrate geophysical, geomagnetic, and ionospheric data for a comprehensive analysis of EQ effects.

- Expand research to include moderate to strong Eqs. ($5 \leq Mw < 7$) in addition to major EQs ($Mw \geq 7$).

7. Conclusion

The EAIM system presents a complex, open network of interactions, particularly when analyzing anomalies potentially linked to seismic activity. While correlations between earthquakes and ionospheric or magnetospheric disturbances have been observed, the lack of comprehensive evidence and the influence of external factors, such as solar activity, pose significant challenges. Moving forward, a cautious and critical approach is needed in interpreting these phenomena. Future research should prioritize interdisciplinary collaboration, improved observational technologies, and a more integrated analysis of both top-down (solar) and bottom-up (EQ) drivers. This holistic approach will enhance our ability to differentiate between EQ-induced effects and other geophysical influences, paving the way for more reliable interpretations.

Acknowledgements

The author acknowledges the use of OpenAI's ChatGPT 3.0 for language polishing of the manuscript.

Thanks

The author extends his gratitude to Professor Leonid Chernogor, Professor Serhiy Shulga, Professor Giovanni Lapenta, and Professor Kwinten Van Weverberg for their insightful discussions and suggestions on related topics. The author expresses gratitude to the academic editor, Professor Ali Ismet Kanli, for his valuable suggestions, which significantly contributed to the enhancement of the manuscript.

Author details

Yiyang Luo
V.N. Karazin Kharkiv National University, Kharkiv, Ukraine

*Address all correspondence to: yiyangluo@163.com

References

[1] Chernogor LF. The earth-atmosphere-geospace system: Main properties and processes. International Journal of Remote Sensing. 2011;**32**(11):3199-3218. DOI: 10.1080/01431161.2010.541510

[2] Pulinets SA, Ouzounov DP, Karelin AV, Davidenko DV. Physical bases of the generation of short-term earthquake precursors: A complex model of ionization-induced geophysical processes in the lithosphere-atmosphere-ionosphere-magnetosphere system. Geomagnetism and Aeronomy. 2015;**55**(4):521-538. DOI: 10.1134/S0016793215040131

[3] Kelley MC. Ionosphere. In: Holton JR, editor. Encyclopedia of Atmospheric Sciences. New York, NY: Academic Press; 2003. pp. 1022-1030. DOI: 10.1016/B0-12-227090-8/00184-6

[4] Marchetti D, Zhu K, Piscini A, et al. Changes in the lithosphere, atmosphere, and ionosphere before and during the Mw = 7.7 Jamaica 2020 earthquake. Remote Sensing of Environment. 2024;**307**:114146. DOI: 10.1016/j.rse.2024.114146

[5] Li Q, Wang W, Wang J, Zhang J, Geng D. Exploring the relationship between InSAR coseismic deformation and earthquake-damaged buildings. Remote Sensing of Environment. 2021;**262**:112508. DOI: 10.1016/j.rse.2021.112508

[6] Anagnostopoulos G. On the origin of ULF magnetic waves before the Taiwan Chi-Chi 1999 earthquake. Frontiers in Earth Science. 2021;**9**:730162. DOI: 10.3389/feart.2021.730162

[7] Shah M, Calabia Aibar A, Tariq MA, Ahmed J, Ahmed A. Possible ionosphere and atmosphere precursory analysis related to Mw > 6.0 earthquakes in Japan. Remote Sensing of Environment. 2020;**239**:111620. DOI: 10.1016/j.rse.2019.111620

[8] Pulinets S, Herrera VMV. Earthquake precursors: The physics, identification, and application. Geosciences. 2024;**14**(8):209. DOI: 10.3390/geosciences14080209

[9] Lizunov G, Skorokhod T, Hayakawa M, Korepanov V. Formation of ionospheric precursors of earthquakes—Probable mechanism and its substantiation. Open Journal of Earthquake Research. 2020;**9**(2):142-169. DOI: 10.4236/ojer.2020.92009

[10] Chernogor LF. Possible generation of quasi-periodic magnetic precursors of earthquakes. Geomagnetism and Aeronomy. 2019;**59**(3):374-382. DOI: 10.1134/S001679321903006X

[11] Sidorin AY. Search for earthquake precursors in multidisciplinary data monitoring of geophysical and biological parameters. Natural Hazards and Earth System Sciences. 2003;**3**:153-158. DOI: 10.5194/nhess-3-153-2003

[12] Karakelian D, Klempeter SL, Fraser-Smith AC, Thompson GA. Ultra-low frequency electromagnetic measurements associated with the 1998 (Mw = 5.1) San Juan Bautista, California earthquake and implications for mechanisms of electromagnetic earthquake precursors. Tectonophysics. 2002;**359**(1-2):65-79. DOI: 10.1016/S0040-1951(02)00439-0

[13] Park SK, Johnson M, Madden JS, Morgan FD, Morrison HF.

Electromagnetic precursors to earthquakes in the ULF band: A review of observations and mechanisms. Reviews of Geophysics. 1993;**31**(2):117-132. DOI: 10.1029/93RG00820

[14] Guglielmi AV. Ultra-low-frequency electromagnetic waves in the Earth's crust and magnetosphere. Physics-Uspekhi. 2007;**50**(12):1197-1216. DOI: 10.1070/PU2007v050n12ABEH006413

[15] Sobolev GA. Low frequency seismic noise before a magnitude 9.0 Tohoku earthquake on March 11, 2011. Izvestiya Physics of the Solid Earth. 2011;**47**:1034-1044. DOI: 10.1134/S1069351311120093

[16] Romanova NV, Pilipenko VA, Stepanova MV. On the magnetic precursor of the Chilean earthquake of February 27, 2010. Geomagnetism and Aeronomy. 2015;**55**(2):219-222. DOI: 10.1134/S0016793215010107

[17] Schekotov A, Fedorov E, Hobara Y, Hayakawa M. ULF magnetic field depression as a possible precursor to the 2011/3.11 Japan earthquake. Journal of Atmospheric Electricity. 2013;**33**(1):41-51. DOI: 10.1541/jae.33.41

[18] Campbell WH. Natural magnetic disturbance fields, not precursors, preceding the Loma Prieta earthquake. Journal of Geophysical Research. 2009;**114**:A05307. DOI: 10.1029/2008JA013932

[19] Smirnova NA, Hayakawa M, Gotoh K. Precursory behavior of fractal characteristics of the ULF electromagnetic fields in seismic active zones before strong earthquakes. Physics and Chemistry of the Earth A/B/C. 2001;**26**(10-12):745-751

[20] Luo Y, Chernogor L, Garmash K, Guo Q, Zheng Y. Seismic-ionospheric effects: Results of radio soundings at oblique incidence. Radio Physics and Radio Astronomy. 2020;**25**(3):218-230

[21] Faheem H, Li X, Zhu W, Ji Y, Feng L, Zhu Y. Refinement of different frequency bands of geomagnetic vertical intensity polarization anomalies before M > 5.5 earthquakes. Sensors (Basel). 2024;**24**(10):3240. DOI: 10.3390/s24103240

[22] Han C, Yan R, Marchetti D, Pu W, Zhima Z, Liu D, et al. Study on electron density anomalies possibly related to earthquakes based on CSES observations. Remote Sensing. 2023;**15**(13):3354. DOI: 10.3390/rs15133354

[23] Chernogor LF. Advanced methods of spectral analysis of quasiperiodic wave-like processes in the ionosphere: Specific features and experimental results. Geomagnetism and Aeronomy. 2008;**48**(5):652-673. DOI: 10.1134/S0016793208050101

[24] Luo Y, Garmash K, Chernogor L, Shulga S. Geomagnetic field fluctuations during Chuysk earthquakes in September–October 2003. Vestnik of KhNU—Radiophysics and Electronics. 2019;**31**:87-104. DOI: 10.26565/2311-0872-2019-31-09

[25] Chernogor L, Garmash K, Holub M, Leus S, Shevelev M, Luo Y. Geomagnetic effect of the Albanian earthquake on November 26, 2019. Vestnik of KhNU—Radiophysics and Electronics. 2020;**32**:44-52. DOI: 10.26565/2311-0872-2020-32-05

[26] Luo Y, Chernogor L, Garmash K. Geomagnetic effect of the Turkish earthquake of January 24, 2020. Radio Physics and Radio Astronomy. 2020;**25**(4):276-289. DOI: 10.15407/rpra25.04.276

[27] Moreno B, Calais E. Evidence of correlation between high frequency

geomagnetic variations and seismicity in the Caribbean. Open Journal of Earthquake Research. 2021;**10**(2):1-10. DOI: 10.4236/ojer.2021.102003

[28] Astafyeva E. Ionospheric detection of natural hazards. Reviews of Geophysics. 2019;**57**:1265-1288. DOI: 10.1029/2019RG000686

[29] Yang M, Zhang X, Ouyang X, Liu J, Qian G, Li T, et al. Polarization method-based research on magnetic field data associated with earthquakes in Northeast Asia recorded by the China Seismo-electromagnetic satellite. Atmosphere. 2023;**14**(10):1555. DOI: 10.3390/atmos14101555

[30] Kulichkov SN, Chunchuzov IP, Bush GA, et al. Physical modeling of long-range infrasonic propagation in the atmosphere. Izvestiya, Atmospheric and Oceanic Physics. 2008;**44**:175-186. DOI: 10.1134/S0001433808020059

[31] Chen H, Han P, Hattori K. Recent advances and challenges in the seismo-electromagnetic study: A brief review. Remote Sensing. 2022;**14**(22):5893. DOI: 10.3390/rs14225893

[32] Le Pichon A, Blanc E, Hauchecorne A. Infrasound Monitoring for Atmospheric Studies. Switzerland: Springer Nature; 2019. DOI: 10.1007/978-1-4020-9508-5

[33] Chernogor L, Garmash K, Guo Q, Luo Y, Rozumenko V, Zheng Y. Oblique-incidence ionospheric radio-sounding: Seismo-ionospheric effects. In: 2020 IEEE Ukrainian Microwave Week (UkrMW) Kharkiv, Ukraine. New York: IEEE; 21-25 Sep 2020. pp. 354-359. DOI: 10.1109/UkrMW49653.2020.9252767

[34] Luo Y, Guo Q, Zheng Y, Garmash K, Chernogor L, Shulga S. HF radio-wave characteristic variations over China during moderate earthquake in Japan on September 5, 2018. Vestnik of KhNU—Radiophysics and Electronics. 2019;**30**:16-26. DOI: 10.26565/2311-0872-2019-30-02

[35] Pulinets S, Budnikov P, Karelin A, Žalohar J. Thermodynamic instability of the atmospheric boundary layer stimulated by tectonic and seismic activity. Journal of Atmospheric and Solar-Terrestrial Physics. 2023;**246**:106050. DOI: 10.1016/j.jastp.2023.106050

[36] Moore GW. Magnetic disturbances preceding the 1964 Alaska earthquake. Nature. 1964;**203**:508-509. DOI: 10.1038/203508b0

[37] Antselevich MG. The influence of the Tashkent earthquake on the Earth's magnetic field and the ionosphere. In: Tashkent Earthquake 26 April 1966. Tashkent: FAN Publishing; 1971. pp. 187-188

[38] Rikitake T. Earthquake precursors in Japan: Precursor time and detectability. Tectonophysics. 1987;**136**(3-4):265-282. DOI: 10.1016/0040-1951(87)90029-1

[39] Shestopalov IP, Soloviev AA, Belov SV, Kuzmin YD. Neutron generation and geomagnetic disturbances in connection with the Chilean earthquake of February 27, 2010 and a volcanic eruption in Iceland in March-April 2010. Geomagnetism and Aeronomy. 2013;**53**(1):124-135. DOI: 10.1134/S0016793213010179

[40] Sobisevich LE, Kanonidi KK, Sobisevich AL. Ultra low-frequency electromagnetic disturbances appearing before strong seismic events. Doklady Earth Sciences. 2009;**429**(5):1549-1552. DOI: 10.1134/S1028334X09090281

[41] Sobisevich LE, Sobisevich AL, Kanonidi KK. Anomalous geomagnetic disturbances induced by catastrophic tsunami producing earthquakes in the region of Indonesia. Geofizicheskiy Zhurnal. 2012;**34**(5):22-37. DOI: 10.24028/gzh.0203-3100.v34i5.2012.116661

[42] Sobisevich LE, Kanonidi KK, Sobisevich AL, Miseyuk OI. Geomagnetic disturbances in the geomagnetic field's variations at stages of preparation and implementation of the Elazig (March 8, 2010) and M 5.3 (January 19, 2011) earthquakes in Turkey. Doklady Earth Sciences. 2013;**449**(1):93-96. DOI: 10.1134/S1028334X13030069

[43] Sobisevich AL, Starostenko VI, Sobisevich LE, Kendzera AV, Shuman VN, Vol'Fman YM, et al. Black Sea earthquakes in late December 2012 and their manifestations in the geomagnetic field. Geofizicheskiy Zhurnal. 2013;**35**(6):54-70. DOI: 10.24028/gzh.0203-3100.v35i6.2013.116455

[44] Surkov VV, Pilipenko VA. Magnetic effects due to earthquakes and underground explosions: A review. Annales de Geophysique. 1997;**40**(2):227-239. DOI: 10.4401/ag-3904

[45] Surkov V, Hayakawa M. Ultra and Extremely Low Frequency Electromagnetic Fields. Tokyo (Japan): Springer; 2014. DOI: 10.1007/978-4-431-54367-1

[46] Jánský J, Pasko VP. Earthquake lights: Mechanism of electrical coupling of Earth's crust to the lower atmosphere. Journal of Geophysical Research – Atmospheres. 2018;**123**:8901-8914. DOI: 10.1029/2018JD028489

[47] Kopytenko YA, Matiashvili TG, Voronov PM, Kopytenko EA, Molchanov OA. Detection of ultra-low-frequency emissions connected with the Spitak earthquake and its aftershock activity, based on geomagnetic pulsations data at Dusheti and Vardzia observatories. Physics of the Earth and Planetary Interiors. 1993;**77**(1-2):85-95. DOI: 10.1016/0031-9201(93)90035-8

[48] Fraser-Smith AC, Bernardi A, Mcgill PR, Ladd ME, Halliwell RA, Villard OG Jr. Low frequency magnetic field measurements near the epicenter of the Ms 7.1 Loma Prieta earthquake. Geophysical Research Letters. 1990;**17**(9):1465-1468. DOI: 10.1029/GL017i009p01465

[49] Molchanov OA, Kopytenko YA, Voronov PM, Kopytenko EA, Matiashvili TG, Fraser-Smith AC, et al. Results of magnetic field measurements near the epicenters of the Spitak (Ms = 6.9) and the Loma Prieta (Ms = 7.1) earthquakes: Comparative analysis. Geophysical Research Letters. 1992;**19**(14):1495-1498. DOI: 10.1029/92GL01152

[50] Fraser-Smith AC, Mcgill PR, Helliwell RA, Villard OG Jr. Ultra-low frequency magnetic field measurements in southern California during the Northridge earthquake of 17 January 1994. Geophysical Research Letters. 1994;**21**(20):2195-2198. DOI: 10.1029/94GL01984

[51] Fraser-Smith AC. Ultralow-frequency magnetic fields preceding large earthquakes. Eos. 2008;**89**(23):211. DOI: 10.1029/2008EO230007

[52] Swati SB, Pundhir D, Sinha AK, Rao KM, Guha A, Hobara Y. Ultra-low frequency (ULF) magnetic field emissions associated with some major earthquakes occurred

in Indian subcontinent. Journal of Atmospheric and Solar-Terrestial Physics. 2020;**211**:105469. DOI: 10.1016/j.jastp.2020.105469

[53] Geller RJ. Earthquake prediction: A critical review. Geophysical Journal International. 1997;**131**(3):425-450. DOI: 10.1111/j.1365-246X.1997.tb06588.x

[54] Hayakawa M, Kawate R, Molchanov OA, Jumoto K. Results of ultra-low-frequency magnetic field measurements during the Guam earthquake of 8 August 1993. Geophysical Research Letters. 1996;**23**(3):241-244. DOI: 10.1029/95GL02863

[55] Berthelier J, Godefroy M, Leblanc F, et al. ICE, the electric field experiment on DEMETER. Planetary and Space Science. 2006;**54**:456-471. DOI: 10.1016/j.pss.2005.10.016

[56] Chmyrev VM, Isaev NV, Bilichenko SV, Stanev G. Observation by space-borne detectors of electric fields and hydromagnetic waves in the ionosphere over an earthquake centre. Physics of the Earth and Planetary Interiors. 1989;**57**:110-114. DOI: 10.1016/0031-9201(89)90220-3

[57] Burmaka VP, Taran VI, Chernogor LF. Wave-like processes in the ionosphere under quiet and disturbed conditions. 1. Kharkov incoherent scatter radar observations. Geomagnetism and Aeronomy. 2006;**46**(2):183-198. DOI: 10.1134/S0016793206020071

[58] Burmaka VP, Taran VI, Chernogor LF. Wave-like processes in the ionosphere under quiet and disturbed conditions. 2. Analysis of observations and simulation. Geomagnetism and Aeronomy. 2006;**46**(2):199-208. DOI: 10.1134/S0016793206020083

[59] Laštovička J, Baše J, Hruska F, et al. Simultaneous infrasonic, seismic, magnetic and ionospheric observations in an earthquake epicentre. Journal of Atmospheric and Terrestrial Physics. 2010;**72**:1231-1240. DOI: 10.1016/j.jastp.2010.08.005

[60] Pulinets SA, Legen'ka AD. Spatial-temporal characteristics of the large-scale disturbances of electron concentration observed in the F-region of the ionosphere before strong earthquakes. Cosmic Research. 2003;**41**:221-229

[61] Liu JY, Chen YI, Pulinets SA, et al. Seismo-ionospheric signatures prior to M≥6.0 Taiwan earthquakes. Geophysical Research Letters. 2000;**27**(19):3113-3116. DOI: 10.1029/2000GL011395

[62] Bolt BA. Seismic air waves from the great 1964 Alaskan earthquake. Nature. 1964;**202**:1095-1096. DOI: 10.1038/2021095a0

[63] Hirshberg J, Currie RG, Breiner S. Long period geomagnetic fluctuations after the 1964 Alaskan earthquake. Earth and Planetary Science Letters. 1967-1968;**3**:426-428. DOI: 10.1016/0012-821X(67)90072-6

[64] Molchanov OA, Mazhaeva OA, Golyavin AN, Hayakawa M. Observation by the Intercosmos-24 satellite of ELF-VLF electromagnetic emissions associated with earthquakes. Annales de Geophysique. 1993;**11**(5):431-440

[65] Hayakawa M, Nickolaenko AP. Variations of atmospheric ELF/VLF radio noise due to seismogenic modifications in tropospheric conductivity. Open Journal of Earthquake Research. 2024;**13**:113-132. DOI: 10.4236/ojer.2024.132005

[66] Ouyang X, Zhang X, Shen X, et al. Disturbance of O+ density before major

earthquake detected by DEMETER satellite. Chinese Journal of Space Science. 2011;**31**:607-617. DOI: 10.11728/cjss2011.05.607

[67] Sytinskii AD. Relationships between Earth's seismicity, solar activity, and atmospheric processes. Leningrad: Gidrometeoizdat; 1987. 97 p. (In Russian)

[68] Odintsov SD, Ivanov-Kholodnyi GS, Georgieva K. Solar activity and global seismicity of the earth. Bulletin of the Russian Academy of Sciences: Physics. 2007;**71**:593-595. DOI: 10.3103/S1062873807040466

[69] Muafiry IN, Heki K. 3-D tomography of the ionospheric anomalies immediately before and after the 2011 Tohoku-Oki (Mw9.0) earthquake. Journal of Geophysical Research: Space Physics. 2020;**125**. DOI: 10.1029/2020JA027993

[70] Heki K. Ionospheric electron enhancement preceding the 2011 Tohoku-Oki earthquake. Geophysical Research Letters. 2011;**38**. DOI: 10.1029/2011GL047908

[71] Kunitsyn VE, Krysanov BY, Vorontsov AM. Acoustic-gravity waves in the Earth's atmosphere generated by surface sources. Moscow University Physics. 2015;**70**:541-548. DOI: 10.3103/S0027134915060120

[72] Hegai V, Zeren Z, Pulinets S. Seismogenic field in the ionosphere before two powerful earthquakes: Possible magnitude and observed ionospheric effects (case study). Atmosphere. 2023;**14**:819. DOI: 10.3390/atmos14050819

[73] Parrot M, Benoist D, Berthelier J, et al. The magnetic field experiment IMSC and its data processing onboard DEMETER: Scientific objectives, description and first results. Planetary and Space Science. 2006;**54**:441-455. DOI: 10.1016/j.pss.2005.10.015

[74] Parrot M, Berthelier J, Lebreton J, et al. Examples of unusual ionospheric observations made by the DEMETER satellite over seismic regions. Physics and Chemistry of the Earth A/B/C. 2006;**31**:486-495. DOI: 10.1016/0031-9201(89)90220-3

[75] Li M, Shen X, Parrot M, et al. Primary joint statistical seismic influence on ionospheric parameters recorded by the CSES and DEMETER satellites. Journal of Geophysical Research, Space Physics. 2020;**125**:1-13. DOI: 10.1029/2020JA028116

[76] Zhang X, Shen X, Parrot M, et al. Phenomena of electrostatic perturbations before strong earthquakes (2005-2010) observed on DEMETER. Natural Hazards and Earth System Sciences. 2012;**12**:75-83. DOI: 10.5194/nhess-12-75-2012, 2012

[77] Zeng ZC, Zhang B, Fang GY, Wang DF, Yin HJ. An analysis of ionospheric variations before the Wenchuan earthquake with DEMETER data. Chinese Journal of Geophysics. 2009;**52**:13-22. DOI: 10.1002/cjg2.1322

[78] Zheng L, Yan R, Parrot M, et al. Statistical research on seismo-ionospheric ion density enhancements observed via DEMETER. Atmosphere. 2022;**13**:1252. DOI: 10.3390/atmos13081252

[79] Yan R, Parrot M, Pinçon JL. Statistical study on variations of the ionospheric ion density observed by DEMETER and related to seismic activities. Journal of Geophysical Research, Space Physics. 2017;**122**:12421-12429. DOI: 10.1002/2017JA024623

[80] Harrison RG, Aplin KL, Rycroft MJ. Atmospheric electricity coupling between earthquake regions and the ionosphere. Journal of Atmospheric and Solar—Terrestrial Physics. 2010;**72**(5-6):376-381. DOI: 10.1016/j.jastp.2009.12.004

[81] Muhammad A, Külahcı F, Birel S. Investigating radon and TEC anomalies relative to earthquakes via AI models. Journal of Atmospheric and Solar—Terrestrial Physics. 2023;**245**:106037. DOI: 10.1016/j.jastp.2023.106037

Chapter 6

Further Studies on Three Large Earthquakes in the Makran Subduction Zone Region: The Crustal Velocity and Source Rupture Models

Parisa Asgharzadeh, Shutian Ma and Darioush Motazedian

Abstract

Three large earthquakes (2011-01-18 Pakistan Mw 7.2; 2013-04-16 IRAN Mw 7.7; and 2013-09-24 Pakistan Mw 7.8) occurred in the Makran subduction zone north region. This chapter introduces an average crustal model retrieved using Rayleigh wave dispersion data in the three earthquake region and their vicinity, the source rupture modeling results, and some geological background features. The centers of the major ruptures of the 2011 Mw 7.2 and the 2013 Mw 7.7 earthquakes were at depth about 65 km; the major rupture of the shallow 2013 Mw 7.8 earthquake was within the upper crust. The focal mechanisms of the two deep earthquakes were normal type; while the shallow earthquake was a left-lateral strike–slip type with a thrust compo nent. The dipping directions of all the three earthquakes were at northwest directions. The rupture process of the Mw 7.2 earthquake was simple; the Mw 7.7 became complex; while the shallow one was the most complex among the three. The sources of the two deep earthquakes were farther than that of the shallow earthquake to the Makran subduction front. All three earthquakes were in the interception region of the Arabian plate, Indian Plate, and the Eur-Asian Plate. The shallow earthquake occurred along a known fault; while the deeper two may occur along a deep, large, potent fault.

Keywords: crustal velocity model, source rupture model, tectonics, Makran subduction zone, Sistan suture zone, grid search method, 2011-01-18 Pakistan earthquake, 2013-04-16 IRAN earthquake, 2013-09-24 Pakistan earthquake

1. Introduction

Three large earthquakes, the January 18, 2011 Pakistan Mw 7.2, the April 16, 2013 IRAN Mw 7.7, and the September 24, 2023 Pakistan Mw 7.8 earthquake, occurred the same tectonic region—Makran subduction zone region in a short time duration. Since the time durations between their occurrences and the differences between their locations are small, we speculated that these three earthquakes were the representations of the same tectonic movements in that region. These three earthquakes have been more or less studied. Some of the available results were summarized in the following paragraphs.

The 2013-04-16 Mw 7.7 earthquake occurred in the Sistan suture zone near the Western end of Makran subduction zone, Iran. The epicenter of this earthquake is about 100 km east-south of the city Khash (**Figures 1** and **2**). It is sometime called Khash, IRAN earthquake [3]. The earthquake affected a mountainous region between the cities of Saravan and Khash in Sistan and Baluchestan provinces, Iran, near the border with Pakistan. The quake lasted approximately 25 seconds and occurred at an intermediate depth of about 50 km within the Arabian Plate.

This earthquake was the largest shaking event to hit this region of IRAN in the past 300 years, matching the magnitude of the 1721 East Azerbaijan Earthquake that resulted in up to 250,000 fatalities, and possibly the Iranian largest in the last 50 years. It followed a magnitude 6.3 earthquake near Bushehr. The town of Mashkel, close to the epicenter, saw around 85 percent of its buildings destroyed, and the earthquake resulted in 41 fatalities and injured more than 180.

The earthquake was characterized by a normal faulting mechanism [4, 5]. It provided a valuable opportunity to study the regional tectonic and fault characteristics [5]. Its focal depth determined by the US geological survey (USGS) is about 60 km; the centroid depth provided by the GCMT group is about 50 km. The focal mechanism is normal faulting; its dipping direction is at northwest direction (about Az 328°). The principal extensive stress axis (T) is at about 337°, as such this earthquake

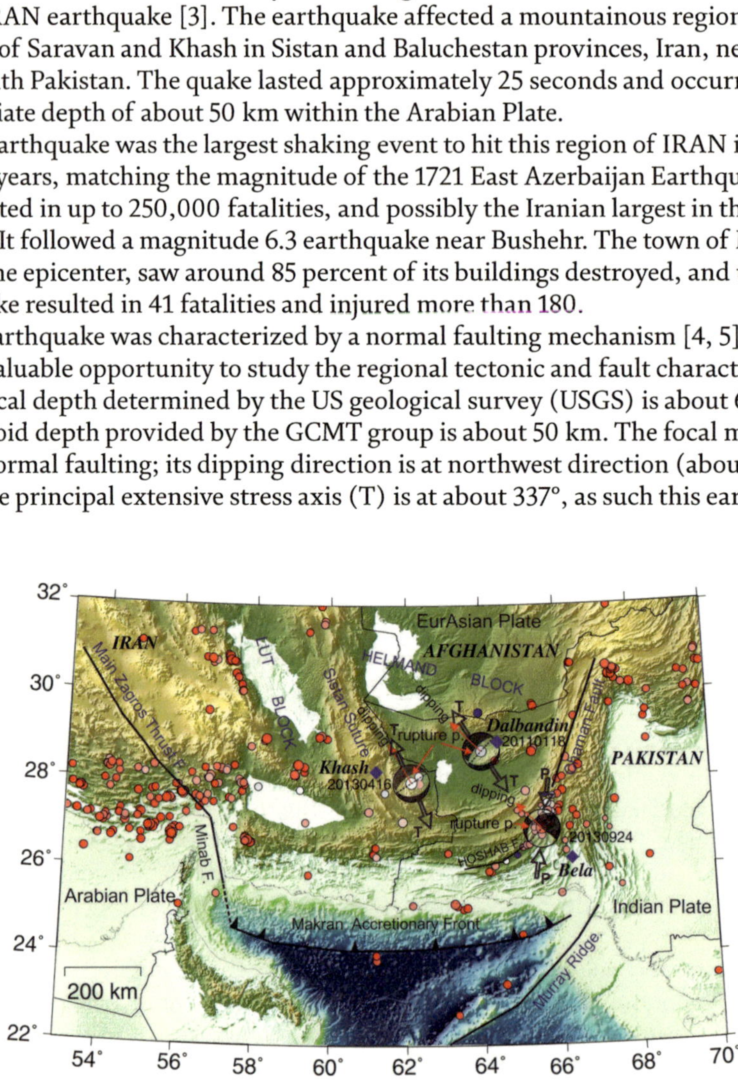

Figure 1.
*Major tectonic features, and the seismicity in the southern IRAN and southwestern Pakistan region from January 2000 to August 2024 with magnitude ≥ 5.0. There are three large earthquakes with magnitude >7.0. They are the 2011-01-18, southwestern Pakistan Mw 7.2; 2013-04-16, southern IRAN Mw 7.7; and 2013-09-24, Pakistan Mw 7.8 earthquakes. The arrows attached each beach-ball (focal mechanism lower half spherical projection) show the extensive (**T**) or compressive (**P**) force direction in the source region of each large earthquake. A visual inspection found that the focal mechanisms of the 2013 IRAN and the 2011 Pakistan were very similar; while the focal mechanism of the 2013-09-24 Pakistan earthquake is different from those of the two. All the dipping directions of the three rupture planes are at northwest. The major tectonic features were referred to the article by Byrne et al. [1].*

DOI: http://dx.doi.org/10.5772/intechopen.1007812

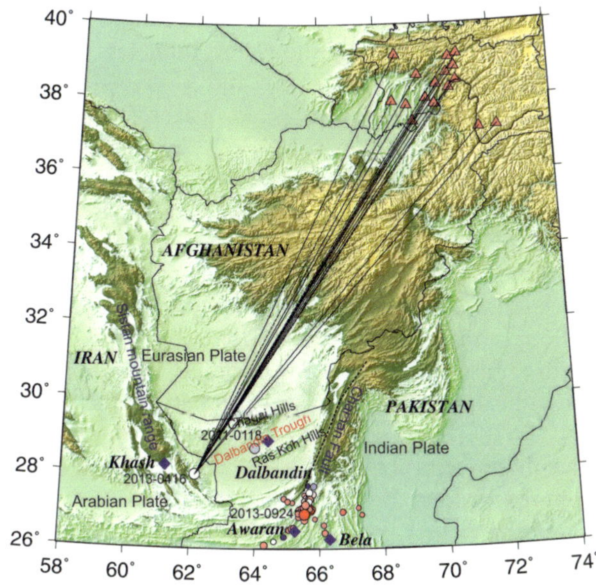

Figure 2.
Rayleigh wave travel paths used to retrieve surface wave dispersion data. The triangles show the locations of the seismic stations that recorded the Rayleigh waves. The epicenters of the three large earthquakes are indicated with three larger solid circles. The smaller circles show the aftershocks occurred within one month following the 2013-0924 Mw 7.8 main shock. The linear trend of the epicenters of those aftershocks approximately shows the strike direction of the 2013-0924 Mw 7.8 earthquake. Between the Chagai Hills and the Ras Koh Hills, the Dalbandin Trough is located there. Along the trough a fault connects to the Chaman fault; the fault is not long enough to reach the epicenter of the 2013-0416 earthquake [2].

is almost pure normal faulting. Scientists paid attention to this earthquake, and some studies on this earthquake have been performed. Its source rupture processes were also studied to some extent (e.g., [3, 5]).

The 2011-01-18 Mw 7.2 earthquake occurred in southwestern Pakistan, specifically in the Dalbandin region of Balochistan. The earthquake was felt over a vast area, from northern Persian Gulf states to Central Asia, and from Tehran to Delhi. Its significant radial impact was highlighted [6]. The 2011 earthquake resulted in relatively minor damage and casualties, due to the earthquake's depth and the low population density around the epicenter [6].

The seismicity in southern Pakistan is influenced by the active plate margins of the Arabian, Eurasian, and Indian plates, with significant seismic potential due to the accumulation of stresses along these fault lines [7]. The Dalbandin region is situated at the intersection of the Indian, Eurasian, and Arabian tectonic plates. This area is characterized by complex tectonics, including active subduction along the Makran subduction zone [6, 8].

The earthquake is mainly linked to the subduction of the Arabian Plate beneath the Eurasian Plate. The Makran-Balochistan volcanic arc at its south suggests that it likely took place within the subducting Arabian-Ormara Plate, which is being forced beneath the Eurasian Plate [6].

The earthquake occurred at an approximate depth of 70 km [6]. This depth suggests an intermediate-focus earthquake, which is characteristic of subduction zones [9].

The 2013-09-24 Mw 7.8 earthquake was also known as the Awaran earthquake. The earthquake primarily resulted from strike–slip activity along the Hoshab fault, which is part of the extensive Chaman fault system that delineates the boundary between the

Indian and Eurasian tectonic plates [10, 11]. The event caused widespread destructions, resulting in 386 deaths and leaving thousands homeless. It also impacted areas in neighboring countries, including Iran, India, and Afghanistan [12].

The earthquake took place in the eastern Makran accretionary wedge. The Hoshab fault, which has an inclination of approximately 60° toward the northwest, exists within a complicated tectonic setting formed by the convergence of the Indian and Arabian plates with the Eurasian plate [13, 14]. The earthquake was caused by displacement along an oblique-strike–slip fault at a depth of approximately 15 km. The dynamics of the rupture were influenced by preexisting geological structures, with the fault rupture alternating between long-term fault lines and secondary branches due to the compressional context of the region [15].

The surface rupture traces displayed significant left-lateral movement along with some vertical shifts, indicating a complex interplay between strike–slip and dip-slip motions [11, 14].

Since earthquake phenomena are very complex, well understanding to the phenomena has not been reached yet. We performed further studies on these three large earthquakes, with more work on the 2013-04-16 Mw 7.7 IRAN earthquake. In this

No	Source	Date	Time	Latitude (°)	Longitude (°)	Depth (km)	Mw	Moment (10^{20}NM)
1	GCMT	2013-04-16	10:44:32	27.890	62.210	50.8	7.7	5.093
	USGS	2013-04-16	10:44:19	27.9704	62.136	63.1	7.7	
2	GCMT	2011-01-18	20:23:31	28.610	63.900	52.3	7.2	0.88
	USGS	2011-01-18	20:23:25	28.6831	63.9948	79.9	7.2	
3	GCMT	2013-09-24	11:30:08	26.700	65.040	12.0	7.8	5.59
	USGS	2013-09-24	11:29:48	26.9109	65.5315	15.5	7.8	

Table 1.
Catalog of the three large earthquakes.

No.	Nodal plane and dipping	Nodal plane	Principal axes			
		strike/dip/ rake (°/°/°)	Name	Az (°)	Plunge (°)	Length (10^{20}NM)
1	NP1	80/35/−72	T	337	11	5.381
	NP2(normal)	*238/56/-102*	N	245	10	−0.578
	Dipping Az	170/*328*	P	112	75	−4.804
2	NP1	77/31/−60	T	326	17	0.924
	NP2(normal)	*224/63/-107*	N	231	15	−0.088
	Dipping Az	167/*314*	P	102	67	−0.836
3	*NP1*(thrust)	*223/39/4*	T	73	36	5.037
	NP2	130/87/129	N	307	39	1.106
	Dipping Az	*313*/220	P	188	31	−6.143

Table 2.
Focal mechanism parameters of the three large earthquakes.

chapter, we summarize the methods used, introduce an average crustal velocity model obtained using Rayleigh wav dispersion data, and the source rupture modeling results for the three large earthquakes.

The catalog and the basic source parameters for the three large earthquakes obtained by the USGS and the GCMT (The Global CMT project, see globalcmt.org) group were retrieved from the the Incorporated Research Institutions for Seismology (IRIS) Web site. **Tables 1** and **2** list the values of source parameters provided by the GCMT group and the USGS.

2. Tectonic background in the three large earthquake region

The epicenters of the three large events are located in the north region of the Makran subduction front (**Figure 1**). The Makran subduction zone is the convergent boundary between the Arabian oceanic plate and the southern margin of the Iranian micro-continent in southeast IRAN and in the south of Pakistan [1], it runs along the southeastern coast of Iran and the southern coast of Pakistan. It is a major control on the seismic hazard of the region [16].

Extensive research has explored the geology, petrology, and geochemistry of the Sistan suture zone (SSZ) [6–9, 17]. The north–south trending SSZ is the result of Sistan Ocean closure and the subsequent collision between the Lut and Afghan/Helmand blocks [11, 13, 14].

The SSZ is subducted beneath the east–west Makran coastal range, corresponding to the northward subduction of the Arabian plate. The 2013-04-16 earthquake occurred at the southern end of the SSZ.

Opposed to the SSZ is the large Chaman fault system. Connected to the fault system is the Hoshab fault, where the 2013-09-24 earthquake occurred.

In the 2011-01-18 earthquake region, the Dalbandin Trough is located there. Along the trough a small fault [2] connects to the Chaman fault.

Overlooking **Figure 1**, we found that the Sistan suture zone, the Chaman fault zone, and the Hoshab fault zone form a large U-shaped structure. The 2013-04-16 Mw 7.7 and the 2013-09-24 earthquakes occurred along the structure; while the 2011-01-18 earthquake occurred within it. The Makran subduction zone region is a natural laboratory for scientists to study subduction zone and its seismicity. The summaries of the tectonic settings have been very well organized, such as, by Bascou and Jouanne [17]; Martin and Kakar [6]; Barnhart et al. [3]; and Burg [18].

3. Method introduction

3.1 Crustal velocity model

Surface waves are formed when P and S waves interact at the Earth's surface. There are two types of surface waves—Rayleigh wave and Love waves. Rayleigh waves are generated by P-waves and SV-waves (S-waves which vibrate along a vertical plane); while Love waves are generated by SH waves (S-waves which vibrate along a horizontal plane). S-wave velocities could be retrieved using the surface wave dispersion data. The dispersion data of Rayleigh waves could be measured using the multiple filter technique (MFT) technique developed by Dziewonski et al. [19], implemented in the computer program package by Herrmann and Ammon [20].

Once the Rayleigh wave dispersion data are measured, the S-wave velocities in the earth through which the Rayleigh waves travel could be extracted. The method starts with an initial crustal model. This model serves as a basis for further refinement in the inversion. The inversion utilizes the observed dispersion data to iteratively update the crustal model. At each step, the predicted dispersion, computed using the current model, is compared to the observed dispersion. The misfit between the observed and the calculated dispersion curves is then minimized by adjusting the model parameters. This iterative process continues until a model that exhibits the best possible agreement with the observed dispersion is achieved. The mathematical background using Rayleigh wave dispersion data to retrieve S-wave velocities has been introduced in literatures (e.g., [21–25]).

3.2 The source rupture modeling method

The procedure to model the rupture process of an earthquake is to set up a Cartesian coordinate system on a selected nodal plane. The selected plane is divided into many small rectangles which are called sub-faults. At a specific station, a synthetic seismogram is the summation of the synthetic seismograms generated by all of the sub-faults. The source time function of each sub-fault (the point source) is usually described by overlapped triangles. The layout of the source rupture model can be found in Hartzell and Heaton [26]. A unit rupture slip vector for each sub-fault can be divided into two orthogonal vector components. At a station, each component of the vector on a sub-fault can be used to calculate a synthetic seismogram called Green's function. Any slip vector on the sub-fault can be obtained by multiplying the two constant components with appropriate coefficients. An inversion aims to obtain the coefficients for all of the sub-faults to establish the rupture model for the earthquake.

For shallow large earthquakes, the aftershock distribution may be used to identify the causative plane. For deep earthquakes, usually not many aftershocks or even no aftershocks following; as such the rupture plane could not be selected using the aftershock distribution.

When the earthquake source is a finite type, the patterns of the seismograms, generated by the source on either of the two nodal planes, at some stations around the epicenter are not the same. When one of the two nodal planes is used as the rupture plane to perform source rupture modeling, the fit between the observed and the synthetic seismograms at the same station is at a certain extent better when the plane is close to the real rupture plane. Accordingly, the average variance between the observed and the synthetic seismograms is smaller, when the assumed plane is close to the real rupture plane. Thus, we could identify the rupture plane by comparing the average variance obtained using the two nodal planes.

Based on the above facts, Ma et al. [25] introduced a grid search procedure to identify the causative plane and at the same time to obtain an optimal rupture initial depth and an average rupture velocity. Once a nodal plane is determined, and the rupture initial depth and the average rupture velocity are obtained, a source rupture model could be established. The procedure to perform a source rupture modeling was introduced (e.g., [25–28]).

4. Seismogram records for the crustal velocity and source rupture modeling

The data used for the crustal velocity models are Rayleigh surface wave vertical records. The waveform records were downloaded from the IRIS database. Only the high-quality Rayleigh wave segments were considered in the analysis to ensure high signal-to-noise ratios. A filtering process using a band-pass filter of a frequency range of 0.02 to 0.2 Hz was applied to the dataset. **Figure 2** shows the Rayleigh wave paths used to retrieve crustal models.

The data used for the source rupture inversions are P-wave segments in the teleseismic waveform vertical records. The waveform records were also retrieved from the IRIS database. Specifically, records within an epicentral distance of 30° to 90° surrounding the epicenter were selected for analysis. Only the high-quality P-wave segments were considered. Many waveform records are available. To save time, only the waveform records, provided by the global seismographic network (GSN), were used for our current work.

5. Crustal velocity models in the three large earthquake region and their vicinity

The vertical components, generated by the 2013-04-16 M_W 7.7 earthquake, were retrieved from the IRIS. The dataset was obtained from approximately 40 stations. Only high-quality data were considered. To facilitate the Green's function calculations for the source rupture modeling, an averaged crustal model was constructed from seventeen crustal models. **Figure 3** shows the average crustal velocity model obtained using Rayleigh-wave dispersion data. The two red lines represent the final average crustal velocity model. The two blue lines represent the initial global velocity model AK135 [29].

Within the depth range of 15 km to 30 km, both the Vs and Vp profiles exhibit a prominent high-velocity zone compared to the global crustal profile. Another zone with contrasting velocity characteristics, potentially a low-velocity zone, is observed at depths shallower than 15 km and deeper than 30 km. This means that the crustal profile within the depth range from 15 to 30 km in this region has more brittle behavior than what is generally expected for that portion of the crust. On the other hand, the crust profile at depth of less than 15 km and the depth deeper than 30 km generally demonstrate weaker rock strata.

Figure 3 also presents the standard deviation associated with the inverted S-wave velocity profile, depicted by the dotted lines. These standard deviations were calculated by analyzing data from all used Rayleigh wave travel paths. The standard deviation for the P-wave velocity is slightly higher than that of the S-wave velocity. This observation aligns with the expected behavior, as P-wave velocity is derived from the inverted S-wave velocity through conversion. Consequently, the inherent uncertainties associated with the S-wave model propagate and potentially amplify when calculating the P-wave model, leading to a higher standard deviation for P-waves.

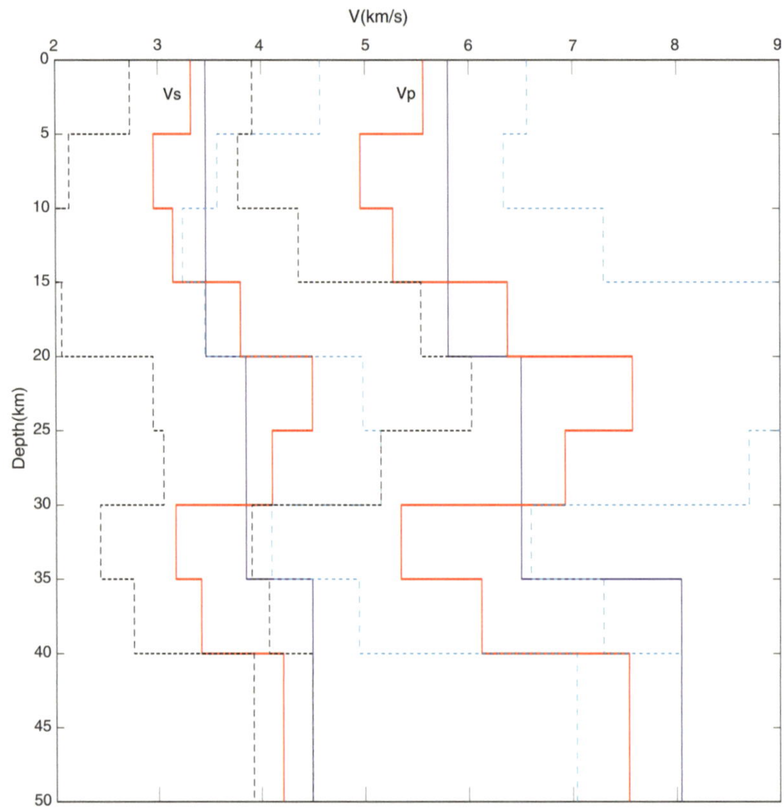

Figure 3.
An average crustal velocity model obtained using Rayleigh wave dispersion data. Red lines demonstrate the developed Vs and Vp models along with standard deviations of the values. Blue line shows the global average velocity profile. (This figure will be re-produced).

6. Source rupture modeling for three large earthquakes

6.1 Source rupture modeling for the 2013-04-16 Mw 7.7 earthquake

6.1.1 Selection of the rupture plane and the determination of the rupture initial depth and the average rupture velocity

Selecting one of the two nodal planes as the rupture plane is necessary when establishing a source rupture model. Typically, the distribution of aftershocks could be used to determine the rupture plane for strike–slip type of earthquakes. For the normal or thrust/reverse slip earthquakes, the linear trend of the aftershocks may not be clear, and as such, the rupture plane may not be able to select from the two nodal planes. In our work, we assumed each of the nodal planes to be a rupture plane and performed the grid search. First, one of the nodal planes is used as the rupture plane; then a grid search scheme is set up.

For the 2013-04-16 Mw 7.7 earthquake, along the depth axis (y-axis), for the NP1 there are 63 grid points (2 km to 64 km); for the NP2 there are 57 grid points (2 km to 58 km). Along the rupture velocity axis (x-axis), there are 16 grid points (from 1.3 km/s to 2.8 km/s). The total grid points are 63 × 16 = 1008 and 57 × 16 = 912; the total inversion number is 1920, and that is, there are 1920 objective function values to be calculated. After the calculations for a grid node were completed, the average variance value for the node was obtained. This value together with the used depth value and the used rupture velocity value were stored in an output data file. Each nodal plane has one output data file for plotting a contour map for the objective function.

The upper panel of **Figure 4** was plotted using the objective function values for NP1 (**Table 2**); while the bottom panel was plotted using the objective function values for

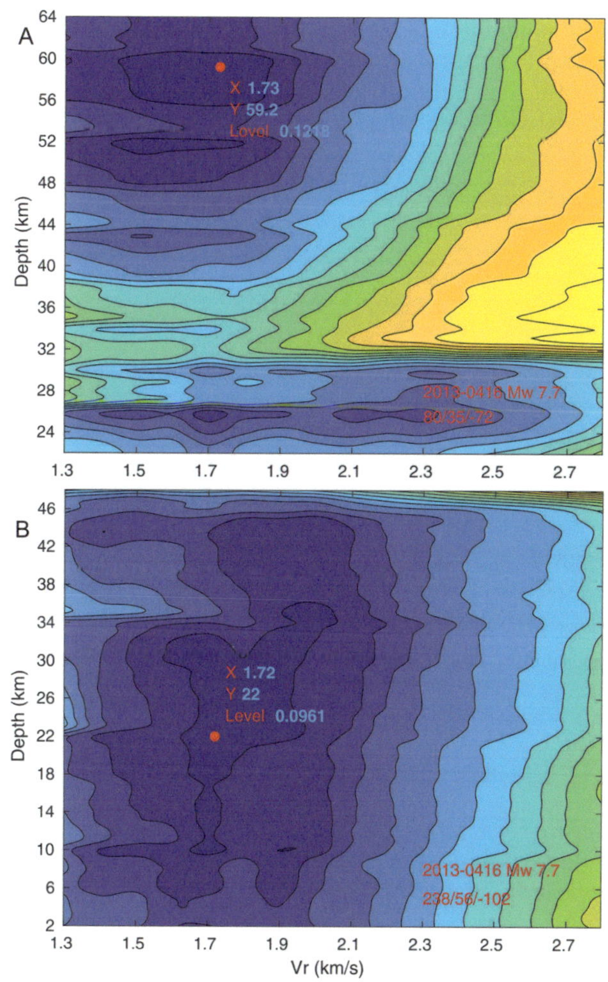

Figure 4.
Contour maps of the objective function values (variance values, the values of the objective function f(x, y); x = vr, rupture velocity; y = rupture initial depth). At a specific point on the maps, a deeper color shows a smaller variance value. A variance value shows the fit between the observed and the synthetic seismograms. The smaller the value is, the better the fit between the observed and the synthetic seismograms. Upper panel, A, when the nodal plane, NP1, strike/80°, dip/36°, rake/−72°, provided by the GCMT, was used as the rupture plane to perform the grid search procedure, the minimum of the variances was approximately at the rupture velocity of 1.72 km/s and the initial depth of 59.2 km; the minimum variance is 0.1218, directly measured on a computer screen. Bottom panel, B, when the nodal plane, NP2, strike/238°, dip/56°, rake/−102°, was used as the rupture plane to perform the grid search procedure, the minimum was at the rupture velocity of 1.72 km/s and the initial depth of 22.0 km; the minimum variance is 0.0961. When the rupture velocity values and the rupture initial depths at the above two minimum points were used in the source rupture inversions, the obtained variances in the inversions may be changed slightly.

NP2. The minimum point on the upper panel was approximately at (1.73 km/s, 59.2 km). At this point, the measured objective function value was 0.1218. On the bottom panel the minimum was approximately at (1.72 km/s, 22 km), and the minimum value was 0.0961. After comparing the two minima, the nodal plane NP2 for the bottom panel was selected to be the rupture plane. Accordingly, the rupture velocity 1.72 km/s and the depth 22 km were selected for the final source rupture modeling for the earthquake.

6.1.2 Source rupture model for the 2013-04-16 Mw 7.7 earthquake

Taking the minimal point (1.72 km/s, 22 km) as the average rupture velocity and the initial rupture depth, a rupture model was obtained for the M_W 7.7 earthquake. **Figure 5** shows the slip distribution. The key features are the center of the major rupture patch was at about 70 km; the secondary patch was within the lower crust;

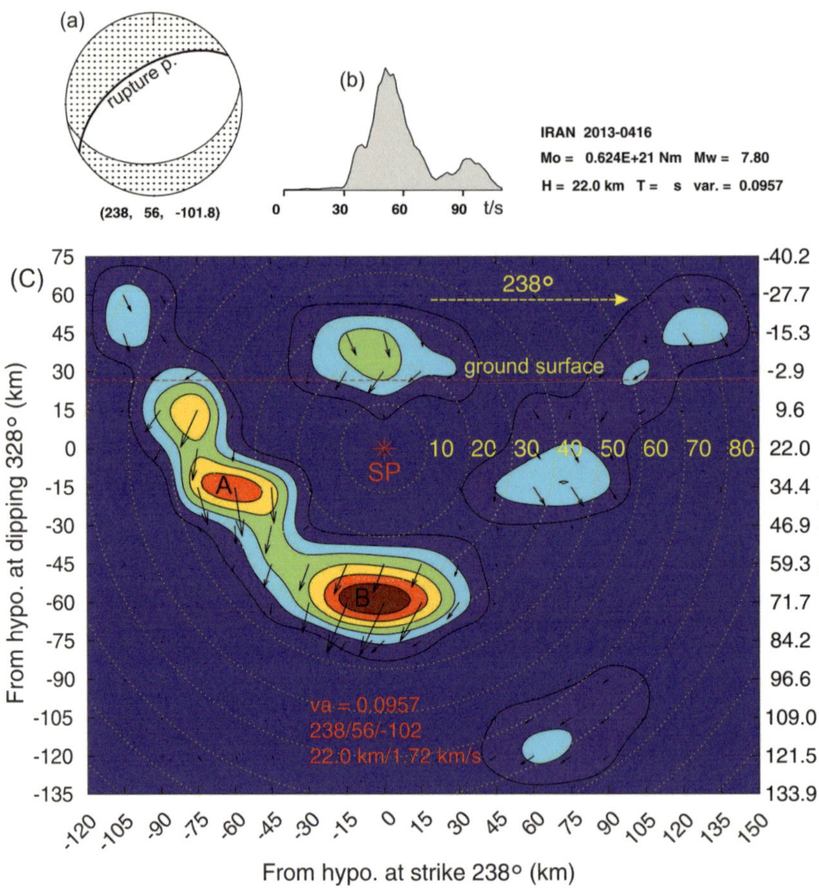

Figure 5.

Rupture distribution obtained using the GCMT nodal plane (strike/238°, dip/56°, and rake/−102°) at the minimum (rupture velocity of 1.72 km/s, initial rupture depth of 22.0 km). The star (*) marked with SP denotes the rupture start point. One large patch indicated with letter B was obtained. The circles represent the rupture fronts at specific time intervals, indicated as 10 (s), 20 (s), and so on. The focal mechanism, the source time function (the moment release rate), and the major source parameters were shown at the top. The solid curve in the beach ball shows the rupture plane. The depth of the major rupture center was at about 70 km; the calculated minimum variance va = 0.0957, slightly smaller than that measured from the contour map (0.0961). The maximum slip in patch B was 2.1 m.

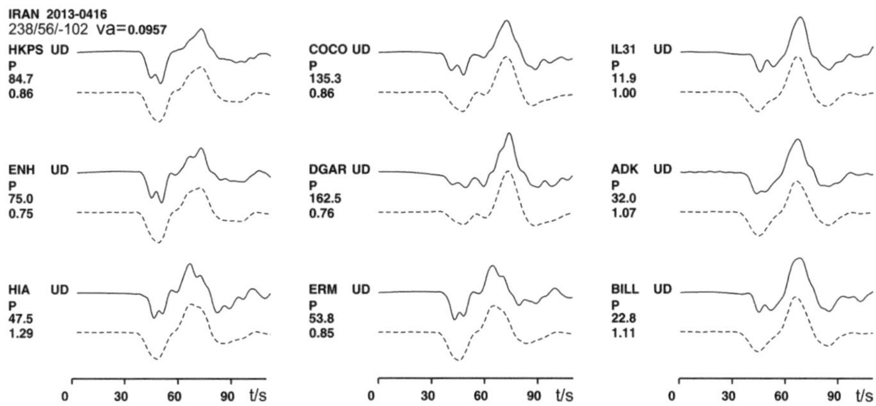

Figure 6.
Comparison between the observed and synthetic seismograms. For each pair the upper trace represents the recorded waveform, whereas the lower trace represents the synthetic waveform generated using the slip distribution depicted in ***Figure 5****. Both the observed and synthetic waveforms were filtered with a band-pass filter, ranging from 0.01 to 0.2 Hz. The symbols and numbers on the left side of each pair indicate the station name, the P-wave vertical component, the station azimuth in degrees, and the ratio between the observed and synthetic maximum amplitudes. The average variance va = 0.0957. Thirty-seven (37) high quality records were used in the inversion. Only nine records were plotted in this figure. The quality of fit between the observed and the synthetic seismograms for the other 28 pairs is similar to that in this figure.*

the rupture reached the ground surface. The variance between the observed and the synthetic seismograms at a station is used as a criterion to judge the fit between the two seismograms. The smaller the variance the better the fit is. **Figure 6** shows nine pairs of comparisons between the observed and synthetic seismograms. Visually the shapes of the observed and the synthetic waveforms are very similar. This indicates that the modeling is good. The average variance va = 0.0957. This number is not exactly the same as that (0.0961) measured directly from the contour map (**Figure 4**). The reason is that the number 0.0957 came from inversion, while the number 0.0961 was the measured value. These two numbers should be very close. Thirty-seven (37) high-quality records were used in the inversion. To save page space only nine records were plotted in **Figure 6**. For the other 28 records, the fit quality is similar.

6.2 Source rupture modeling for the 2011-01-18 Mw 7.2 earthquake

6.2.1 Focal depth determination for the 2011-01-18 earthquake

The focal depth of this Mw 7.2 earthquake, provided by the USGS was 79.9 km; the centroid depth provided by the global CMT group was 52.3 km. When we analyzed the P-wave segments, we found that the depth phase sP was very well developed. Using the time difference between sP phase and its reference phase, the direct P wave, the focal depth can be calculated. The measured sP–P time from the P wave segment recorded at station HIA is 29.4 s. Using this value and the procedure, developed by Motazedian and Ma [30], the calculated focal depth is 81.4 km.

6.2.2 Selection of the rupture plane; and the determination of the initial rupture depth and the average rupture velocity

Using the same procedure for the 2013-04-16 earthquake, the causative plane was selected for the 20,121-01-18 earthquake. The selected nodal plane is the NP2

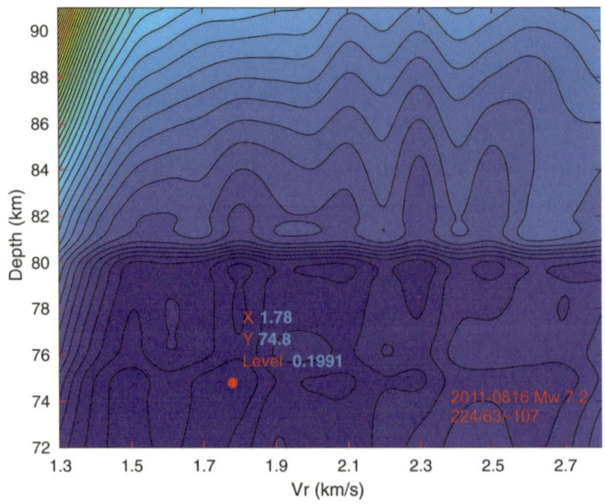

Figure 7.
Contour map of the objective function. The nodal plane, strike/224°, dip/63°, and rake/−107°, provided by the GCMT, was used as the rupture plane to perform the grid search procedure. The measured minimal point was at (1.78 km/s, 74.8 km). At this point the measured average variance va was 0.1991.

in **Table 2** (strike/224°; dip/63°; and rake/−107°). **Figure 7** was generated using the depth range from 70 km to 91 km; the rupture velocity ranges from 1.3 km/s to 2.8 km/s. The pattern on the contour map is complex; the measured optimal lowest point was at (1.78 km/s, 74.8 km).

6.2.3 Source rupture model for the 2011-01-18 Mw 7.2 earthquake

Using the obtained average rupture velocity (1.78 km/s) and the initial rupture depth (74.8 km), a rupture model was obtained for the M_W 7.2 earthquake. **Figure 8** shows the rupture distribution. The key features are the center of the major rupture patch was at depth about 60 km. The shapes of the source time function, panel (b) in **Figure 8** show that the source rupture process was simple and completed in a short time. **Figure 9** shows six pairs of waveform comparison. It could be found that the fit between the observed and the synthetic waveforms is generally good. Along the traces there are two large phases. The first one is the directive P phase; while the second is the depth phase sP. The time difference between these two phases can be used to calculate focal depth with a small uncertainty.

6.3 Source rupture modeling for the 2013-09-24 Mw 7.8 earthquake

6.3.1 The determination of the rupture initial depth and the average rupture propagation velocity

The causative plane of this earthquake could be easily determined. The aftershock distribution of this earthquake (**Figures 1** and **2**) indicated that the nodal plane NP1 in **Table 1** was the rupture plane.

Since this earthquake was a crustal event, the depths range in the grid search for the optimal rupture initial depth and the average rupture velocity was taken from 2 km to

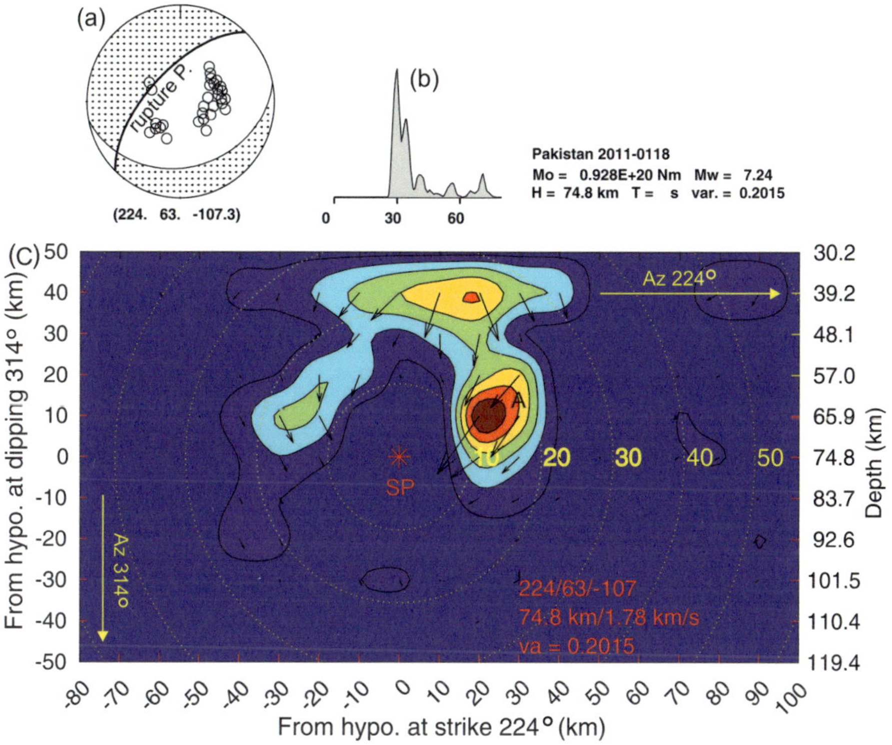

Figure 8.
Rupture distribution obtained using the GCMT nodal plane (strike/224°, dip/63°, rake/−107°) at the minimal point (rupture velocity of 1.78 km/s, initial rupture depth of 74.8 km) on the grid search contour map. The star (*) marked with SP denotes the rupture start point. One large patch, indicated with letter A, was obtained. The depth of the major rupture center was at about 70 km. The calculated minimal variance va = 0.2015; slightly larger than that measured from the contour map (0.1991). The ruptures were limited within the earth.

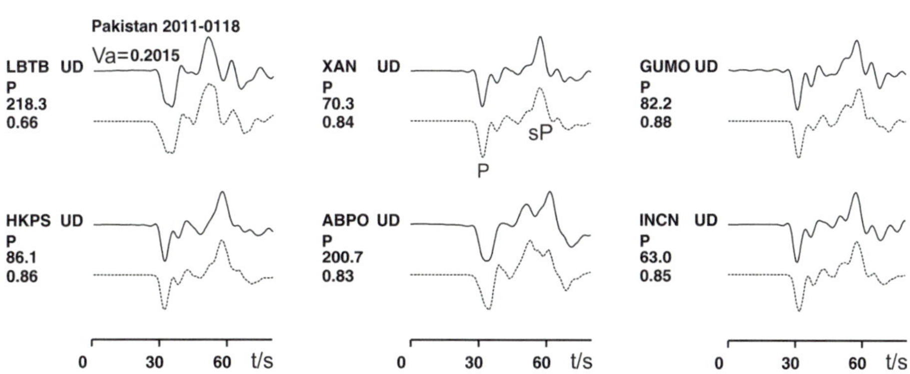

Figure 9.
Comparison between the observed and synthetic seismograms. For each pair the upper trace represents the recorded waveform, whereas the lower trace represents the synthetic waveform generated using the slip distribution depicted in ***Figure 8****. Both the observed and synthetic waveforms were filtered with a band-pass filter, ranging from 0.01 to 0.2 Hz. The average variance va = 0.2015. Thirty (30) vertical waveform records were used in the inversion. Only six pairs were plotted in this figure. The quality of fit between the observed and the synthetic seismograms for the other 24 pairs is similar to that in this figure. Along the traces there are two large phases. The first one is the directive P phase; while the second is the depth phase sP.*

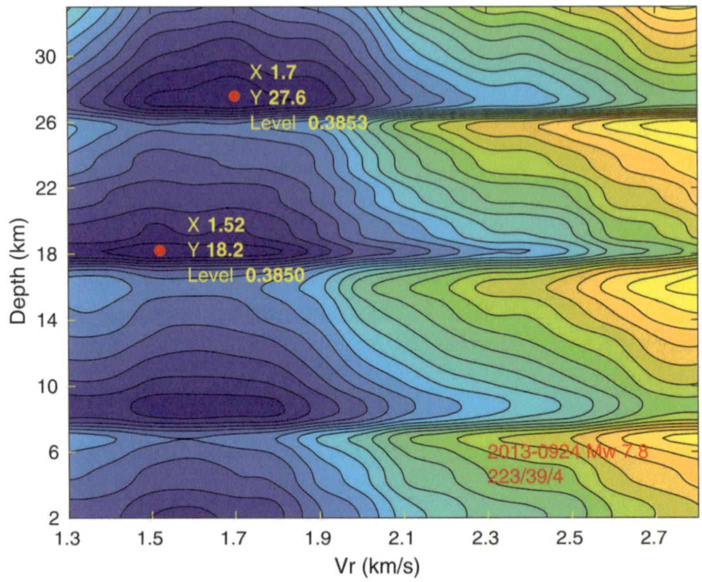

Figure 10.
Contour map of the objective function for the 2013-09-24 Mw 7.8 earthquake. The nodal plane, strike/223°, dip/39°, and rake/4° (when the rake's value is positive, the source rupture has a thrust component), provided by the GCMT, was used as the rupture plane to perform the grid search procedure. There are two the minimal points. The point with va = 0.3850, formed by the rupture velocity of 1.52 km/s and the initial depth of 18.2 km, was used in the inversion. The local minimum point with va = 0.3853, formed by the rupture velocity of 1.70 km/s and the initial depth of 27.6 km, was also tested. The obtained rupture model seems not to be acceptable.

33 km, with a 1-km grid interval; the average rupture velocity range was taken from 1.3 km/s to 2.8 km/s; and the nodal plane (strike/223°; dip/39°; and rake/4°) was taken as the rupture plane, and the grid search was performed (**Figure 10**). It was found that there are two minimal points at (1.52 km/s, 18.2 km) and (1.70 km/s, 27.6 km).

6.3.2 Source rupture model for the 2013-09-24 Mw 7.8 earthquake

Taking the two minimal points to perform the source rupture inversions, it was found that when the point (1.70 km/s, 27.6 km) was used, only one large rupture patch was obtained. After referring to the preliminary source rupture modeling results by the USGS, we thought only one large rupture patch for this large crustal event is not reasonable. So the inversion results using the point (1.52 km/s, 18.2 km) were presented. **Figure 11** shows that there were a large rupture patch and a smaller patch in the upper crust. The ruptures broke through the ground surface of the earth.

Figure 12 shows the six pairs of waveform comparison. The complexity of the waveform shapes shows that the source rupture process was complex. As such, the fit between observed and the synthetic seismograms is not as good as those for the other two earthquakes. Since the major features were modeled, the obtained rupture model may in some extend show the real rupture process of this Mw 7.8 earthquake.

7. Discussion and conclusions

Three types of the P-waves are the major contents in a P-wave segment recorded at a tele-seismic station. (1) Direct P-wave, traveled from an earthquake source to the

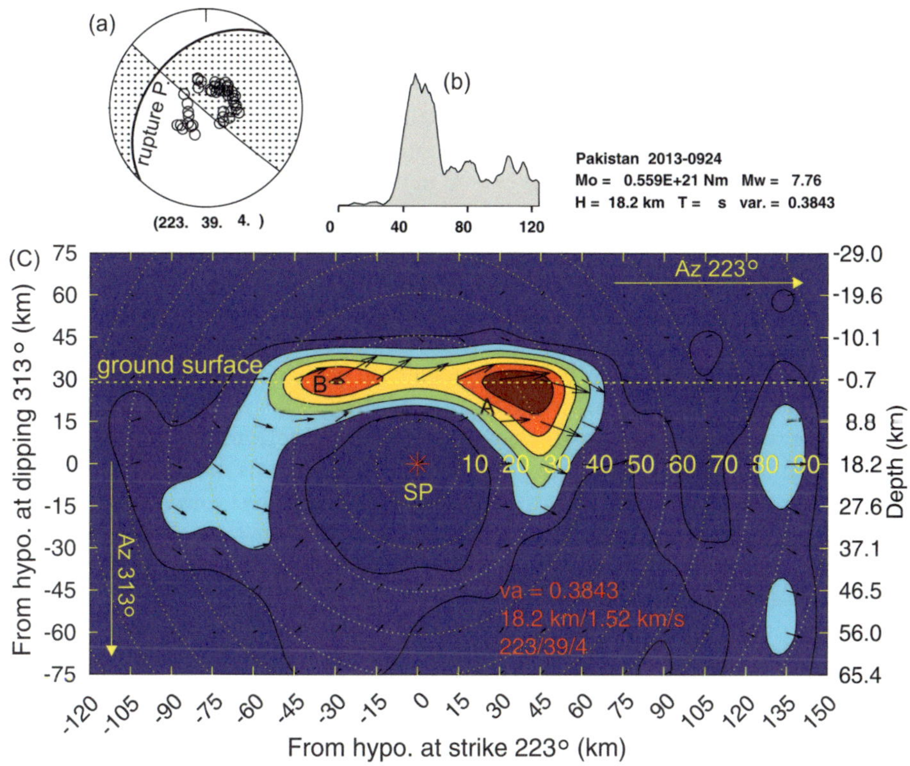

Figure 11.
Rupture distribution obtained using the GCMT nodal plane (strike/223°, dip/39°, rake/4°) at the minima (1.52 km/s, 18.2 km). The star (*) marked with SP denotes the rupture start point. The major ruptures occurred in the upper crust; the calculated minimal variance va = 0.3843; slightly smaller than that measured from the contour map (0.3850). The ruptures ran through the ground surface, meaning that the inversion program does not have the ability to limit the ruptures within the crust. This phenomenon is similar to that an earthquake's focal depth is in the air, when locating a shallow earthquake.

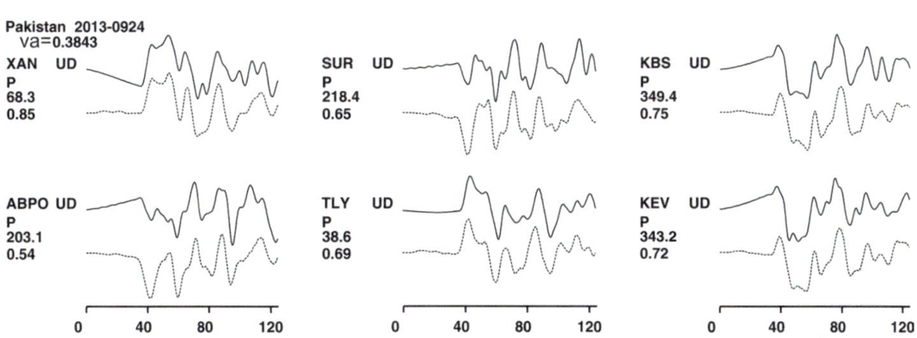

Figure 12.
Comparison between the observed and synthetic seismograms. For each pair the upper trace represents the recorded waveform, whereas the lower trace was generated using the slip distribution depicted in ***Figure 11****. Both the observed and synthetic waveforms were filtered with a band-pass filter, ranging from 0.01 to 0.2 Hz. The average variance va = 0.3843. Forty-three (43) vertical waveform records were used in the inversion. Only 6 records were plotted in this figure. The quality of fit between the observed and the synthetic seismograms for the other 37 pairs is similar to that in this figure. Since the shapes of the waveforms are complex, the fit is not as good as those for the other two earthquakes. As the major features are caught, the modeled ruptures may indicate to some extent the real ruptures of this earthquake.*

station; (2) pP wave, reflected P-wave at the surface of the earth above the source, then traveled to the station; and (3) sP waves, the S-wave converted to P-wave at the surface above the source, then traveled to the station. The amplitudes of the pP and sP in the P-wave segment are related to the source magnitude, and the focal mechanism, and the structures of the crust in the source region. When the synthetic seismograms (Green's functions) are calculated, a reasonable crustal model in the source region is required.

For shallow large earthquakes, the aftershock distribution may be used to identify which nodal plane is the causative plane. For deep earthquakes, usually not many aftershocks or even no aftershocks follow the main shock. In **Figure 1**, we can see that some aftershocks distributed around the epicenter of the 2013-04-16 Mw 7.7 earthquake; but the pattern of those epicenters did not form a linear trend. For the 2011-2101-18 Mw 7.2 earthquake, basically no aftershocks occurred around its epicenter. When the earthquake source is a finite type, the patterns of the seismograms, generated on either of the two nodal planes, recorded at some stations around the epicenter are not the same. When we treat one of the two nodal planes as the rupture plane, the fit between the observed and the synthetic seismograms at the same station is at a certain extent better when the assumed nodal plane is close to the real rupture plane. Accordingly, the average variance, the amplitude difference between the observed and the synthetic seismograms at a station, is smaller (**Figure 4**).

Using the above procedure, the source rupture plane could be identified. At the same time, the initial rupture depth and the average rupture propagation velocity are also estimated. In our previous source rupture modeling work, the initial rupture depth value was the focal depth provided by the USGS; the rupture velocity value was from a guess or trail-and-error. As such, the rupture velocity value used in the inversion may not be reasonable. Up to date, other people who perform the source rupture modeling may still use the practice as we did before.

In our knowledge, other scientists have performed the source rupture modeling for the 2013 Mw 7.7 earthquake. Barnhart et al. [3] concluded that the earthquake ruptured at least half of the subducted oceanic lithosphere, potentially including the entire oceanic crust. The depth of the major rupture patch we obtained (**Figure 5**) is from about 60 to 85 km, similar to that Barnhart et al. [3] obtained. In our model, the rupture reached the ground surface. This modeling result is consistent with the observed ground fissure. Figures in the report (Zare et al. [31]) clearly show the surface fissure in the Gol-Pichak village (nearby Gosht city) created during the 2013-04-16 earthquake.

The preliminary source rupture inversion conducted by the USGS shows that they used a focal depth of 60 km, and the major rupture they obtained was from depth about 40 km to 90 km. Walpersdorf et al. [5] used interferograms, coseismic GPS velocity field to combine with teleseismic P-wave data to model source fault parameters. They found that the Saravan causative fault is a north-dipping normal fault with a dip of about 67°. The major rupture patch they obtained was from a depth of 30 km to 65 km. In our modeling, the causative plane has a northwest dipping (**Figure 1**). The ruptures distributed from the ground surface to the depth of about 85 km.

For the 2011-01-18 Mw 7.2 earthquake, we have not spent enough time to study its source rupture process. We also have not found the source rupture modeling results performed by other people. We checked the USGS Web site and found no preliminary source rupture modeling results there. Our results show that the depth of the major rupture of this earthquake is on the same order of that the major rupture of the 2013 Mw 7.7 earthquake displayed. The identified causative planes of these two earthquakes have similar orientations (**Figure 1**).

For the 2013-09-24 Mw 7.8 earthquake, the key rupture feature is that the major ruptures occurred within the upper crust. Its ruptures also broke the surface of the ground. Bascou and Jouanne [10] stated that: our simulations indicate that after-slip along the basal dĕcollement of the accretionary prism is required to explain the observed post-seismic deformation. This basal dĕcollement seems to be located at 18 km depth with a 7° northward dip, which is consistent with the geometry proposed by Grando and Mclay [32] and Burg et al. [18]. Coincidentally, in the grid search procedure the optimal initial rupture depth we obtained was 18.2 km (**Figure 10**).

For the source rupture modeling for the 2013 Mw 7.7 IRAN earthquake, we conclude that the nodal plane (strike/ 238°, dip/56°, and rake/ -102°) was the causative fault plane. It was dipping at Az 328°; the rupture distributed from the ground surface to as deep as about 85 km; the major rupture patch occurred between depth about 60 km to 85 km. The maximum slip was about 2 m to 4 m, depending on the used sub-fault dimension. The fault formed by the earthquake may cut through the Arabian oceanic lithosphere slab.

For the 2011 Mw 7.2 Pakistan earthquake, we conclude that the nodal plane (strike/ 224°, dip/63°, and rake/ -107°) was the causative fault plane. It was dipping at Az 314°; the major rupture patch occurred between depth about 55 km to 75 km. The maximum slip was about 1.2 m when the used sub-fault dimension was 10 km × 10 km. The fault formed by the earthquake may also cut through the Arabian oceanic lithosphere slab.

For the 2013 Mw 7.8 Pakistan earthquake, the nodal plane (strike/ 223°, dip/39°, and rake/4°) was the causative plane; the fault orientation was at 313°. The major ruptures occurred within the upper crust. The rupture fault was a left-lateral strike–slip type with a small thrust component.

Acknowledgements

This research was made possible through the support of the Natural Sciences and Engineering Research Council of Canada under the Discovery Grant program. We would like to acknowledge the following programs and tools in our data processing and figure preparation: SAC2000, Rdseed, geotool, MATLAB, and Generic Mapping Tools (GMT).

Data sources

In this study, the seismograms, GCMT, and USGS nodal plane solutions were obtained from the Incorporated Research Institutions for Seismology (IRIS) database, accessed at http://www.iris.edu

Author details

Parisa Asgharzadeh, Shutian Ma* and Darioush Motazedian
Carleton University, Ottawa, Canada

*Address all correspondence to: shutian33@yahoo.ca

References

[1] Byrne DE, Sykes LR, Davis DM. Great thrust earthquakes and aseismic slip along the plate boundary of the Makran subduction zone. Journal of Geophysical Research. 1992;**97**:449-478. DOI: 10.1029/91JB02165

[2] Ul-Hadi S, Khan SD, Owen LA, Khan AS, Hedrick KA, Caffee MW. Slip-rates along the Chaman fault: Implication for transient strain accumulation and strain partitioning along the western Indian plate margin. Tectonophysics. 2013;**608**:389-400

[3] Barnhart WD, Hayes GP, Samsonov SV, Fielding EJ, Seidman LE. Breaking the oceanic lithosphere of a subducting slab: The 2013 Khash, Iran earthquake. Geophysical Research Letters. 2014;**41**:32-36. DOI: 10.1002/2013GL058096

[4] Musavi Z, Amiri M, Walpersdorf A, Khorrami F, Pathier E, Samsonov S, et al. Constraints on the 2013 Saravan intraslab earthquake using permanent GNSS, InSAR and seismic data. Geophysical Journal International. 2024;**239**:155-172. DOI: 10.1093/gji/ggae248. Advance Access publication 2024 August 10 GJI Geodynamics and Tectonics

[5] Walpersdorf A, Amiri M, Pathier E, Mousavi Z, Khorrami F, Samsonov S. Slip model of the 2013 April 16 mw 7.7 Saravan intra-slab earthquake (Makran subduction zone) derived from InSAR, GPS, and Teleseismic P-wave modeling show affiliations. 2022. In: EGU22, the 24th EGU General Assembly, Held 23-27 May 2022 in Vienna, Austria and Online. Available from: https://egu22.eu/id.EGU22-7507

[6] Martin SS, Kakar DM. The 19 January 2011 mw 7.2 Dalbandin earthquake, Balochistan. Bulletin of the Seismological Society. 2012;**102**(4):1810-1819

[7] Khan MJ, Lamessa G, Ali M, Khan M. An updated catalog of earthquakes for southern Pakistan and surrounding offshore region. Arabian Journal for Science and Engineer. 2021;**47**:601-617

[8] Shafiq-Ur-Rehman A, Azeem T, Khairy A, Nasir A. Parameterization of 18th January 2011 earthquake in Dalbadin region, Southwest Pakistan. NRIAG Journal of Astronomy and Geophysic. 2013;**2**(2):203-211

[9] Ray S, Chakrabarti SK, Chakrabarti S. A study of the behavior of the terminator time shifts using multiple VLF propagation paths during the Pakistan earthquake (M = 7.2) of 18 January 2011. Natural Hazards and Earth System Science. 2013;**13**(6):1501-1506

[10] Bascou P, Jouanne F. Quantification and modelling of post-seismic deformation following the 24 September 2013 mw 7.7 earthquake in the Makran region, Pakistan. Geophysical Journal International. 2022;**233**(2):999-1017

[11] Mahmood I, Iqbal M, Rizwan Mughal M, Shahzad MI, Tariq S, Ullah K. An investigation of 2013 M w 7.7 Awaran earthquake, Pakistan. International Letters of Natural Science. 2016;**61**:36-42

[12] Raja I, Iqbal J, Jadoon I, Desk S. The Balochistan earthquake 2013: Emergence of a new island in the arabian sea. Geology, Environmental Science. 2017;**2**(1):153-164

[13] Barnhart W, Hayes GP, Briggs RW, Gold RD, Bilham R. Ball-and-socket tectonic rotation during the 2013 Mw7.7 Balochistan earthquake. United States

Geological Survey, Cooperative Institute for Research in Environmental Sciences—Earth and Planetary Science Letters. 2014;**403**:210-216

[14] Harrington J, Wang T, Feng GTG, Akoglu AM, Jónsson S, Motagh M. Coseismic surface displacements from optical and SAR image offset tracking, fault modeling and geomorphological analysis of the sept. 24th, 2013 M7.7 Balochistan earthquake. In: 2014. EGU General Assembly 2014, Held 27 April—2 May, 2014 in Vienna

[15] Vallage A, Klinger Y, Lacassin R, Delorme A, Pierrot-Deseilligny M. Geological structures control on earthquake ruptures: The Mw7.7, 2013, Balochistan earthquake, Pakistan. Geophysical Research Letters. 2016;**43**(19):10, 155-10, 163

[16] Musson RMW. Subduction in Western Makran: A historian's contribution. Journal of the Geological Society. 2009;**166**:387-391

[17] Khan MJ. Modelling of seismicity in southern Pakistan using GIS techniques. Earth Science Informatics. 2020;**13**(4):1327-1340

[18] Burg JP, Dolati A, Bernoulli D, Smit J. Structural style of the Makran tertiary accretionary complex in SE-Iran. In: Al Hosani K, Roure F, Ellison R, Lokier S, editors. Lithosphere Dynamics and Sedimentary Basins: The Arabian Plate and Analogues. Heidelberg: Springer Verlag; 2013. pp. 239-259

[19] Dziewonski A, Bloch S, Landisman M. A technique for the analysis of transient seismic signals. Bulletin of the Seismological Society of America. 1969;**59**:427-444

[20] Herrmann R, Ammon C. Computer Programs in Seismology. Missouri, USA: Saint Louis University; 2002; version 3.30

[21] Båth M. Spectral Analysis in Geophysics. Amsterdam: Elsevier; 1974. p. 563

[22] Rodi WL, Glover P, Li TMC, Alexander SS. A fast, accurate method for computing group-velocity partial derivatives for Rayleigh and love modes. Bulletin of the Seismological Society of America. 1975;**65**:1105-1114

[23] Badal J, Corchete V, Payo G, Canas JA, Pujades L, Serón FJ. Processing and inversion of long-period surface wave data collected in the Iberian Peninsula. Geophysical Journal International. 1990;**1990**(100):193-202

[24] Kafka AL, Reiter EC. Dispersion of Rg waves in southeastern Maine: Evidence for lateral anisotropy in the shallow crust. Bulletin of the Seismological Society of America. 1987;**77**(3):925-941

[25] Ma S, Asgharzadeh P, Motazedian D. The source rupture models and Seismogenic structures of the Iran 2017 MW 7.3 earthquake. In: Salazar W, editor. Earthquake Ground Motion. London, UK: IntechOpen; 2024. DOI: 10.5772/intechopen.1003138

[26] Hartzell SH, Heaton TH. Inversion of strong ground motion and teleseismic waveform data for the fault rupture history of the 1979 Imperial Valley, California, earthquake. Bulletin of the Seismological Society of America. 1983;**73**:1553-1585

[27] Bjork A, Lawson CL, Hanson RJ. Solving least squares problems. Mathematics of Computation (Philadelphia, PA: Society for Industrial and Applied Mathematics). 1995, 1976;**30**(135):665-666

[28] Yagi Y, Mikumo T, Pacheco J, Reyes G. Source rupture process of the

Tecomán, Colima, Mexico earthquake of 22 January 2003, determined by joint inversion of teleseismic body-wave and near-source data. Bulletin of the Seismological Society of America. 2004;**94**(5):1795-1807. DOI: 10.1785/012003095

[29] Kennett BLN, Engdahl ER, Buland R. Constraints on seismic velocities in the earth from travel times. Geophysical Journal International. 1995;**122**:108-124

[30] Motazedian D, Ma S. Source parameter studies on the 8 January 2017 mw 6.1 resolute, Nunavut, Canada, earthquake. Seismological Research Letters. 2018;**89**(3):1030-1039. DOI: 10.1785/0220170260

[31] Zare M, Ansari A, Heydari H, Shahvar M, Daneshdust M, Mahdian M, et al. Reconnaissance report on two Iran, Makran earthquakes; 16 April 2013, Mw7. 8, Gosht (Saravan) and 11 may 2013 Irar (Goharan), Bashagard, SE of Iran. In: Technical Report June 2013. 2013. DOI: 10.13140/RG.2.2.34981.63200

[32] Grando G, McClay K. Morphotectonics domains and structural styles in the Makran accretionary prism, offshore Iran. Sedimentary Geology. 2007;**196**(1-4):157-179. DOI: 10.1016/j.sedgeo.2006.05.030

Chapter 7

Geotribometamorphism in the Paleoseismic Zones of the Earth's Crust

Evgenia Kozhoukharova

Abstract

Geotribometamorphism represents the transformation of rocks in the seismic zones of the Earth's crust, which takes place under the influence of friction. It differs in spread, place in geological structures, time, and thermodynamic parameters from regional and dynamic metamorphisms. Geotribometamorphism gives a new face to earthquake events by pointing out not only the destructive but also the creative activity of seismic energy. The chapter offers a brief overview of geotribological zones of different scale, also adding pressure zones. Deformation and recrystallization processes are considered. In addition to the well-known pseudotachylites, high-thermobaric eclogites, garnet lherzolites, and others belong to the formations of the seismic zones. Geotribometamorphism provides a new solution to some hotly debated problems in metamorphic petrology such as the location and genesis of eclogites, thus indicating the energy connection between geophysical and petro-mineralogical processes. The geotribometamorphic rocks established in the metamorphic complex of the Rhodope massif mark paleoseismic zones.

Keywords: geotribometamorphism, seismotectonic geotribological zones, eclogitization, friction, deformation, seismic waves

1. Introduction

The earthquakes are constant events in the past and present geologic life of the planet. Such have happened in past eras as well, but few of them have left traces of their activity in paleoseismotectonic zones. They take place at different depths in the Earth's crust, where rock blocks or layers move, rub against each other, receive deformations, and generate energy, which spreads as seismic waves. We know the consequences of modern events only from the surface disturbances of the relief that is only one echo of the processes in the hypocentre of the event. Destruction, melting, and crystallization take place in the hypocentre, which remain invisible to us. These changes become available for observation and study only when, after millions of years, as a result of uplift and erosion, these places come to the surface and the old paleoseismic zones show their face through the metamorphic rocks in them. Thanks to mostly Precambrian metamorphic complexes, we have the opportunity to know what happened long ago deep in the paleohypocentre of the earthquake events.

A rupture zone is tectonic one when it is in a state of rest, but when movement with sliding and friction begin between the rock blocks or layers, it becomes a seismic zone, which produces friction energy and seismic waves. The updated formulation "Geotribometamorphism in seismic zones of the Earth's crust", announced at a National Conference in 2023 [1], reveals the specific relationship between rock friction processes and seismicity.

Until now, pseudotachylites have been traditionally interpreted as an indicator of high-velocity slip >10 m/s and friction products formed in the shallow seismic zones at a depth of up to 5 km and therefore as a fossil remnant of paleoseismic events [2]. Petrographers long ago described [3] eclogites as rocks composed of high-thermobaric minerals, but they search for place of formation by calculating the temperature along geothermal gradient and the pressure along lithostatic (confine) pressure. This leads to the conclusion that eclogites are formed in sub-crustal and even mantle depths or in subduction zones under conditions of T > 800°C and P - 10-12 kbar. However, this contradicts the geological facts because eclogites are found in the middle levels of the crust where the temperatures are 460–600°C and the pressure is 4–6 kbar.

Geotribometamorphism provides a more plausible explanation for eclogite genesis. The main factor for the transformation of rocks is the kinetic energy arising from friction in the geotribological (seismic) zones. Seismic waves propagating outside of the geotribological zone also have power and influence on the rocks transformation, leaving microtraces in some metamorphic structures, as well as the concentric hills around meteorite craters. Seismic waves should also be considered among the factors of geotribometamorphism.

Geotribometamorphism takes place precisely in seismic zones. It is of theoretical importance because it provides a new perspective in the understanding of deep crustal processes. "It appears to be no clearly defined differences between the processes or products of dynamic and those of low-grade regional metamorphism" [4]. This characteristic allows too wide limits of its manifestations in terms of place and action, but covers only the destructive processes. In contrast geotribometamorphism is precisely defined in place, these are seismotectonic zones of friction and cover a full cycle of destructive and constructive recrystallization ending in new consolidation and new minerals and rocks. Geotribometamorphism demonstrates the energetic connection between seismotectonics and petrological-mineralogical processes.

The main goal of the chapter is to draw the attention to geotribometamorphism, presenting an overview of the most essential features of it. Geotribometamorphism provides satisfactory solutions for the origin of eclogites. It is reasonable to distinguish it as an independent type in the Earth's crust.

2. Methods for studying geotribometamorphism

2.1 Tribological methods

Geotribometamorphism is the object of study of geotribology, which is a branch of overall tribology and an interdisciplinary trend among Earth's sciences. The geotribology studies the transformation of rocks and minerals in the zones of friction in the Earth's crust under influence of kinetic energy. Geotribological zones are equal in characteristics and importance to seismic zones.

In the aspect of mechanophysics, geotribological zones in the Earth's crust are zones of friction, which in development follow the principles of the tribology. Friction is a widespread phenomenon in nature. Due to the great economic importance of friction in technical and construction sites, the science *Tribology* was founded by Jost [5] in 1966. The processes of friction became the subject of intensive, multifaceted research of friction in technical systems, tribochemical reactions in the mineral lattice [6], deformations [7, 8], and thermal processes [9] on the frictional contact. The role of contact is highlighted as a matter of particular importance for transformations [10]. Scholz [11] points out the main dependencies between sliding speed, rock type, and temperature of the friction.

The tribological configurations are also found in geological structures so the principles of friction are applicable of geological processes. The geotribological processes are manifested at different scale levels, and their complete characterization requires the use of the knowledge of many related sciences. So the geotribology occupies an interdisciplinary position between geophysics, tectonics, petrology, mineralogy, inorganic chemistry, and solid state physics. In the present chapter, we consider only the geotribological zones on the example of the amphibolite facies metamorphic complexes in the Rhodope massif.

2.2 Geophysical methods

Geophysical methods [12] include field and laboratory studies—geological investigation of the active or fossil faults, physical and chemical features obtained by experiments, as well as the seismological estimation from seismic waveforms. Earthquake dynamic rupture can now be modelled using simulations, which should be useful for understanding earthquake rupture dynamics [12]. Now micro-earthquakes are experimentally simulated in order to trace the dependences between sliding speed, rock type, temperature of environment, and degree of friction destruction [11]. It has been established that at a high sliding speed, the friction decreases [13]. The successive transformation of rocks passing through grain fragmentation, powdery mylonitization [14], and viscosity [15] until the appearance of the first drops and thin layers of melt [16, 17] is sufficiently well experimentally documented. Smith et al. [18] claim that pseudotachylites are the only indisputable geological records of rocks from seismic zones which we dispute because eclogites are also geological records of ancient seismic events. The pseudotachylites form at >5 km depth [18] and the eclogites at a depth of 15-20 km.

Friction melt is implied in a variety of processes as seismic slip. It would be interesting for seismological "to obtain a generalization of rate-and-state friction laws nowadays implemented in earthquake fault models [18] not only for the steady-state condition under constant slip rate but for the weakening transient and possibly for the whole cycle of seismic rupture" [18]. An extensive overview of seismic zones including meteorite craters, the processes of disintegration, and melting of rocks which is controlled by the mechanical properties of the rock's mineral composition is [2, 19] illustrated in tables and graphs. Plastic and rupture deformations are particularly expressive around meteorite craters [2], but they have an analogous expression at the micro-level as well.

An important element of geophysical practice is the study of seismic waves propagated beyond the seismic zone into surrounding rocks. The speed of seismic waves depends on the composition of the rocks, but also on the degree of their homogeneity and viscoelasticity [20, 21] which determines the general anisotropy [22] of the environment.

2.3 Geological and petrological methods

Geological field and laboratory microscopic and geochemical methods provide the information of the natural geotribological zones that are considered over a wide scale range from the crystal lattice to planetary dimensions. Geotribological or seismological zones in a state of rest are common tectonic structure, but even the slightest movement causes them to overturn into normal seismic zones.

The rich variety of macro- and micropetrographic structures provides information on the degree of deformation and rheological properties of the metamorphic rock types. The mineral composition of rocks is indicators of temperature-pressure conditions during their formation.

Products of geotribometamorphism: eclogites, garnet lherzolites, phengite, kyanite schists, and calciphyres have been described in detail in the petrological literature as high-pressure and ultra-high-pressure rocks. The place of their genesis until now is not associated with seismic friction zones, but mostly with deep zones of regional metamorphism. Only pseudotachylites were recognized by petrographers as rocks formed in frictional melting zones [23] without being associated with seismic zones.

3. Friction and movement

3.1 Friction

Friction, as a widely manifested process in nature and technical systems, is the subject of research from the tribology. Thermodynamic processes are also studied [24, 25]. Any friction generates energy that deforms and breaks the contact walls of the bodies and produces seismic waves propagating beyond the contact. The tribosystems exchange materials, energy, and information with the environment and between components into interface, and they have the ability to "forget" disturbances by building new self-organized dissipative structures [26].

The friction manifests itself in three main forms: a. sliding; b. rotation and rolling; c. impact and pressing.

a. Sliding occurs when two bodies move relative to a common contact surface. The driving forces are parallel to the contact (**Figure 1a**). The contact surfaces are rough due to numerous unevenly distributed protrusions and depressions called asperities. Therefore, the bodies touch only in some protrusions.

b. Rotation and rolling tribological structures have a more diverse configuration. The contact can be a point, line or circular surface (**Figure 1b**), and the driving forces—parallel or perpendicular to the contact surface.

c. Impact and pressing differ in the duration of the contact—instantaneous in the impact and long-term in the pressing, but in both variants the forces are perpendicular to the contact surface (**Figure 1c**).

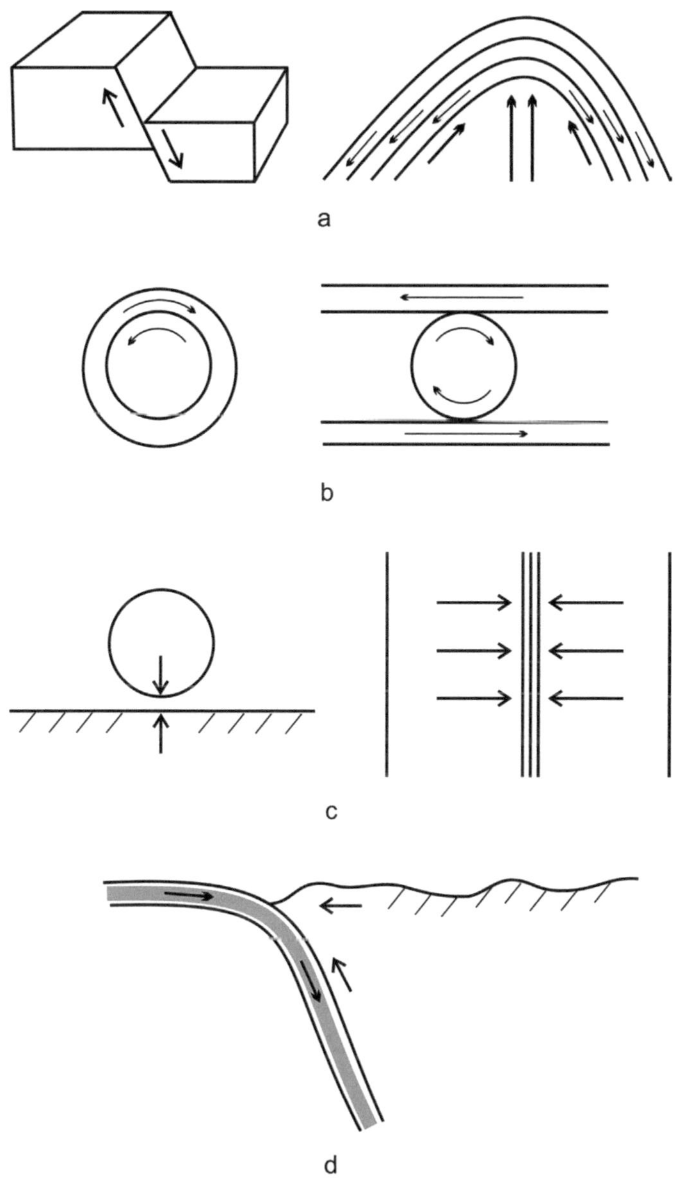

Figure 1.
Geotribological structures: a. sliding—between block and layers; b. rotation and rolling; c. impact and pressing; d. subduction zone.

The subduction zone can be considered as a combined structure, where mutually perpendicular compressive forces and parallel sliding forces act simultaneously on the contact between the two bodies (**Figure 1d**).

3.2 Movement

Movement is a major factor in the creation and development of a geotribological zone. The geotribological zone exists only during periods of movement of the two contacting bodies when producing energy and seismic waves and in periods of rest it is a simple tectonic zone.

Two important movement characteristics are: a. the initial drive moment and b. the stick–slip phenomenon.

3.2.1 Initial drive moment: Stage of preliminary displacement

A body is at rest and motion less only when no tangential force is applied to it, only its own gravitational force acts. But even after applying a tangential force F, the body is not immediately moved due to resistance T, because the tangential force is lower than the maximum frictional resistance $F < T$. Only when the increasing tangential force F exceeds the resistance T and the condition $F > T$ is fulfilled, the body is moved. The period after application of a tangential force and before beginning of the macroscopic (visible) movement of a body is the stage of *preliminary displacement*. During this stage, the frictional contact between the two bodies undergoes various changes: micromovements take place, during which the asperities are broken, and energy is accumulated and compacted, which is reflected in an increase in the coefficient of friction. When the accumulated energy grows sufficiently and the magnitude of the tangential force exceeds that of the so-called static force, the actual sliding of the body begins suddenly "with a jump". This is a strong energy pulse of a sudden release of the accumulated energy, causing deformation damage of the geotribological system, which is especially clearly illustrated in earthquakes.

The coefficient of friction (COF) illustrates also the change in density and energy accumulation in the stage of preliminary displacement. The onset of body motion has been the subject of special study [27–29], which is a starting point toward a deeper understanding of the processes preceding earthquakes [30].

3.2.2 The stick-slip phenomenon

This is a physical characteristic expressing the discontinuity of motion. The modern physical studies of the body motion establish that it is always discontinuous, even though it apparently in macro-terms appears smooth and constant [31]. The stick-slip phenomenon is the spontaneous jerking motion that can occur, while two objects are sliding over each other. Each stick-slip pulse includes the stage of prelaminar displacement and the moment of motion initiation. In this sense, the stick-slip phenomenon is a frequently repetition of this combination. Stick-slip frictional behaviour is encountered over a wide range of length scales from the atomic up to the tectonic, and there is no single underlying physical mechanism responsible for all manifestation of vibrations [32]. The stick-slip phenomenon expresses the unevenness and periodic stopping of the movement as well as the instabilities of the macro-motion [33]. The behaviour of seismically-active faults is also explained using stick-slip model with earthquakes being generated during the period of rapid slip [34].

4. Geotribological structures in rock complexes

The Earth's crust is a living organism in a state of eternal movement. Tectonic processes like earthquakes, folding, faulting, thrusting, and collisions cause dislocation and provoke formation of shear zones of friction and seismic.

4.1 Structure, shapes, and dimensions of the geotribological zones

In physical terms, they are simple tribological zones (**Figure 2**) of: a. sliding, b. rotation and rolling; c. impact and pressing. The forces have a different arrangement in the case of sliding, and they are parallel to the friction surface and perpendicular to impact and pressure (**Figure 1**).

The geotribological zones manifest various scale lengths: a. transcontinental megazones extending over 1000 km; b. regional macrozones with tens to hundreds of kilometres long; c. mesozones with several metre and decimetre lengths, observed on rock outcrops; d. microzones with microscopic dimensions; e. submicro (nano) zones in the crystal lattice, possible to establish by TEM.

4.2 Location and formation of geotribological zones

Two main types are distinguished: *a. block shear geotribological zones*, preceded by the appearance of cracks due to a change in stress and upon reaching critical values, the cracks pass into fault zones (**Figure 1a**). In some cases, secondary shear zones subparallel to the main ones also occur; *b. interlayer geotribozones,* formed along lithological contacts between rock layers that usually occupy flanks of fold structures (**Figure 1a**): anticlinals and synclinals.

The formation of zones follows two paths: a. spontaneous occurrence of a crack in a single rock block followed by horizontal, vertical, or oblique frictional displacement; b. joining of different two blocks moving against each other.

4.3 Main types of geotribological structures

4.3.1 Geotribostructures of sliding

The sliding between two rock bodies takes place along a contact surface to which the tangential forces are parallel (**Figure 1**). The contact can be: a. fault and according to the geometry of structures are named: normal fault, reverse fault, oblique fault or strike-slip fault, transform fault, horst, and graben, b. lithological boundary between

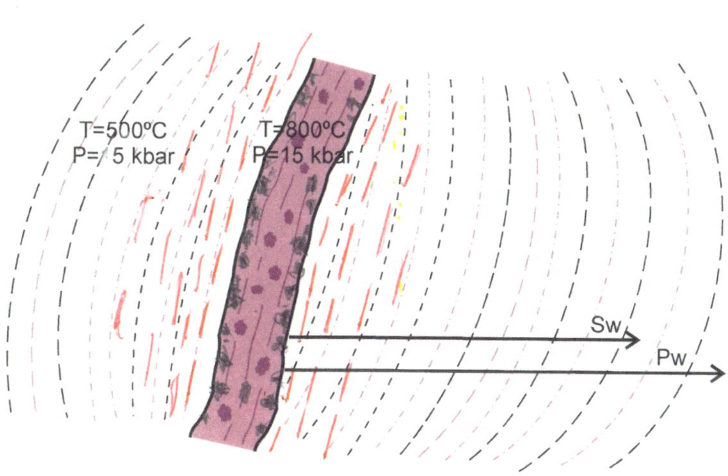

Figure 2.
Geotribological zone: inner part—melt with fragments, outer part—subparallel cracks and seismic waves.

layers of different petrographic composition and rheological properties of the rocks: density, hardness, elasticity, and plasticity. This type of sliding along broken lithological contacts is the characteristic of layered rock complexes in a tilted position or on the flanks of fold structures, the core of which rises; c. thrust structures—sliding of a package of rocks on a subhorizontal surface; d. dislocations in the crystal lattice of minerals at the submicro (nano)-level. In parallel sliding structures, frontal or convergent contact and rear divergent contact are distinguished.

4.3.2 Geotribostructures of rotation and rolling

1. The rolling of pieces of rock and mineral, except in a surface setting, is realized in the interspace, where fragments of rocks and minerals roll, rub, and hit each other, multiplying the frictional surfaces and the amount of generated energy (**Figure 1b** and **2**).

2. Macro- and megarotational structures. Large rock blocks, usually of old consolidated metamorphic complexes, surrounded by a circular belt of younger and relatively more ductile rocks, perform rotational movements. Similar structures have been described in many places in the Precambrian metamorphic terrains in northern Asia and southern Africa. The distribution of forces is tangential parallel to the circular friction surface.

4.3.3 Geotribostructures of impact and pressing

Impact or pressing friction is produced when two bodies in opposite motion meet (**Figure 1c**). The kinetic forces are perpendicular to the frictional contact. The contact of the two bodies in impact structures is instantaneous or short-acting, while in pressing structures it is long-lasting. Of essential importance for the energetic impact between the two bodies is the spatial configuration of the structure, the ratio in the size of the bodies, the speed of movement and whether one body is stationary and the other is falling or striking.

Meteorite structures—the most spectacular and destructive zones, are due to the impact of cosmic bodies fallen on the Earth's. The joining structures of encounter and pressing between two opposing blocks are known as sutures. The contact surface is figuratively defined as a "seam" or "adhesion".

4.3.4 Subduction zone

It is a mega geotribological seismic structure, often of a planetary nature. It is formed at convergent contacts between two sliding tectonic plates, when they meet and one, usually the heavier one, is pushed under the other and slides along an inclined surface (**Figure 1d**). The distribution of forces at the moment of the encounter and the subsequent action is a combination between perpendicular to the contact and at the same time tangential—parallel to the inclined surface.

Three types of subduction zones are distinguished:

4.3.4.1 Oceanic-oceanic subduction

An oceanic plate is tucked under another one due to the appearance of a transform fault on the ocean floor. As a result of the friction between the two plates, oceanic volcanic island arcs arise.

4.3.4.2 Continental-continental subduction

A continental plate turns under another continental one. The famous example is the Alpo-Himalayan system, along which many local subduction zones were formed: Arabian plate under Anadola Massif, the Indian plate under the Himalayas and others. Everywhere this kind of subduction marks the most active orogeny on the planet.

4.3.4.3 Oceanic-continental subduction

An oceanic plate sinks under continent—for example, Nazca plate under the Andes. A combination of orogeny and volcanism is observed in the section of the Chilean sector of the Andes. In this case, two options are possible: a. the oceanic plate sinks under the continent plate dragging sediments from the ocean floor; b. obduction of scraped fragments of the oceanic plate onto the continental surface. A second case is feasible when the oceanic plate is two-layered, covered with a softer material that is scraped off like a grater. Abducted serpentinite fragments in ophiolite associations, extracted from the serpentinized coat of ultrabasic oceanic plates, illustrate precisely such cases [35]. In the friction zone between the continental and oceanic plates, foci of melting occur. The melt penetrates through channels in the continental rocks, forming intrusive bodies and reaching the surface effuses on it.

5. Construction of geotribological zone

The geotribological zone is a tightly closed mobile zone with an evolving structure. It starts from a crack, along which its two walls are closely adjoined, but in the process of their destruction, the area expands (**Figure 2**). It is logical to assume that the deep zones are closed, unlike the surface ones, which are open. An inner and an outer part can be distinguished.

1. The inner, real part of the zone is the interface, filled with fragmented material from the collapsed walls. The surface of the walls is always rough, covered with unevenly distributed asperites—protrusions and depressions. Areas with a denser presence of asperities form contact spots. The actual contact between the two walls occurs only at the tops of the asperities. The friction surfaces are constantly developing—some of the asperities are broken, erased, and smoothed, new smaller or larger ones appear. Depending on the rheological properties of the bodies, the friction surfaces become smoother or rougher. The hard pieces of rock scratch, gouge, and destroy the walls, while the soft minerals are ground to a powdery mylonites.

2. The outer part of the zone is determined by the distance to which the impact of the kinetic energy of seismic waves reaches. This influence is manifested by the appearance of a bundle of sub-parallel thin cracks that expand the volume of the outer zone, a phenomenon known as dilatancy. Banded metamorphic textures parallel to the main geotribological zonal were observed [36].

6. Deformation of rock bodies of the geotribiological zone

The deformation of bodies is the visualized energy of motion. It is the result of body motion and contact interaction between them. The friction of the walls generates energy that changes the thermodynamic situation and performs work, manifested in the deformation of rocks. The type and intensity of deformation depend on: a. intensity of the energy impact; b. matter characteristics of the interacting bodies; and c. environmental conditions.

Deformations can be considered as external and internal.

1. External deformations are expressed in changes in the shape and volume of the body into rock fragments, sand, mylonite, or melt.

2. Internal deformations are manifested in damage to the crystal lattice of minerals. Under the energy impact, the particles begin to vibrate more intensively, increasing their amplitude and at critical values to break bonds and become migrating atoms. Defects such as vacancies, dislocations, and bulk voids appear.

3. Deformations and influence of rheological properties of rocks.

The geotribological zones can arise and develop among homogeneous rock or among rocks of different petrographic composition and rheological properties. The walls of the zone are subjected to friction and destruction, so interface is filled with different sizes of rock and mineral fragments that roll, slide, and hit each other, receiving different deformations and generating additional energy. Harder and stronger rock or mineral pieces scratch, gouge, and press the softer ones, contributing to their deformation and disintegration. Garnets, quartz, pyroxenes, and rutile will play the role of damage to feldspars and calcite. In a variegated metamorphic complex subjected to interlayer friction, some of the rocks—schists, marbles, serpentinites—will react plastically thinning or accumulating, forming small folds and lenses, while the harder amphibolites will fracture and split into various boudinage structures. The degree of plasticity of a rock mainly depends on its mineral composition.

6.1 Deformation and contact geometry: Distribution of forces and energy

The nature and intensity of the deformations are directly dependent on the geometry of the contact, that is, from the spatial relationships between the rubbing bodies and the distribution of forces between them. Spatial relationships, combined with the rheological properties of frictional bodies, create numerous combinations in nature, of which will be mentioned here some of the most common cases.

6.1.1 Sliding deformations: Convergent and divergent contacts

When one body slides over another in opposite motion, convergent and divergent parts of the general contact are distinguished. The convergent is the frontal, forepart of the meeting of the two bodies, and the divergent—the rearguard. Two combinations are possible:

a. a softer body is sliding on a harder base (**Figure 3a**). The harder body attacks and damages the upper one. At the convergent frontal part, the resistance T is greatest due to the effect of preliminary displacement and stick-slip phenomenon. Consequently, the kinetic energy in the upper softer and plastic body compressed in its frontal part causes mass accumulation there. In the divergent rearguard part, a tear of the upper body will occur. Similar relations are found in some thrusts, where a rock mass accumulates in the frontal part, and in the rear part, the thrust is torn into separate clips.

b. a harder body slides over a softer one, making indentations and furrows, a case often described in tribology. In the convergent part, the harder body scrapes thin plates from the softer base according to the principle of the grater (**Figure 3b**). The accumulated mass in some cases is transferred to the upper body, a case described [34] during the obduction of the serpentinites on the continental margins.

6.1.2 Deformations during rolling and rotation

In the space of the geotribological zone, there are multiple rolling and hitting rock fragments and mineral grains that mutually damaging each other. Metamorphic rocks offer a variety of deformation structures. Softer and plastic rocks acquire a lenticular shape. Serpentinites form elongate boudins or flattened lenses, while amphibolites break into parallelepiped boudins when stretched. Particularly illustrative are the microscopic deformation structures described in the petrographic articles and atlases. It is worth mentioning the well-known spiral deformation structure in garnet, the so-called snowball, which shows internal deformations of rolling garnet, which has relatively preserved its integrity.

6.1.3 Impact and pressing deformations

In this type of friction, the reaction forces have opposite directions, and the damage of the bodies in impact and pressure interaction is much more pronounced than in sliding bodies of the same mass [37].

a. Impact of a falling strong body on a stationary platform causes a successively elastic, plastic and destructive reaction on both striking bodies, but usually stronger for the struck body [7]. If the two bodies are of the same hardness, the reaction is elastic, the striking body bounces back and the depression formed at the place of impact is restored. When the falling body is harder than the platform, it has a highly pronounced plastic or destructive effect on the surface, but the body waves penetrate deep and disorder the atomic arrangement in the crystal lattice. In case of uniform low hardness, the inverter hurts the surface, but it flattens itself [38]. Tiessen et al. [39] proposed the so-called magma-plasma model of an impact between two bodies, with a strong effect reaching

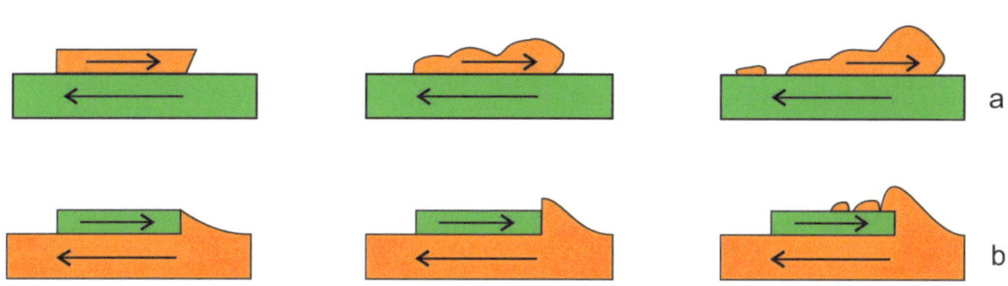

Figure 3.
Friction between bodies with different rheological properties: orange—soft and green—hard matter.

the complete disintegration of matter, became a starting point for the study of physico-chemical processes during friction [7, 40].

Meteoric structures are the most spectacular examples of impact friction. Due to the enormous energy of the impact, the temperature and pressure reach extreme values, the rocks from the target are scattered, turned into plasma, and vaporized. The surrounding rocks undergo alternating elastic-plastic deformations that remain fixed as raised ridges at the impact site or as concentric hills around the crater (reminiscent of frozen water waves), which associate with pseudotachylite dykes developed in gravity slip zones [2]. The concentric hills and cracks filled with melt testify to the power of the seismic waves, which caused plastic and rupture deformations during their propagation. They are like visualized waves that are analogous in appearance to the rhythmic streaks of compressions and expansions of atoms caused by P-waves.

b. Pressing friction presents two bodies in long-term contact, corresponding almost to their surface, the driving forces are mutually perpendicular (**Figure 3c**). Geotribological examples are the so-called sutures or tectonic "seams", representing lines of contact between big blocks. One of the most representative planetary sutures is the Ural chain connecting the European and Asian continental plates. The recrystallized products adhering the blocks also have colossal dimensions, for example, the eclogite massifs in the Southern Urals covering tens of kilometres.

c. Subduction zones, being more complex configurations between impact, pushing, and sliding with dip at an angle (**Figure 1d**), show a variety of deformations. Macrodeformations are expressed in orogeny over convergent contacts. Convergent contact, with uneven pressure between the plates, sometimes forms a *deformation arc*. A clear one is well expressed on the geographical maps between the Pacific Nazca Plate subducting under the Chilean Andes. Similar deformation arcs expressed on the relief are observed in several regions along the Alpo-Himalayan system (European Alps, Nepal, Arabian plate), which require special study in this aspect.

6.2 Hierarchy of deformation and autonomy of structural elements

Geological systems offer a favourable opportunity to observe and study frictional processes at different levels—from planetary scales to nanolevels in the crystal lattice. Physical and geotribological studies of deformations on bodies find that the degree of damage appears different, depending on the scale of observation. A body viewed on a macro-scale appears consolidated and unified, but on a meso-scale it shows plastic deformations, and on a micro-level—multiple fractures and dislocations. For example, a granitogneiss massif is depicted on the geological maps as a large consolidated body, upon direct observation of the outcrops, various deformations are established: folds, shear zones, microscopic microfolds, foliation, and cataclasis of the minerals. And on the TEM photos dislocations of the crystal lattice are visible. Leonov [41] describing the internal deformations of a granite massif from the Caucasus came to the conclusion that the rock mass behaves as a granular medium.

The rank hierarchy of deformations is obvious. The most significant damages have developed at the atomic and micro-level, at the meso-one they are only plastic while at a higher level they are compensated, so that the entire body has preserved its integrity and shape. Deformation considered in this aspect also shows that the individual parts of the bodies exhibit relatively independent stress behaviour.

DOI: http://dx.doi.org/10.5772/intechopen.1008370

7. Thermodynamic development of the geotribological zone

Individual sections of the Earth's crust for some time interval can be in tectonic rest. Then, the substance processes are controlled by the geothermal gradient and the lithostatic (confine) pressure, and the physicochemical reactions are in equilibrium. This is the thermodynamic state of regional metamorphism for which Escola's facies scheme is valid.

The occurrence of a seismic event creates a geotribological zone of friction, in which a new thermodynamic system is established. The friction mode is controlled by the sliding speed. At low-sliding rates creep as in low-slope strata frictional resistance between fault rocks is much higher than the shear stress and at high-sliding rates and vice versa—at a high-sliding speed, such as during earthquakes, the frictional resistance decreases [13]. The seismic energy provokes rapid, even instantaneous, extreme increase in temperature (flash heating) and pressure, causing disintegration and melt of the rocks. The energy in the geotribological zone is generated due to the breaking of the billions of crystal chemical bonds. The temperature in the tribological zone is always higher than that in the environment (**Figure 2**).

The secondary thermodynamic system established during earthquakes starts with an external energy pulse. It is short-lived as long as the seismic event lasts and is limited in the space of the geotribological zone. The new thermodynamic system is submerged in the primary large one of the environment and is not in equilibrium with it. The temperature gradient between the environment and the geotribological zone is very high, which actually turns it into a closed or semiclosed thermodynamic system. The life of this local thermodynamic system in the geotribological zone is short-lived, but traces of its previous existence remain in the recrystallized geotribological products that mark the paleoseismic zone.

8. Model of the geotribometamorphism along the geotribozone

8.1 The processes inside the geotribozone

At the interface of the geotribological zone, various deformation and physicochemical processes take place simultaneously. Three stages of development of processes are distinguished: a. destruction and melting, b. culmination of activity, and c. crystallization and consolidation (**Figure 4**) [42].

8.1.1 Stage of destruction and melting

The first stage starts with elastic and plastic deformation, followed by brecciation, cataclasis, and mylonitization. Point, linear, and planar defects emerge also in the mineral lattice. The inner dislocations and translation of the crystal structure lead to fracture and fragmentation of minerals. The comminution process is controlled by the mechanical properties of the minerals being crushed. For rocks and minerals, these properties include the yield strength, shear yield strength, fracture toughness, and thermal conductivity [2, 19]. In an advanced stage of deformation, the crystal chemical bonds are torn to complete decomposition of minerals, reaching to molecular and atomic level up to exoemission of ions and electrons and melt formation. The slope of the rising line depends on the speed of the process—in a high-magnitude seismic event, the line rises steeply almost vertically, while in a slow interlayer creep, it will

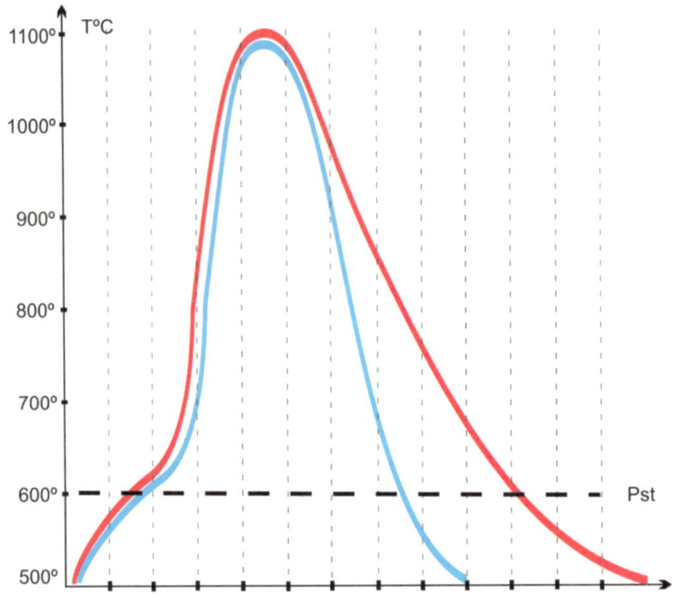

Figure 4.
Diagram of the development of the geotribological process—the red line of temperature, the blue line of pressure. Dashed line—pseudotachylite level.

be slightly inclined and step-like. Simultaneously to the deformation, a number of physicochemical endothermic reactions take place [7]. They begin with dehydration of the hydrous minerals such as serpentine and amphiboles. Oxidation reactions, diffusion of electrons and components, phase transitions in quartz, restructuring of the crystal lattice, for example, graphite-diamond and others take place in the crushed rock material. Some authors assume that before the onset of melting, the destroyed material of interface can form a solid lubricant [16]. The appearance of a frictional melt is a crucial moment for the tribological and seismic process [16]. Melting begins with the appearance of first drops, followed by thin layers that settle irregularly in the interstices between the aspirites [43].

Friction melting is depending on the temperature and the properties of minerals which melt in order corresponding to their melting points. Feldspars and biotite are the first to be melted [T ~ 600°C], followed by quartz, Fe-Mg silicates [T ~ 1100°C], and lastly the oxides [19]. As long as the melting of the minerals lasts, the increase in temperature in the system decreases or stops (**Figure 4**). Partial melting of the interface results in a silicate fluid containing mineral and rock fragments, from which, after cooling and crystallization, pseudotachylites appear.

Steady-state friction in the presence of melt is described by a system of coupled equations involving viscous shear, temperature-dependent viscosity, thermal balance of shear heating, diffusion, latent heat, and extrusion of melt under an applied normal stress [44].

8.1.2 Stage of culmination of energy activity

The second stage is the time of maximal activity of geotriboprocesses when specific thermodynamic conditions in the tribozone are established. The rocks from the interface are completely melted, and they roughly correspond to the walls of the geotribological zone. If it is developed among gneisses, the melt would correspond to the chemistry of granitic magma, or if the zone is among amphibolites and serpentinites—to a gabbro and peridotite magma composition, respectively, which is consistent with the experimental results [17, 43–45].

A part of the kinetic energy transforms to thermal one. The internal energy, entropy, and enthalpy increase, and the free energy decreases. The amounted effect of tribo- and mechanical-chemical processes is realization of extremely high temperature and pressure only in the narrow closed space of the geotribozone. It is possible for a brief moment to get to temperature "explosion" and during a second or microsecond the temperature to increase more than 1000°C, but beyond boundaries of tribozone they decrease rapidly [9]. This fact allows the tribozone to be considered as a closed or semi-closed thermodynamic system comparable to an autoclave, in which chemical reactions attain a local equilibrium.

8.1.3 Stage of crystallization and consolidation

At the third stage, the seismic event and the friction are over, so is the energy supply and the melt goes into cooling conditions. In the shallow levels of the Earth's crust due to rapid cooling of the melts, glassy pseudotachylites may occur but in deeper levels of amphibolite facies at 15–20 km and at ambient temperature of 480–600°C, the melt will crystallize into medium- to coarse-grained rock. Temperature and pressure begin to decrease with a speed dependent on the depth level of the geotribozone, although in any case much slower than the rapid or flash increase during the progressive stage of the tribological process. Chemical reactions are exothermic, simultaneously with energy dissipation to the environment, which changes the thermodynamic system into an open one. The process ends with the development of the so-called dissipative or protective structures [26]. The newly crystallized rocks are nearly isochemical to the host destroyed rocks. In a magnesian basic substrate, the first minerals will be high temperature olivines and pyroxenes, but the possible presence of rutile, spinel, and garnet in the mineral association indicates high-pressure conditions.

The final geotribometamorphic products (rocks) will correspond to the TP conditions in the seismic zone but also to the depth of crystallization. In surface levels up to 1–2 km, the rocks will be unwelded or compacted with fine-grained solder tectonic breccias. Below them, at a depth of about 3–5 km, the pseudotachylites appear—a mixture of fragments and a glass matrix, increasing in quantity with depth. Epidote-chlorite amphibolites, followed downward by garnet amphibolites and eclogites, mark an increase of melt crystallization temperature with depth (**Figure 5**), corresponding to a gabbro composition. At 15–20 km—the emblematic eclogites with a mineral association of omphacite, garnet, and rutile appear. In the case of very high pressures, a microdiamond is present in a garnet. A similar picture was observed in the metamorphic complex of the Rhodope massif, Bulgaria.

The onset of tribometamorphism is triggered by an external factor, such as a seismic event, while the third stage is a self-developing process [46].

8.2 The processes outside the geotribozone under the influence of seismic waves

The indirect influence of seismic waves on relief and building in the epicentre is well known. Action of seismic waves on rock matter and especially on the structure of minerals is a question of primary importance for geotribometamorphism which, however, has not yet received the necessary attention in the geological, tribological,

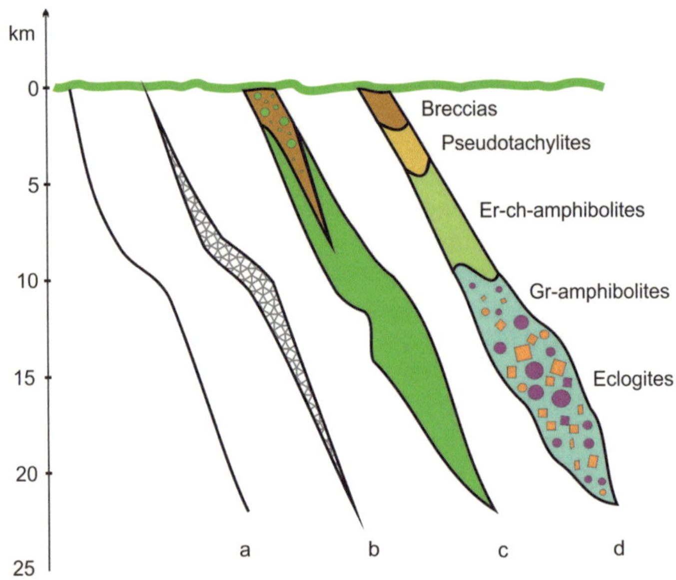

Figure 5.
Sketch of the growth of a geotribological zone from a crack, through destruction to consolidation by crystallization of the fragmented and molten material.

and geophysical literature. It is known from geophysics that every friction in its three modifications—sliding, rolling, and impact, produces two types of seismic waves moving through solid rocks (**Figure 2**). The first one—longitudinal P-wave provokes intense vibration, rarefaction, and condensation of the particles, forming rhythmic zones of dilatation and compression in the direction of wave propagation. The second type—transverse S-wave is slower than the P-wave, and the particles fluctuate in two directions, perpendicular to the first one and to the direction of propagation. It is logical to assume that the multiple consecutive energy wave pulses are superposed, creating a complex interference pattern with a rhythmic character. However, until now there is a lack of descriptions in the literature of a fixed deformation effect of seismic waves on the structure of minerals.

The principles of the deformation hierarchy suggest that the influence of seismic waves can also be expected at a lower level, for example, at the meso- and micro-level in the crystal lattice. In Section 9—eclogitization in the geotribological zone, a rare case of rhythmic textures that could be explained by the action of seismic waves will be considered.

9. Geotribometamorphism and eclogitization

9.1 Eclogitization in geotribological zones

Understanding geotribometamorphism enables us to find a more convincing solution to the genesis of eclogitization, an important and not fully resolved problem of metamorphic petrology. Eclogitization *in sensu lato* is a process of recrystallization of minerals at high temperature and pressure, a typical product of which are eclogites. According to the dominant hypothesis, eclogites are formed in subduction

DOI: http://dx.doi.org/10.5772/intechopen.1008370

or subcrustal zones at a depth of 150–200 km, and they were tectonically exhumed to higher levels in the crust where we find them today. This hypothesis is based on the traditionally followed in petrology principle that temperature and pressure increase with depth, according to the geothermal gradient and lithostatic (confine) pressure. The hypothesis is contradictory in some points. Pilshin himself [47] notes on the inconclusiveness of the "subduction-exhumation" mechanism. The idea of upward escape of heavy eclogites through continental rocks contradicts the law of gravity. Furthermore, depth is not the only factor controlling temperature and pressure values. According to the geotribometamorphic concept, the eclogites, found in the middle levels of the Earth's crust, are formed *in situ* in seismic zones, where, thanks to the friction, causing high temperatures and pressure, they are created. It is possible for eclogites to form in subcrustal and subduction zones, but they remain there in their place of formation.

The eclogites found in the Rhodope metamorphic complex are most often present as thin layers with a thickness of centimetres to 1–2 m between isochemical amphibolite beds, which are in steep contact with dense strong gneisses. The geological setting indicates that the contact is an old consolidated strike-slip zone and eclogites are formed *in situ* in seismic geotribological friction zones. More rarely, eclogites are found in cracks up to several centimetres [48] thick, cutting gabbronorites. The mineral composition of the eclogites in the Rhodope massif is represented by almandine garnet, omphacite, rutile, and magnetite. Almost everywhere they are being replaced by amphibole and plagioclase.

The eclogitization in *senso lato* can develop not only in a basic substance, but also in various petrographic ones that have included into a geotribological zone. Geotribometamorphic products are also garnet-lherzolites arising on serpentinites, garnet-phengite, or garnet-kyanite schists with microdiamonds—on metapelites, and calciferous skarns—on carbonate rocks. They are heterofacial to these of the country rocks. The eclogitized rocks are almost isochemical to those containing them, with minor deviations—for example, the eclogites formed among amphibolites are enriched in Ti, manifested in an increase in the rutile content to 3–4%. Rutile, a typical high-pressure mineral, has the ability to extract titanium atoms from the surrounding rocks, necessary for its crystallization.

Eclogitization may be synchronous with Precambrian regional metamorphism, but it may naturally be later throughout Phanerozoic time. Eclogite products mark paleoseismic zones in the Earth's crust.

9.2 Eclogitization under the influence of seismic waves

The influence of seismic waves outside the geotribological zone is manifested in a rare case of banded eclogitization on serpentinites that are in steep tectonic contact with leptite gneisses [36]. Bands of garnet-lherzolites, consisting of garnet, enstatite, diopside, olivine and spinel, form thin (1-2 cm) parallel stripes, alternating with serpentinites into peripheral zones of a serpentinite bodies at the contact with gneisses (**Figure 6**). The temperature/pressure conditions of the garnet-lherzolite are T = 800°C; P = 10–15 kbar. The background regional metamorphism of the country rocks is typical medium pressure amphibolite facies: T = 480–540°C; P = 4–5 kbar. The myrmekite-like symplectites built up of diopside, spinel, enstatite, magnetite, and actinolite indicate a rapid recrystallization.

The striated texture of the eclogitized serpentinites is of a rhythmic character. The texture shows closest analogy to the alternating spaces of dilation and concentration

Figure 6.
Samples of eclogitized serpentinites: the dark stripes—serpentinite, the light stripes—garnet lherzolite. Avren village, Krumovgrad region, Southern Bulgaria.

due to the rarefaction and densification of atoms that the longitudinal P-seismic wave causes, knowledge long proven by geophysics. This analogy is complemented by the stick-slip phenomenon and particularly expressive by the meteorite hill rings. The banded texture of eclogitized serpentinites manifested at the meso-level is similar with this picture. Therefore, tying rhythmic striated eclogitization to the action of seismic waves is the most suggestive solution. For the moment a hypothetical idea is briefly presented:

The longitudinal P-seismic wave causing dilation, when critical values are reached, the weak van der Waals bonds between the layers in the serpentinite lattice can be broken. The released particles, in the front of the pulsating seismic wave, form striated spaces of viscous or liquid matter. It extracts additional calcium atoms from the surrounding environment, necessary for the construction of pyroxene and garnet in the lherzolites. After termination of the seismic event and energy supply, the chaotic mass tends to order. The duration of crystallization of the molten matter is dependent on the temperature difference with the host rocks.

The current knowledge of the effect of mechanical body waves on the structure of minerals is scarce. The idea of the wave action on minerals is developed very slowly. Heinicke [6] studying the impact of a metal ball on a softer surface observes the formation of a thin layer of plastic flow below the surface of impact. Brantut et al. [15] performed stick-slip experiment of natural antigorite during laboratory earthquakes and observed a rapid coseismic slip. It caused a significant overheating of asperities along the sliding surface, sufficient to produce amorphization and melting. Furthermore, they registered a layer of the order of 1 to 3 μm of foam-like matrix in the internal structure of antigorite. Serpentine, because of its layered structure, favours interlayer slip [49]. The serpentine is unstable to temperature and pressure, loses H_2O in conditions of prograde metamorphism, and is successively replaced by talc, chlorite, olivine, and at high pressure—enstatite.

All these phenomena suggest to the possibility of seismic waves to disorder and destruction of the internal structure of some minerals outside the frictional surface, which can be explained by the action of seismic waves.

DOI: http://dx.doi.org/10.5772/intechopen.1008370

10. Discussion

Understanding geotribometamorphism requires a more accurate combination of knowledge from tribology, solid-state physics, and geophysics with geological and petrological-mineralogical data obtained through field, microscopic, and geochemical studies.

Experimental geophysical tests simulating a seismic event have greatly contributed to clarifying the deformation effects on the rocks from the initial phases of the process developing in the shallow levels of the crust. But they cannot reproduce the extremely high TP realized in the hypocentre, as well as the extended geological time required for the crystallization of the melt at depth. And here again the necessity of the traditional field observations on the spatial relations between the rocks to identify the old already consolidated seismotectonic zones stands out. Unfortunately, modern geologic field research is greatly reduced, and even tectonicists rely on the geochemical composition of a few rock samples to obtain definitions of geodynamic zones in the Earth's crust.

The impact of seismic waves on the mineral structure about which there is very little information needs special attention and a new type of research.

Hierarchical deformations and relative autonomy in the behaviour of building elements have not been sufficiently addressed in the geological literature and interpretations. If a researcher localizes his observations only to some rock outcrops and microscopic preparations, he will note numerous meso- and small-scale drag folds, cataclasis, and mylonitization, and he can easily reach the incorrect conclusion that the entire rock complex is a system of thrusts. In fact, however, micro- and meso-level deformations have not disturbed the stratigraphic sequence that allows us to state with confidence that the Rhodope Massif represents a unified stratigraphic system from which we can draw examples for a reliable interpretation regarding geotribometamorphism.

The scenario presented here for formation of eclogites in seismic zones does not reject the possibility of their creation in deep subcrustal levels as well as in subduction zones. But they remain there at the slave of crystallization and are not exhumed as migrating bodies up into the crust. It is against physical principles for heavy eclogite bodies to rise to mid-crustal levels. The presence of eclogitized serpentinites among medium-facies metamorphites is another evidence for the possibility of high-thermobaric parageneses to be created in the amphibolite facies environment.

Eclogitization on serpentinites outside the friction zone raises the question of the impact of seismic waves directly on the structure of minerals, which has not been discussed until now in geological and geophysical literature.

11. Conclusion

- Geotribometamorphism means transformation of rocks under the influence of friction energy in the seismic zones of the Earth's crust. Its identification as an independent type of metamorphism different from the regional one gives a new face of seismic processes, where the destructive and creative actions are in a balanced unity;

- Geotribometamorphism provides a new more plausible theoretical solution to the genesis *in situ* of high-thermobaric eclogite rocks and the evolution of the thermodynamic environment in the crust during earthquakes;

- The triad *movement-friction-seismicity* is the main energy factor and connection between tectonoseismic and petrological-mineralogical-geochemical processes.
- Eclogitization on serpentinites outside the friction zone raises the question of the impact of seismic waves directly on the structure of minerals, which has not been discussed until now in geological literature.
- Eclogitization *in sensu lato* develops on different petrographic varieties in seismic zones in the Earth's crust.

Author details

Evgenia Kozhoukharova
Geological Institute, Sofia, Bulgaria

*Address all correspondence to: ekozhoukharova@abv.bg

References

[1] Kozhoukharova E. Geotribometamorphism in seismic zones of the Earth's crust. Review of the Bulgarian Geological Society. 2023;**84**(3):113-116

[2] Spray JG. Frictional melting processes in planetary materials: From-hypervelocity impact to earthquakes. Annual Review of Earth and Planetary Sciences. 2010;**38**:221-254

[3] Escola P. On the eclogites of Norway. Vidensk. Meddr. Dansk. Naturh Foren KL. 1921;**8**:118

[4] Spry A. Metamorphic textures. Pergamon Press Oxfort; 1969. 350 pp

[5] Fitch J. Interview with luminary professor H. Peter Jost – The man who gave birth to the word "tribology". In: Machinery Lubrication. Nonia Corporation, Tusla; 2006; 1 pp

[6] Heinicke G. Tribochemistry. Berlin: Academie Verlag; 1984; 350 pp

[7] Stachowiak GW, Batchelor AW. Engineering Tribology. Butterworth-Heinemann. 1st ed. Perth, Australia: University of Western Australia; 1993. 871 pp

[8] Bhushan B. Introduction to Tribology. New York: Wiley & Sons; 2013; 711 pp

[9] Chichinadze AV. Fundamentals of Tribology (Friction, Wear, Lubrication). Moscow: Mashinostroenie Publ. House; 2001; 664 pp (in Russian)

[10] Manolov N. Tribology and Interdiscipline. Sofia: TEMTO Publ. House; 2003; 387 pp. (in Bulgarian)

[11] Scholz CH. Earhquakes and friction laws. Nature. 1998;**391**(1):37-42

[12] Fukuyama E. Fault-zone. In: Properties and the Earthquake Rupture Dinamics. International Geophysics Series 94. London, UK: Elsevier Acad. Press; 2009; 304 pp

[13] Chen J, Rumpel AW. Progressive flash heating and evolution of high velocity rock friction. Journal of Geophysical Research: Solid Earth. 2014;**119**:3182-3200

[14] Hung CC, Kuo LW, Spagnuolo E, Wang CC, Di Toro G, Wu WJ, et al. Grain fragmentation and frictional melting during initial experimental deformation and implications for seismic slip at shallow depths. Journal of Geophysical Research: Solid Earth. 2019;**124**(11):11150-11169. DOI: 10.1029/2019J8017905

[15] Brantut N, Passelègue FX, Deldicque D, Rouzaud J-N, Schubnel A. Dynamic weakening and amorphization in serpentinite during laboratory earthquakes. Geology. 2016;**44**(8):607-610

[16] Jacobson S, Hogmark S. On the crucial importance of tribologically induced surface modification. In: Nicas GK, editor. Friction and Lubrication. 5. Tribofilms. 2010. pp. 197-226

[17] Niemeijer A, Di Toro G, Nielsen S, Di Felice F. Frictional melting of gabbro under extreme experimental conditions of normal stress, acceleration and sliding velocity. Journal of Geophysical Research. 2011;**116**(B 07404):1-18. DOI: 10.1029/2010Jb008181

[18] Smith SAF, Di Toro G, Kim S, Ree JH, Nielsen S, Billi A, et al. Coseismic recrystallization during

shallow earthquake slip. Geology. 2013;**41**:63-69

[19] Spray JG. Evidence for melt lubrication during large earthquakes. Geophisical Research Letters. 2005: 1-5. L07301;**32**. DOI: 10.1029/2004GL022293

[20] Colombaro I, Guistt A, Mainardi F. On transient waves in linear viscoelasticity. Wave Motion. 2017;**74C**:191-212. DOI: 10.1016/j.wavemoti.2017.07.008

[21] Pshenichnov S, Ivanov R, Datcheva M. Transient wave propagation in functionally graded viscoelastic structures. Mathematics. 2022;**10**(23):4505, 2-18. DOI: 10.3390.math10234505

[22] Alashloo S, Moussavi Y, Yasir Bashir Y. Seismic wave propagation and imaging in anisotropic media: A review. Journal of Review in Science and Engineering (University Sains Malaysia, P. Pinang, Malaysia). 2021;**2021**:1-18

[23] Wintsch RP. The possible effects of deformation on chemical processes in metamorphic fault zones. In: Thompson DB, Rubie DC, editors. Metamorphic Reactions. Kinetics, Textures, and Deformation. Springer–Verlag; 1985. pp. 251-268

[24] Banjac M, Vencl A, Otovich S. Friction and wear processes. Thermodynamic approach. Tribology in Industry. 2014;**36**(4):341-347

[25] Fedorov S. Structural-energy interpretation of the friction. In: Chowdhury MA, editor. Friction, Lubrication and Wear. Vol. 2019. London, UK: IntechOpen; 2019. pp. 35-62. DOI: 10.5772/Intechopen.86123

[26] Nicolis G, Prigogin I. Self-Organization in Nonequilibrium Systems: From Dissipative Structures to Order through Fluctuations. New York: Wiley; 1977; 206 pp

[27] Rubinstein SM, Cohen G, Fineberg J. Detachment fronts and the onset of dynamic friction. Nature. 2004;**430**:1005-1009

[28] Sleep NH. Real contact and evolution laws for rate and state friction. Geochemistry, Geophysics Geosystems. 2006;**7**(8):1-1 9, Paper Q08012. DOI: 10.1029/2005GCOO1187

[29] Braun DM, Barel I, Urbakh M. Dynamic transition from static to kinetic friction. Physical Review Letter. 2009;**103**:19 ; Paper 194301. DOI: 10.1103/PhysRevLett.103.194301

[30] Tullis TE, Goldsby D. 2003. Laboratory experiments on fault shear resistance relevant to coseismic earthquake slip. SCEC Annual Report. Available from: https//files.scec.org/s3fs-public/report/2003/03084_report.pdf

[31] Popp K, Stelter P. Stick-slip vibration and chaos. Philosophical Transaction: Physical Sciences and Engineering. 1990:80-105

[32] Berman AD, Dicker WA, Israelachvili JN. Origin and characterization of different stick- slip mechanism. Langmuir. 1996;**12**(19):4559-4563

[33] Manini N, Braun OM, Vanossi A. Nanotribology: Nonlinear mechanisms of friction. In: Gnecco E, Mayer E, editors. Fundamental of Friction and Wear on the Nanoscale. Swizerland: Springer Cham; 2015. pp. 175-208

[34] Scholz CH. The Mechanics of Earthquakes and Faulting. 2nd ed. Cambridge University Press; 2002. pp. 81-84

[35] Kozhoukharova E. Obducted serpentinites in the Rhodope massif and their stratigraphic, paleotectonic and metamorphic significance. Review of the Bulgarian Geological Society. 2019;**80**(3):127-135

[36] Kozhoukharova E. Eclogitized layered serpentinites in the east Rhodope block-Comptes. Sofia, Bulgaria: Rendus Acad. Bulg. Sci. 1996;**49**(6):69-71

[37] Strong WJ. Impact Mechanic. Cambridge, UK: Cambridge University Press; 2000; 280 pp

[38] Stachowiack G, Batchelor AW. Engineering Tribology. 3th ed. Elsevier; 2011; 832 pp

[39] Tiessen PA, Mayer K, Heinicke G. Grudlagen der Tribochemie. In: Abhandlungen der Deutschen. Vol. 1. Akademie der Wissenschaften zu Berlin/Klasse für Chemie, Geologie und Biologie; 1966; 194 pp

[40] Tole I, Habermehl-Cwirzen K, Cwirzen A. Mechanochemical activation of natural clay minerals an alternative to produce sustainable cementitous binders – Review. Mineralogy and Petrology. 2019;**113**:449-462

[41] Leonov MG. Mechanics of granular media. Five short novellas about geologies. Novella four. Granites and oil. In: Degtyarev KE, editor. Proceedings of the Geological Institute, Russ. Acad. of Sci. Vol. 636. Wienna: Springer; 2020. pp. 127-135. (in Russian)

[42] Kozhoukharova E. Geotribology, a new view on the tectonometamotphism. Rev. Bulg. Geol. Soc. 2016;**77**(2-3):51-60

[43] Nielsen S, Di Toro G, Griffith WA. Friction and roughness of a melting rock surface. Geophysical Journal International. 2010;**182**:299-310. DOI: 10.1111/j.1365-246X.2010.04607.x

[44] Nielsen S, Di Toro G, Hirose T, Schimamoto T. Frictional melt and seismic slip. Journal of Geophysical Research. 2008;**113**:1-20. B01308. DOI: 10.1029/2007JBOO5122

[45] Del Gaudio P, Di Toro G, Han R, Hirose T, Nielsen S, Schimamoto T, et al. Frictional melting of peridotite and seismic slip. Journal of Geophysical Research. 2009;**114**:1-19. DOI: 10.1029/2008JBOO5990

[46] Assenova E. Complexity of contact systems: Self-organization and synergetic approach in tribology. In: Papers of 15th International Conference on Tribology Kraguejevac. Serbia: House University of Kraguejevac; 17-19 May 2017. pp. 53-58

[47] Pilchin A. The role of serpentinization in exhumation of high – To ultra-high pressure metamorphic rocks. Earth and Planet Science Letters. 2005;**237**:815-528

[48] Nenova P, Zidarov N. Eclogites from Maleshevska Mountain, SW Bulgaria. In: Papers of Conference 60 Annual of Geology in Sofia University. Sofia Bulgaria: House "St. Kliment"; 2008. pp. 109-114

[49] Evans BW, Hattory A, Baronnet A. Serpentinite: What, why, where? Elements. 2013;**9**:99-106

Chapter 8

Perspective Chapter: Socio-Economic Impacts of Earthquakes – A Survey of the Planning Literature

David Emanuel Andersson and Oliver F. Shyr

Abstract

This article provides a brief review of recent studies on the impact of earthquakes. We begin by discussing the literature based on topics, year of publication, and analyzed regions. The studies are then categorized into six key areas: 1) planning for earthquakes: theory and applications; 2) aggregate economic impacts; 3) aggregate social impacts; 4) earthquake effects on house prices; 5) case studies of specific earthquakes and their socio-economic impacts; and 6) simulations of potential future impacts. In conclusion, we summarize major findings and suggest future research directions, particularly focusing on the application of artificial intelligence and simulation software for predicting potential damages from future earthquakes.

Keywords: earthquake, urban planning, house price, socio-economic impact, hedonic model, simulation

1. Introduction

There are an enormous number of studies that deal with engineering aspects of constructing buildings that are able to withstand earthquakes of various magnitudes. In comparison, studies that explicitly address the implications of seismic risks for large-scale urban and regional plans are much rarer. This chapter aims to provide an overview of some of the most useful studies for planning practitioners in the world's seismically active areas.

Table 1 provides an overview of the studies covered by this chapter. It is evident that most researches have been conducted in regions that are themselves affected by earthquakes, notably Turkey, Pacific Asia, and the west coast of the Americas. We have chosen to arrange the studies according to generality and scope. The first studies that we introduce are the most theoretical and are of the greatest general interest. We then turn to aggregate economic and social impacts of earthquakes in specific places, but still with implications that are clearly applicable across space and time. We then look at more micro-focused economic studies concerned with price effects of earthquake risks, before turning to case studies of the effects of specific earthquakes in

Type of study	Reference	Analyzed region(s)
Planning theory and applications	Kunreuther [1]	California; Turkey
	Bostenaru Dan et al. [2]	World (theoretical paper)
	Momenian and Zekavat [3]	World (theoretical paper)
	Kii and Doi [4]	World (theoretical paper)
	Sengezer and Koc [5]	Turkey
	Mirmoghtadaee et al. [6]	Iran
Aggregate economic impacts	Freeman et al. [7]	Argentina; Honduras; Nicaragua
	Cardona et al. [8]	19 countries in the Americas
	Alam and Ali [9]	Turkey
	Dincer et al. [10]	Turkey
	Tatar et al. [11]	Turkey
Aggregate social impacts	Aldrich [12]	Japan; New Zealand; Thailand
	He et al. [13]	World
	Palm [14]	California
	Cofini et al. [15]	L'Aquila, Italy
	Chen [16]	Taiwan
	Lian et al. [17]	China
	Shrestha [18]	Nepal
Impacts on house prices	Murdoch et al. [19]	San Francisco Bay Area
	Chen and Chang [20]	Central Taiwan
	Azegami et al. [21]	Fukushima, Japan
	Tu et al. [22]	Taichung, Taiwan
General-interest case studies	Trifunac and Todorovska [23]	Los Angeles, California
	Hung and Chen [24]	Taipei City, Taiwan
	Liu et al. [25]	Wenchuan County, Sichuan
	Balachandran [26]	Bhuj, Gujarat
	Mamula-Seadon [27]	Canterbury, New Zealand
	Di Ludovico et al. [28]	L'Aquila, Italy
	Kondo and Lizarralde [29]	Eastern Japan
	Kodag et al. [30]	Pune, Maharashtra
	Mavroulis et al. [31]	Southeastern Turkey
Future-oriented simulations	Mestav Sarica et al. [32]	Istanbul; Jakarta; Manila
	Satir et al. [33]	Turkey

Table 1.
Planning-related studies of the social or economic impacts of real or potential earthquakes.

specific places. The penultimate section concerns itself with more speculative analyses of future risks, while the final section offers some tentative conclusions.

2. Planning for earthquakes: Theory and applications

Some of the planning-related literature on earthquakes is more general, in that it deals with principles that may be useful in all seismically active regions. For example, Kunreuther [1] notes that major earthquakes have shed light on the desirability of public-private partnerships in disaster management. Events like California's Northridge earthquake resulted in significant financial losses, costing the insurance industry US$28 billion and the government an additional US$17.6 billion. Similarly, the 1999 Kocaeli earthquake in Turkey, which caused over 17,000 fatalities and disrupted the economy, underscored the urgent need for improved building standards, prompting action not only from the Turkish government, but also from developers and the general public.

Bostenaru Dan et al. [2] argue that most studies on human-related aspects of earthquakes have been carried out in the field of earthquake engineering, which has traditionally focused on structural engineering, while neglecting the role of urban planning in earthquake risk mitigation. This focus has thus downplayed the need for interventions at the city level rather than at the level of individual buildings. A more aggregate urban planning approach is better able to address the multifaceted impacts of earthquakes in urban areas, which necessitate approaches such as multicriteria decision-making. Often, such approaches also require wider public participation than smaller-scale property development.

Momenian and Zekavat [3] propose an earthquake-focused urban design strategy that emphasizes building cities in safe zones that are less vulnerable to earthquake-related disruptions. The main benefits are infrastructure resilience, simplification of repair and reconstruction processes, and fewer social problems. Such "earthquake-conscious" urban design makes use of interdisciplinary tools, such as micro-zonation maps and evaluations of post-earthquake fire hazards, in order to arrive at a comprehensive design strategy. The main problem that the design strategy addresses is the post-quake recovery process. The most important component is increasing the survival probabilities of buildings and infrastructure, which implies that they can withstand minor shocks and avoid collapse during major earthquakes.

Kii and Doi [4] use a more economic type of reasoning by explicitly introducing trade-offs between disaster prevention and other objectives. Population concentration in seismically active metropolitan areas unavoidably poses a trade-off between earthquake disaster mitigation and economic efficiency. This trade-off implies that planners and other policymakers need a coherent assessment method that balances these factors while considering intertemporal fairness. One implication is that while decentralized land-use planning is difficult to justify when assuming a risk-neutral, utilitarian welfare function, the introduction of intertemporal fairness as a normative criterion provides the justification for such decentralization.

Sengezer and Koç [5] argue that there is a difference between developed and developing countries in their ability to implement plans that reduce the adverse consequences of seismic events. To illustrate this difference, the paper considers urbanization and planning policies in the context of the 1992 Erzincan and 1999 Kocaeli-Gölcük earthquakes in Turkey. Important planning-related variables include building heights, geological conditions, and construction periods. In the case of Turkey, the values of these variables imply ineffective planning principles from a risk reduction perspective.

Although the study of Mirmoghtadaee et al. [6] is framed as a case study of the Iranian city of Karaj, it reads more like a proposal for how to use general planning principles when dealing with the type of uncertainty that is associated with seismic events. It uses the strategic choice approach to facilitate decision-making under uncertainties related to fault line determination, buffer zones, and setbacks. If the strategic choice approach is combined with risk-based planning, then common earthquake-related "wicked problems," such as insufficient knowledge and unpredictable policy outcomes, become more manageable. The strategic choice approach may thus enable planners to deal with the complexities of fault zones, with reductions in the numbers of lives lost and buildings damaged, in spite of ambiguous information sets. Another aspect of this approach is the gradual implementation of urban planning restrictions in line with the development of geological data. One of the benefits of this type of planning is that it simultaneously addresses the risk of natural disasters and the uncertainties associated with human responses to disasters.

3. Aggregate economic impacts

When assessing the consequences of seismic events, there is the immediate impact of lives lost and buildings destroyed. However, there is also the medium-term impact of an earthquake in its role as an exogenous shock to the regional economy. Depending on how good engineers, planners, and policymakers have been at mitigating the disruptiveness of natural disasters, these economic impacts range in severity from moderate to substantial.

In a World Bank study, Freeman et al. [7] show how countries can prepare for earthquakes and other natural disasters by estimating annual expected macroeconomic impacts. By way of example, the estimated annualized loss is only 0.025% of the total capital stock in Argentina, whereas in Honduras, the annualized loss is closer to 0.5%. Failure to fund probabilistic annual losses of a country's capital stock can be expected to exacerbate its future economic performance, including by stalling efforts at poverty reduction. For this reason, risk management should be an integrated part of development planning, so that enough resources are available to fund post-earthquake reconstruction. Thus, risk financing is an essential complement to planners' typical concerns, which are risk identification and risk mitigation.

Cardona et al. [8] introduce the Disaster Deficit Index (DDI), which assesses a country's macroeconomic and financial risks due to catastrophic events such as major earthquakes. The DDI is not limited to estimating the loss due to disasters, but also compares this with preparedness in terms of availability of funds for post-disaster recovery. Using insurance industry conventions, the index thus calculates costs in a way that accounts for country-specific resource limitations. The study then estimates DDIs for 19 countries in Latin America and the Caribbean. While both Honduras and Mexico have high exposures to natural disasters, Honduras is in a much more precarious situation due to its low economic capacity. In the case of Honduras, the annual cost of expected future disasters corresponds to 30% of its capital investment.

The real estate industry is more affected by earthquakes than any other industry. Alam and Ali [9] show that aggregate losses to the real estate industry of a major earthquake can be expected to be around US$27.4 billion in the case of Turkey, as compared with US$9.5 billion for the next nine industries combined. Using a dynamic input-output model together with a non-linear optimization technique, it becomes possible to analyze the sectoral composition of an earthquake's impact, as well as industry-specific risks, recovery times, and adaptability.

Dincer et al. [10] use an approach focused more on direct observation than on official statistics to analyze the overall economic impacts of major earthquakes in Turkey. Using field visits and interviews, this study identifies significant labor shortages that are more serious in government-related industries than elsewhere, as well as severe depletion of physical capital in earthquake-affected areas, particularly affecting small businesses. Due to the lack of a comprehensive disaster waste disposal plan, debris management was one of the least satisfactory aspects of post-quake recovery processes.

Tatar et al. [11] use yet another approach to analyze the overall impact of a major earthquake in Turkey. Using geographic information systems (GIS) and remote sensing, they found that 55% of commercial properties in the Antakya and Defne districts of Hatay province, following the Kahramanmaraş earthquake in 2023. The findings underscore the vulnerability of commercial real estate to seismic disruptions and the need for targeted recovery and resilience strategies. Geospatial technologies are often particularly effective at assessing earthquake-related economic losses, apart from their benefits related to disaster response efforts and strategic planning.

4. Aggregate social impacts

Apart from direct economic impacts such as the destruction of valuable real estate, natural disasters such as earthquakes may also have impacts that are more difficult to quantify in economic terms, such as impacts on mortality, social capital, migration, and crime. Hence, planning that mitigates the risk of catastrophic events may not only save lives and future revenue streams, but also prevent various types of social disruption.

Aldrich [12] contends that social capital is critical for post-disaster resilience. They analyze the role of social networks in dealing with post-disaster recovery in three distinct locations: Bangkok, the Tohoku region of Japan, and Christchurch in New Zealand. In this context, it is helpful to distinguish between the bonding, bridging, and linking types of social capital. Social capital may aid the recovery process by keeping people in disaster-affected areas and thereby being available for recovery efforts as volunteers. Hence, seismically active areas benefit disproportionately from high levels of social capital.

He et al. [13] examine the relationship between urbanization and earthquake risk, focusing on changes in population and urbanized land area in earthquake-prone zones from 1990 to 2015 and their impact on fatalities. Over two-thirds of population growth and nearly three-quarters of earthquake-related deaths occurred in developing countries with moderate urbanization ratios of between 20 and 60%. Fatality rates tend to increase in the proximity of a 45% urbanization ratio, suggesting that urbanization may lead to more resilient infrastructures. A global perspective shows that more than half of those affected by increased earthquake risk due to urbanization live in South Asia.

Turning to a more micro-based analysis of general social impacts on individual behavior in the face of earthquake risk, Palm [14] looks at the impact of publicly available information on decision-making. Consumer protection laws assume that people want to avoid risks and that they act in response to information. In California, however, the mandated disclosure of environmental hazards such as being in a "surface fault rupture district" has had negligible impact on homebuyers' decisions. Two factors that may have caused this lack of responsiveness among affected individuals are the lack of stringent disclosure requirements and the unrealistic assumption that the availability of information will affect individual responses in predictable ways.

Another way to look at the overall non-economic impacts of earthquakes is to assess the subjective quality of life among a group of people in the aftermath of an earthquake. This is the objective of Cofini et al. [15], which assessed the quality of life and psychological well-being of residents of L'Aquila, Italy, after the April 6 earthquake in 2009. A random sample of 281 individuals in temporary housing was surveyed in 2010. The results indicated that women had worse health outcomes than men and additionally that older cohorts exhibited worse outcomes such as higher levels of depression and depressed levels of self-control. The timing of the survey indicates that such adverse outcomes can be expected to persist for at least 18 months after a catastrophic event.

Chen [16] focuses on how place attachment and risk perception may affect individuals' willingness to live in areas at high risk of catastrophic seismic events. One finding is that behavioral responses to imperfect information, such as fatalism and optimism bias, may lead people to underestimate the risk of injury or property damage. In the case of Taiwan, apparently stringent land-use regulations in fault zones may have done little to discourage new housing construction, which may give the false impression that these areas are safe among residents with high place attachment. Thus, land-use planning in high-risk areas would benefit from including information about residents' risk perception and adaptive behavior as regulatory criteria.

Lian et al. [17] examine the impact of training on earthquake preparedness among rural residents in China. Using survey data on rural residents and applying probit and Poisson models, they find that disaster prevention and mitigation training significantly influence farmers' earthquake preparedness. Specifically, farmers who received training were more than 20% more likely to adopt earthquake preparedness behaviors than others. This implies that they engage in 0.75 more preparedness measures than other farmers.

Earthquakes tend to have more severe human consequences in low-income countries. Shrestha [18] found that the 2015 Kathmandu earthquakes in Nepal caused unusually much damage. The May 12 quake caused many buildings to collapse that had withstood the April 25 quake. This study, which was based on field observation and expert interviews, resulted in an estimated 70% of urban and 95% of rural dwellings in the affected area being irredeemably damaged. In addition, these earthquakes triggered ecological reactions such as dry landslides and disruptions of animal habitats, leading to increasing numbers of animal raids of human settlements. Thus, earthquakes tend to have disproportionate social impacts in less developed areas.

5. Earthquakes and house prices

So far, we have been looking at studies that have attempted to quantify or analyze overall economic or social tolls in the aftermath of a major earthquake. But the perceived effects of potential future earthquakes can also be measured as price effects on land or real estate markets. The perceived risk of future earthquakes often increases in the immediate aftermath of a major earthquake. In general, housing and other real estate located in high-risk areas tend to sell at a discount, and this discount tends to be affected by relevant recent events. Analyses of price impacts illuminate the combined economic effects of risk perception and risk aversion.

Murdoch et al. [19] are an early study that uses a large data set to analyze the impact of the Loma Prieta earthquake on housing prices in the San Francisco Bay Area, controlling for factors, such as location-specific risk and the timing of the earthquake. As expected, the Loma Prieta earthquake led to a decrease in property values within the affected area. Moreover, buyers appeared to factor in other earthquake risk measures, resulting in a discernible price-distance gradient.

Chen and Chang [20] examine the impact of land-use planning on property transactions within and near Taiwan's Chelungpu fault zone. Properties that were built between 1995 and 2008 were built further away from the fault zone, while more recent properties built between 2008 and 2014 were closer to or within the fault zone. Interestingly, more recent buildings in riskier locations commanded higher market prices than older buildings elsewhere. This seemingly paradoxical effect is likely due to more stringent building codes after 2000 as well as specific fault zone regulations that mandate lower building heights and less intensive use of land, all of which mitigate the consequences of future seismic events.

Azegami et al. [21] focus on the Great East Japan Earthquake that combined a severe seismic event with an associated tsunami and nuclear power station accident at Fukushima. The overall effect was a spatial redistribution of relative land prices. Iwaki City, which is more than 120 km from Fukushima, experienced an abnormal increase in land prices. Conversely, housing in reconstruction areas traded at a substantial discount. These price effects started to gradually diminish 4 years after the earthquake.

The study by Tu et al. [22] is the first to simultaneously estimate the risk of earthquakes and the risk of flooding. They do this by estimating a hedonic price function for the Taichung metropolitan area in Taiwan. This area was exposed to the 1999 Jiji earthquake as well as typhoon-induced flooding on a regular basis. Geographically weighted regression estimates show that the effects of elevated earthquake and flooding risks are similar in magnitude, with price effects that are as substantial as neighborhood income and school quality.

6. Case studies of specific earthquakes and their socio-economic impacts

Most planning-related studies of earthquakes are case studies that look at a variety of impacts in a specific seismically active region. The implication is that they often mix impacts that are of general interest with more area-specific considerations.

Many of the early impact studies analyze seismically active parts of California. Trifunac and Todorovska [23] is one such study, focusing on the distribution of property damage associated with the 1971 San Fernando and 1994 Northridge earthquakes. Both earthquakes occurred on blind thrust faults under the densely populated San Fernando Valley. In both events, the areas with the most severe damage overlapped, suggesting that specific local soil and geological properties contributed to the recurrence of damage. The main takeaway from this study is that there are still many factors that are insufficiently understood and there is thus a need for further exploration of what makes two seemingly similar areas dissimilar in their earthquake vulnerability. The aim should be to improve seismic hazard forecast and to create more accurate micro-zonation maps.

Like California, Taiwan is situated on the Ring of Fire, which is the world's most seismically active corridor, spanning the first island chain of Pacific Asia and the west coast of the Americas. Hung and Chen [24] propose a seismic risk-benefit method that they apply to Taiwan's capital city, Taipei. Using land-use maps, surveys, and the HAZ-Taiwan earthquake loss estimation system, they analyze three hypothetical earthquakes to estimate casualties and economic losses, including their respective spatial distributions. Vulnerability was assessed using cluster analysis, and a risk-benefit ratio was calculated to evaluate seismic risk profiles for different land-use plans. The aim of the paper is to help the urban planners of Taipei City make informed decisions regarding earthquake risk.

Liu et al. [25] examine Wenzhou Town in Wenchuan County, which was severely damaged by the 2008 Sichuan earthquake in central China. The reconstruction effort was channeled through unique institutional arrangements, which accelerated the reconstruction of the settlement, including dwellings and public spaces. While the reconstructed town is considered safer than its predecessor, the top-down approach introduced certain challenges that are discussed in the article.

Balachandran [26] and Mamula-Seadon [27] both mentioned that academic research in disaster risk reduction is shifting from basic to applied, implementation-oriented research, with a focus on evidence-based, multistakeholder collaboration, and multidisciplinary approaches.

Di Ludovico et al. [28] analyze the impact of the 2009 earthquake in central Italy on urban systems and mobility. They propose a multidisciplinary approach that integrates urban design with sustainable mobility strategies. The Strategic Projects aimed at reconnecting fragmented urban areas through multimodal sustainable mobility. They found that post-quake reconstruction offers a unique chance to develop a more efficient urban transport model using advanced technologies and innovative urban design.

Kondo and Lizarralde [29] study the effects of post-disaster urban planning and risk mitigation on internal migration after the 2011 Great East Japan Earthquake and Tsunami. Using GIS spatial analysis to track urban footprint changes, they found that government-led efforts to provide safe housing in over 20 cities led to significant unintended consequences: rapid urban sprawl, increased risks of landslides, vulnerabilities in mountainous areas, and enhancing urban fragmentation. They emphasize the need for incremental, adaptive actions in disaster recovery. And they suggest that planners and local authorities should be mindful of the cascading effects of centralized planning decisions.

Kodag, et al. [30] explored the relationship between spatial planning and urban community resilience to earthquakes and floods in Pune, Maharashtra, India. Their study finds that economically weaker households are more vulnerable to disasters due to their spatial locations and limited access to resources, which reduces overall resilience in the city.

Mavroulis, et al. [31] argued that the rush for rapid recovery resulted in critical errors in disaster management planning for the devastating earthquakes that occurred in southeastern Turkey on 6 February 2023. From the post-event field surveys, they found that disposal sites in the most affected provinces were unsuitable for safe earthquake debris management. Moreover, these sites were located near densely populated areas and close to surface water bodies, posing significant environmental and public health risks. Their paper proposes measures for effective debris management, emphasizing the need to avoid similar challenges and threats in future disasters.

7. Simulations of potential future impacts

Sarica et al. [34] focused on forecasting the dynamics of built-up areas at risk in three Asian megacities—Jakarta, Metro Manila, and Istanbul—using a spatiotemporal approach that considers seismic hazards. Landsat Thematic Mapper images from 1995 to 2016 (Metro Manila) and 1995 to 2018 (Jakarta and Istanbul) were analyzed to determine past urban growth. The SLEUTH (slope, land use, exclusion, urbanization, transportation, and hill shade) urban growth model, a cellular automaton-based model, was then used to project urban expansion to 2030. Seismic hazard maps for 10% and 2% probabilities of exceedance were overlaid with built-up area maps. The findings reveal significant increases in urban areas exposed to high seismic risk: in Metro Manila, a 65% increase in areas subjected to Modified Mercalli Intensities (MMI) VIII and IX over 35 years; in Jakarta, a 79% increase at MMI VIII; and in Istanbul, a 54% increase at MMI VIII. For a 2% exceedance probability, the urban areas at MMI IX have increased by 75% in Jakarta, 65% in Metro Manila, and 49% in

Istanbul. Their study demonstrates that urban growth modeling is valuable for assessing and planning the exposure of built-up areas to seismic hazards in megacities.

Satir et al. [33] used the Van 2011 earthquake as a case to analyze its impact on urban development using the Cellular Automata (CA)-MARKOV model. By simulating the 2018 urban development without the earthquake and comparing it with the actual built-up areas, their study found that the earthquake significantly accelerated urban expansion, especially toward the suburb. Additionally, more buildings were constructed on unstable land due to the earthquake, causing urban development to shift from flat land to slopes and move away from roads and existing settlements.

8. Conclusions

8.1 Major findings and methodology

- Disaster risk reduction (DRR) provides a comprehensive framework for controlling earthquake damages by relocation of human settlement through urban planning
- Urban sprawl has been found to be significant resulting from post-earthquake relocation
- A successful recovery plan with effective design of reconstruction could prevent poverty and slow down economic development for the area that suffered from earthquake
- Qualitative interview is often used for the questionnaire survey related to respondents' attitude toward earthquake
- Geographic information systems (GIS) techniques are widely applied for measuring the distance from properties to fault lines that cause earthquake
- Rush and poorly executed reconstruction projects have shown to be ineffective for recovery
- Hedonic price models are often applied to evaluate the impact of earthquake on property values. The functional forms include simple linear, double-log linear, Box-Cox non-linear, and geographically weighted regression
- A review of national and regional plans for resilience strategies is necessary to ensure the effectiveness of any reconstruction plan
- A successful public-private partnership program is very crucial for identifying urgent needs during the post-disaster reconstruction in the US
- Individuals' willingness to live in high-earthquake-risk areas is affected by place attachment and risk perception through bounded rationality. Education and earthquake drill may be the keys to change resident's risk perception
- Earthquake drill practices in China help to enhance farmer's knowledge of evacuation in rural area

- A cellular automaton-based urban growth simulation model called SLEUTH was used to project urban expansion up to 2030 after major disasters, such as earthquake and typhoon in Asian cities

- The strategic choice approach (SCA) can applied, to facilitate decision-making amidst uncertainties related to fault trace determination, fault buffer zones, and fault setbacks

- Geographic information systems (GIS) software are useful tools to monitor land-use changes after earthquake resulting from relocation plans

- A trade-off between earthquake disaster risk and economic efficiency is a crucial question in urban planning. As a result, a national land-use structure that balances risk and efficiency should be formulated by considering intertemporal fairness

- Planning for resilient cities may be achieved by recreating traditional city images after disasters and by shaping post-disaster citizenry

- Earthquake-focused urban design aims to build cities in safer zones, reducing vulnerability to severe earthquakes and their risks.

- Economically weaker households are more vulnerable to disasters due to their spatial locations and limited access to resources, which reduce overall resilience in the city

- Social capital helps keep communities together, facilitates the mobilization of groups, and provides informal insurance when formal resources are unavailable.

9. Future directions of research

Many studies focus on single earthquake event on one region. Cross-sectional studies for comparison of DRR strategies may be useful to provide more detailed guidance for decision makers in urban planning.

The evaluation of potential property damage by earthquake remains difficult. Therefore, an option approach regarding insurance policy for earthquake damage shall be investigated so that the financial feasibility for reconstruction plan can be validated.

Artificial Intelligence (AI) is now being applied to forecast the paths of Typhoon and Hurricane. Nowadays, more and more countries are becoming devoted to developing early warning systems for earthquake. However, simulation of earthquake through AI for pre-disaster evacuation and dispatching of rescue teams after earthquake have not yet been studied.

Many studies focus on the change for hierarchical structure of urban planning system to adopt post-disaster management. However, a detail guidance/handbook regarding how to establish quantitative measures to evaluate the effectiveness of reconstruction project is absent in these studies.

Acknowledgements

Most of the papers were collected by Shao-Jun Chiang. We thank him for his efforts.

Conflict of interest

None.

Author details

David Emanuel Andersson[1] and Oliver F. Shyr[2]*

1 IBMBA Program, National Sun Yat-Sen University, Taiwan

2 Department of Urban Planning, National Cheng-kung University, Taiwan

*Address all correspondence to: oliver.shyr@gmail.com

References

[1] Kunreuther H. Public-private partnerships for reducing seismic risk losses. In: Kleindorfer PR, Sertel MR, editors. Mitigation and Financing of Seismic Risks. Dordrecht, Netherlands: Springer Science+Business Media; 2001. pp. 73-99. DOI: 10.1007/978-0-7923-7098-7_5

[2] Bostenaru, Dan M, Armaş I, Goretti A. Earthquake hazard impact and urban planning: An introduction. In: Bostenaru Dan M, Armaş I, Goretti A, editors. Earthquake Hazard Impact and Urban Planning. Environmental Hazards. Dordrecht, Netherlands: Springer Science+Business Media; 2014. pp. 1-12. DOI: 10.1007/978-94-007-7981-5_1

[3] Momenian A, Zekavat M. Seismic approach to urban design. Research Journal of Recent Sciences. 2015;**4**(8):72-76. Available from: https://www.isca.in/rjrs/v4/i8/13.ISCA-RJRS-2013-947.pdf

[4] Kii M, Doi K. Earthquake risk and intertemporal fairness: An economic assessment of the national land-use structure. Transport Policy. 2020;**87**(March):77-83. DOI: 10.1016/j.tranpol.2018.08.009

[5] Sengezer B, Koç E. A critical analysis of earthquakes and urban planning in Turkey. Disasters. 2005;**29**(2):171-194. DOI: 10.1111/j.0361-3666.2005.00279.x

[6] Mirmoghtadaee M, Abdi MH, Eslamlou MS. Managing uncertainties in fault avoidance zone of urban areas: A case study of Karaj, Iran. International Journal of Disaster Resilience in the Built Environment. 2024;**14**:1-14. DOI: 10.1108/IJDRBE-07-2023-0088

[7] Freeman PK, Martin LA, Mechler R, Warner K, Hausmann P. Catastrophes and Development: Integrating Natural Catastrophes into Development Planning. Dordrecht, Netherlands: Springer Science+Business Media; 2002

[8] Cardona OD, Ordaz MG, Marulanda MC, Carreño ML, Barbat AH. Disaster risk from a macroeconomic perspective: A metric for fiscal vulnerability evaluation. Disasters. 2010;**34**(4):1064-1083. DOI: 10.1111/j.1467-7717.2010.01183.x

[9] Alam A, Ali Y. Studying the effects of Türkiye earthquake disaster and its impact on real estate industry: A risk analysis based on input-output & non-linear optimization models. International Journal of Disaster Risk Reduction. 2023;**96**(October):103920. DOI: 10.1016/j.ijdrr.2023.103920

[10] Dinçer AE, Dincer NN, Tekin-Koru A, Yaşar B, Yılmaz Z. The impact of Kahramanmaraş (2023) earthquakes: A comparative case study for Adıyaman and Malatya. International Journal of Disaster Risk Reduction. 2024;**110**(August):104647. DOI: 10.1016/j.ijdrr.2024.104647

[11] Tatar CO, Cabuk SN, Ozturk Y, Senyel Kurkcuoglu MA, Ozenen-Kavlak M, Bilge Ozturk G, et al. Impacts of earthquake damage on commercial life: A RS and GIS-based case study of Kahramanmaraş earthquakes. International Journal of Disaster Risk Reduction. 2024;**107**(June):104464. DOI: 10.1016/j.ijdrr.2024.104464

[12] Aldrich DP. The importance of social capital in building community resilience. In: Yan W, Galloway W, editors. Rethinking Resilience, Adaptation, and Transformation in a Time of Change. Change. Cham, Switzerland: Springer

International Publishing; 2017. pp. 357-364. DOI: 10.1007/978-3-319-50171-0_23

[13] He C, Huang Q, Bai X, Robinson DT, Shi P, Dou Y, et al. A global analysis of the relationship between urbanization and fatalities in earthquake-prone areas. International Journal of Disaster Risk Science. 2021;**12**(4):805-820. DOI: 10.1007/s13753-021-00385-z

[14] Palm RI. Public response to earthquake hazard information. Annals of the Association of American Geographers. 1981;**71**(3):389-399. DOI: 10.2307/2562898

[15] Cofini V, Carbonelli A, Cecilia MR, Di Orio F. Quality of life, psychological wellbeing and resilience: A survey on the Italian population living in a new lodging after the earthquake of April 2009. Annali di Igiene: Medicina Preventiva e di Comunità. 2014;**26**(1):46-51. DOI: 10.7416/ai.2014.1957

[16] Chen T-L. Structural analysis of how place attachment and risk perceptions affect the willingness to live in an earthquake-prone area. Disaster Prevention and Management: An International Journal. 2020;**29**(4):557-573. DOI: 10.1108/DPM-08-2018-0249

[17] Lian P, Zhuo Z, Qi Y, Xu D, Deng X. The impacts of training on farmers' preparedness behaviors of earthquake disaster: Evidence from earthquake-prone settlements in rural China. Agriculture. 2021;**11**(8):726. DOI: 10.3390/agriculture11080726

[18] Shrestha US. Earthquake mitigation and its effect on eco-environment and social development: A case study from Tamakoshi river basin of Central Mountain region, Nepal. In: Li A, Deng W, Zhao W, editors. Land Cover Change and Its Eco-environmental Responses in Nepal. Singapore: Springer Nature Singapore; 2021. pp. 445-464. DOI: 10.1007/978-981-10-2890-8_20

[19] Murdoch JC, Singh H, Thayer MA. The impact of natural hazards on housing values: The Loma Prieta earthquake. Real Estate Economics. 1993;**21**(2):167-184. DOI: 10.1111/1540-6229.00606

[20] Chen T-L, Chang H-S. The effects of zoning regulations along fault zone areas on land development and property values after the 921 Chi-Chi earthquake in Taiwan. Sustainability. 2018;**10**(4):1175. DOI: 10.3390/su10041175

[21] Azegami T, Takahashi A, Yasufuku K, Otsuka N, Miyagawa T, Abe H. Analysis of matters affecting land price fluctuations in the great East Japan earthquakc. International Journal of GEOMATE. 2022;**22**(89):55-64. DOI: 10.21660/2022.89.gxi314

[22] Tu CH, Andersson DE, Shyr OF, Lin PH. Earthquake risk, flooding risk and housing prices: Evidence from Taichung, Taiwan. Applied Spatial Analysis and Policy. 2023;**16**(2):923-938. DOI: 10.1007/s12061-023-09513-2

[23] Trifunac MD, Todorovska MI. 1971 San Fernando and 1994 Northridge, California, earthquakes: Did the zones with severely damaged buildings reoccur? Soil Dynamics and Earthquake Engineering. 2004;**24**(3):225-239. DOI: 10.1016/j.soildyn.2003.11.006

[24] Hung H-C, Chen L-C. The application of seismic risk-benefit analysis to land use planning in Taipei City. Disasters. 2007;**31**(3):256-276. DOI: 10.1111/j.1467-7717.2007.01008.x

[25] Liu L, Lin Y, Wang S. Urban design for post-earthquake reconstruction: A case study of Wenchuan County, China. Habitat International.

2014;**41**(January):290-299. DOI: 10.1016/j.habitatint.2013.09.001

[26] Balachandran BR. The reconstruction of Bhuj: Reflecting on the planning process. In: Banba M, Shaw R, editors. Land Use Management in Disaster Risk Reduction: Practice and Cases from a Global Perspective. Tokyo, Japan: Springer Japan; 2017. pp. 31-61. DOI: 10.1007/978-4-431-56442-3_4

[27] Mamula-Seadon L. Integrated land use planning in New Zealand and Canterbury earthquakes. In: Banba M, Shaw R, editors. Land Use Management in Disaster Risk Reduction: Practice and Cases from a Global Perspective. Tokyo, Japan: Springer Japan; 2017. 2017. pp. 107-136. DOI: 10.1007/978-4-431-56442-3_7

[28] Di Ludovico D, D'Ovidio G, Santilli D. Post-earthquake reconstruction as an opportunity for a sustainable reorganisation of transport and urban structure. Cities. 2020;**96**(January):102447. DOI: 10.1016/j.cities.2019.102447

[29] Kondo T, Lizarralde G. Maladaptation, fragmentation, and other secondary effects of centralized post-disaster urban planning: The case of the 2011 "cascading" disaster in Japan. International Journal of Disaster Risk Reduction. 2021;**58**(May):102219. DOI: 10.1016/j.ijdrr.2021.102219

[30] Kodag S, Mani SK, Balamurugan G, Bera S. Earthquake and flood resilience through spatial planning in the complex urban system. Progress in Disaster Science. 2022;**14**(April):100219. DOI: 10.1016/j.pdisas.2022.100219

[31] Mavroulis S, Mavrouli M, Vassilakis E, Argyropoulos I, Carydis P, Lekkas E. Debris management in Turkey provinces affected by the 6 February 2023 earthquakes: Challenges during recovery and potential health and environmental risks. Applied Sciences. 2023;**13**(15):8823. DOI: 10.3390/app13158823

[32] Mestav Sarica G, Zhu T, Pan T-C. Estimation of future population exposure to seismic hazard: A case study of Asian megacities. Environmental Research Letters. 2020;**15**(9):0940b5. DOI: 10.1088/1748-9326/ababc7

[33] Satir O, Kemec S, Yeler O, Akin A, Bostan P, Mirici ME. Simulating the impact of natural disasters on urban development in a sample of earthquake. Natural Hazards. 2023;**116**(February):3839-3855. DOI: 10.1007/s11069-023-05838-w

[34] Sarica GM, Zhu T, Pan TC. Spatio-temporal dynamics in seismic exposure of Asian megacities: Past, present and future. Environmental Research Letters. 2020;**15**(9):094092. DOI: 10.1088/1748-9326/ababc7

Chapter 9

Perspective Chapter: Statistical Seismology

Şakir Şahin

Abstract

Seismology, known as the science of earthquakes, is a branch of geophysics that studies earthquakes, seismic wave propagation within the earth, measurement instruments (seismographs) and seismic methods, evaluation of data, and other earthquake-related issues. Statistics is the science of collecting data accurately, learning from data, and transforming data and observations into information. Statistical seismology is the subject of transforming earthquake data into information by analyzing statistical methods, observing, predicting, and interpreting. It is the application of stochastic modeling into seismology. Even if all seismogenic structures and features that can cause an earthquake in a seismotectonic region are known, an earthquake is a natural event that occurs stochastically (randomly) in terms of its occurrence in temporal and spatial. Several probability models are used based on the random nature of the earthquake. Best-fit probability models have always been an important subject in statistical seismology. In this chapter, these subjects of statistical seismology are discussed.

Keywords: earthquake, physical, seismology, seismicity, stochastic, statistic

1. Introduction

Statistical seismology is a subject that reveals the effects of earthquakes using physical and statistical models. These models mainly use the parameters of earthquakes. All earthquakes that occurred in the historical and instrumental periods are included in the statistical seismology. It contains the studies of seismicity, prediction of earthquake occurrence, descriptive models, temporal and spatial distributions of earthquakes, seismic risk, earthquake's hazards and density, and socio-economic effects of earthquakes. For this, it uses all seismic data including historical and instrumental (digital) period data. The digital data has contributed greatly to the development of statistical seismology. The digital data acquisition used in statistical analysis has been begun with the instrumental period from 1900s and has accelerated especially in the last fifty years. With the growth of seismic networks and the increase in the number of data, very important developments have been reached in statistical seismology in recent years. By statistically examining earthquakes, it has become possible to make predictions for both the past and the future.

Statistical seismology has developed rapidly in the last few years and in this subject, a lot of studies have been realized [1–5]. In particular, the following issues stand

IntechOpen

out in statistical seismology: Relationships between seismicity and physics, quality data, statistical analysis of earthquakes, seismic probability estimation and seismic hazard assessment based on seismicity, and alternative data sets [5]. In recent years, many statistical methods have been developed to determine time-varying earthquake activities in the past and future seismic activity. These are Gutenberg-Richter (G-R), Båth, Omori, ETAS, Kegan-Jackson, EEPAS models. The advantage of using these methods, also defined as conditional intensity models, is that earthquake activity analysis can be realized statistically using probabilistic methods and the results can be determined accurately within an acceptable limit range. With the rapid development of observation technologies, the results become more correct as the number of data increases. Artificial intelligence applications will also great contributions to the development of the seismological data set. In addition to the development of the seismological data, GPS observation of surface displacement, InSAR observation of co-seismic deformation, and ionospheric observations provide big data support for statistical seismology.

Earthquake prediction is one of the most challenging problems in Geophysics. Currently, it is thought that individual earthquakes cannot be predicted due to the inability to fully observe the fundamental processes of the system and also due to natural randomness. Therefore, there is more focus on probability predictions, and reference models are used to create reliable earthquake prediction methods. This situation is not only limited to stress and fracture processes at the earthquake source but also requires the development and application of statistical methods to test data-based physical hypotheses. Ultimately, statistical prediction models need to be tested and applied to determine seismicity.

2. Seismicity

Seismicity is the determination of earthquake occurrences, earthquake potential, mechanisms and magnitude in a defined geographical region. Seismicity is also the statistical study of the distribution of earthquakes as temporal and spatial. The aim of seismicity studies is to determine the earthquake that will occur in the future and its damage by using data from past earthquakes in a certain period of time taking into account the tectonic structure. For this purpose, earthquake parameters such as date, time, epicenter coordinates (latitude-longitude) focal depth, and magnitude values are used. It also involves compiling and using statistical data such as the number of earthquakes, tectonic activity, and intensity-epicenter distance based on these parameters and using them as a whole (**Figure 1**).

After the first scientific classification in the field of seismicity was made by Aristotle, John Michell [7] realized publications stating that earthquakes were related to wave movements, and in 1840 Von Hoff published an earthquake catalog (the parameters of earthquake's date, time, epicenter coordinates, magnitude, depths, and locations) covering the whole world [8]. After the great Naples earthquake in 1857, Robert Mallet from Ireland conducted the first field study on earthquakes and prepared a damage map of the region [9]. Oldham [10] and Montessus de Ballore [11] expressed seismicity with epicenter maps indicating the historical occurrence of earthquakes and their magnitudes in temporal and spatial. Following these studies, the concept of seismicity which can be generally defined as the temporal and spatial distribution of earthquakes has been examined from different aspects by many researchers. These include the occurrence number and magnitude relationship,

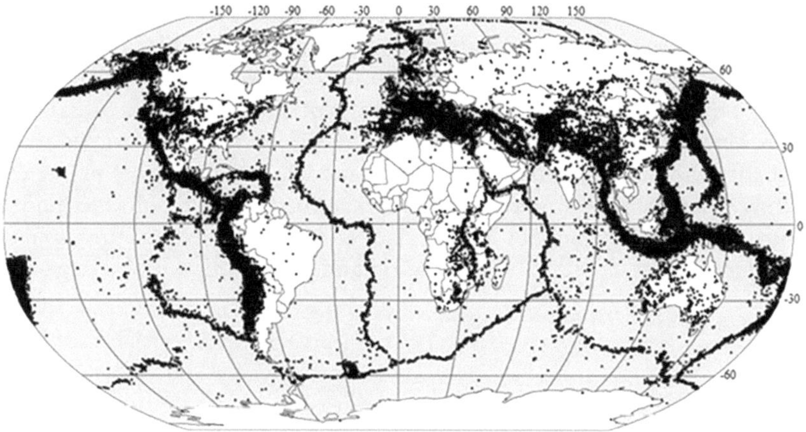

Figure 1.
The earthquake epicentral distribution around the world. The relationship between the epicenter distribution of earthquakes and tectonic units is clearly seen in this world seismicity map. Earthquakes are arranged according to plate boundaries and along tectonic zones [6].

tectonic structure, seismic energy, and seismic regime depending on the spatial and temporal distribution of earthquakes. In these seismicity studies, the tectonic structure and earthquake distributions are clearly revealed. Determining the relationships between earthquakes and tectonic structures in the studying region is very important in terms of seismicity studies.

3. Stochastic and physical models

Although modeling earthquake occurrences temporal and spatial was only an idea in the 1980s, today earthquake predictions can be made using fast and powerful computers with qualified and large-sized datasets. High-quality seismic data, especially those obtained using widespread blasting data, are of great importance in the development of statistical seismology. Statistical seismology is the application of stochastic modeling to seismology. In other words, stochastic modeling explains the principles and goals of statistical seismology.

Stochastic models are systems that include randomness. This randomness may be in the parameters, dynamics, and inputs of seismological calculations. Therefore, the results obtained will also have a similar randomness. In such a process, if the same calculations are repeated under the same conditions, the same results may not be obtained. But it can be calculated in which range or in which distribution the results are to be and what the probability of which results are to be occurring. In stochastic models, instead of knowing the future position of the system, it is estimated and probabilistic.

The physical model attempts to understand and predict the process entirely, while the stochastic model accepts that, at least for practical purposes, the physical process contains some irregularities (randomness) and is, therefore, a random process. The main reason for identifying uncertainties is to build them into the model, and only in this way one can determine the diversity of predicted results. The resulting stochastic model derives properties of physical events that are suitable and acceptable for measurement.

Stochastic does not mean unphysical. Because in the stochastic model, some conditions of the process behave randomly. It does not mean that it lacks physical content. The idea put forward by Jeffreys [12], a pioneer in statistics as well as geophysics, is that every physical theory involves estimating not only the relevant quantities but also their uncertainties. In the terminology of Vere-Jones [13], it is argued that every physical theory should be based on a stochastic model. Physics should not be excluded when adding the need to estimate uncertainties to the theory. Stochastic models of earthquakes aim to bring together limited data and limited physical theory in direct relation to questions such as rupture initiation and its evolution into a large-scale earthquake.

4. Earthquake occurrence models

The occurrence of earthquakes is random and independent of each other in time. However, in the future earthquakes can be statistically predicted by taking past data. Stochastic models are used to determine the probability of earthquake occurrence due to these uncertainties. Many stochastic models have been developed for the prediction of earthquake occurrence and earthquake hazard. The most commonly used stochastic models are Poisson and Markov models.

4.1 Poisson model

The probability of earthquakes occurring completely independently of each other is given by the Poisson distribution. In this model, events occur completely by chance and the probability of an occurrence is given by the Eq. (1) which is the probability of n earthquakes greater than M_{min} occurring in each time interval δt;

$$P(n, \delta t) = e^{-\lambda \delta t} (\lambda \delta t)^n / n! \tag{1}$$

Here λ is the number of earthquakes occurring per unit time. Accordingly, the real distribution of the annual number of earthquakes can be calculated. The theoretical Poisson distribution is compared with the data obtained as a result of this calculation. For $\delta t = 1$ it is given by the Poisson distribution relation as:

$$P(n, 1) = e^{-\lambda} (\lambda)^n / n! \tag{2}$$

The real distribution is given by the equation:

$$P(n) = n_i / \sum n_i \tag{3}$$

Here n_i is the annual number of occurrences of i earthquakes. In the Poisson model, earthquake occurrences are independent of each other. Energy accumulates slowly and is suddenly released. Foreshocks or aftershocks cannot be considered independent of the main shock. For this reason, it is stated that the Poisson Model is valid for large earthquakes, but when small local areas are taken into consideration, they do not fit the Poisson model due to the scarcity of data [14]. Therefore, the probability of an occurrence is not exactly the same for any interval δt along the time axis. Poisson and real distribution distributions of earthquakes that occurred within the time period T are given in **Figure 2**.

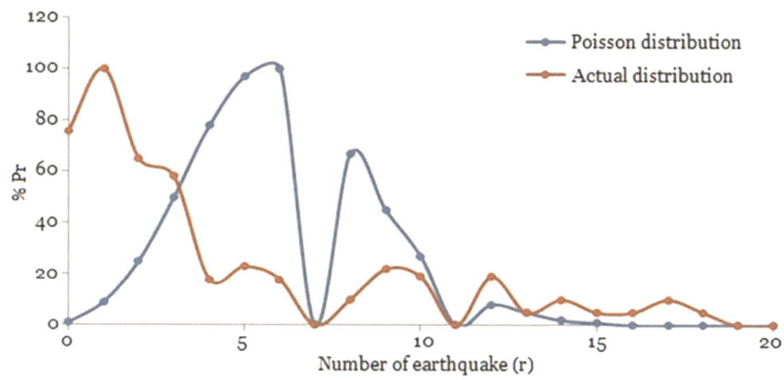

Figure 2.
The distributions of Poisson and actual of earthquakes that occurred in the interval time T.

4.2 Markov model

The Markov Model follows the elastic rebound theory, which assumes that future earthquakes are dependent on past earthquakes. If the conditional probability of $x(t)$, which is a random process with discrete parameters (t = 0,1,2, ...), depends only on the valuc $x(t\text{-}1)$ one time unit ago, such random processes are called first-order Markov chain. Mathematically, this is expressed as follows:

$$P\{x(t)\|x(1),x(2),\ \dots,x(t-1)\} = p\{x(t)\|x(t-1)\} \tag{4}$$

The processes of the Markov Model are applied in certain n steps. The transition probabilities from one state to another in these processes can be defined by a step matrix called the transition matrix and denoted [p] as:

$$[P] = \begin{bmatrix} p(0,0) & p(0,1) \\ p(1,0) & p(1,1) \end{bmatrix} \tag{5}$$

In this matrix, the probability of an earthquake occurring in the current time unit is $p(0,1)$ while it is known that there was an earthquake in the previous time unit with $p(1,1)$. $p(1,0)$ can be defined similarly. $p(0,0)$ represents the situation of no earthquake. In this model, it is assumed that the probability of more than one earthquake occurring within the selected time unit is very small. If the initial state of the system is defined by the probability row vector $\Pi(0)$, the next state of the system after n steps is found as follows;

$$\Pi(n) = \Pi(0)[p]^n\ n = 0,1,2,\ \dots. \tag{6}$$

Here $[p]^n$ refers to the n-step transition matrix [15].

$$[p]^n = \frac{1}{2-p(1,1)-p(0,0)}\begin{bmatrix} 1-p(1,1) & 1-p(0,0) \\ 1-p(1,1) & 1-p(0,0) \end{bmatrix} + \frac{[p(1,1)+p(0,0)-1]^n}{2-p(1,1)-p(0,0)} \tag{7}$$

This matrix contains the n-step transition probabilities between defined states of the system. The first and second parts of the matrix are the stationary and transient components, respectively. For bigger values of n, the transient component of the

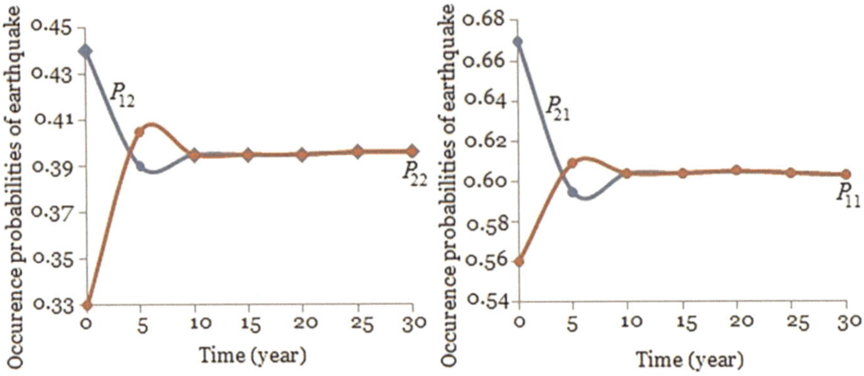

Figure 3.
The probability of earthquakes of a certain magnitude occurring (P_{12}, P_{22}) as 40% and the probability of not occurring (P_{21}, P_{11}) as 60% is determined by using the Markov model.

matrix approaches zero and in the limit state the probabilities become equal to the values obtained from the stationary matrix. Each point represents the status of the process. $p(1,1)$ is the probability that an earthquake occurring at the moment in the region where earthquakes are known to have occurred in the previous time period. $p(0,1)$ is the probability of an earthquake occurring at the moment in the region where it is known that there was no earthquake in the previous time unit.

NxN dimensional transition probability matrix *Pij* (*i* is the current state and *j* is the post-transition state);

$$[P_{ij}] = \begin{bmatrix} P_{11} & P_{12} \dots & P_{1N} \\ P_{21} & P_{22} \dots & P_{2N} \\ \dots & & \\ P_{N1} & P_{N2} \dots & P_{NN} \end{bmatrix}, 0 \le P_{ij} \le 1, 1 \le i,j \le N \tag{8}$$

The transition probability matrix of a Markov process can be graphically represented by a transition diagram (**Figure 3**).

5. Descriptive models in seismology

Descriptive models in seismology can be defined as the Gutenberg-Richter (G-R) relationship, Båth Model, Omori Model and Etas, Kagan-Jackson, and Eepas Models. One of these, the modified Omori law, uses the Gutenberg-Richter (G-R) relationship to determine the magnitude of aftershocks while estimating the number of possible aftershocks. These models are used to examine seismic activity called the main shock-aftershock. A study to calculate the probability of an aftershock that includes the time after the main shock requires the identification of specific examples in the calculations. In order to determine the occurrence of aftershocks, it is necessary to determine when and where major aftershocks may occur, the effects of major aftershocks in the region where they occur, and then to reveal the signs of a possible earthquake. When an earthquake occurs, many smaller earthquakes occur after the first earthquake in regions close to the epicenter of the earthquake. The first earthquake is defined as the main shock, and subsequent smaller earthquakes are called aftershocks. This seismic

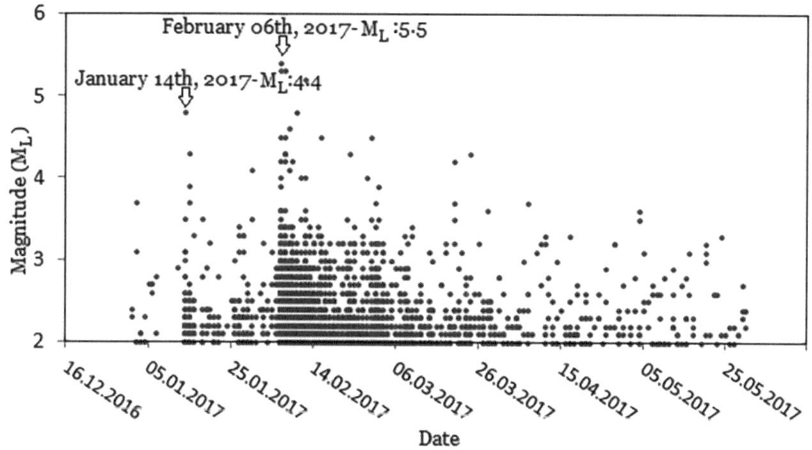

Figure 4.
The main shock-aftershock distributions that occurred in Çanakkale-Ayvacık between January 1, 2017 and May 30, 2017. Here it can be seen that the seismic activity continued with the first shock on January 14, 2017 (M_L = 4.8) and the second shock on February 6, 2017 (M_L = 5.4) and ended over time.

activity occurrence is called the main shock-aftershock sequence. The aftershock distribution almost points to the focal region of the main shock. Earthquakes that occur after the main shock on the fault where the main shock occurred are called aftershocks and they end over time (**Figure 4**).

Descriptive models explain the physical event used to consist of the basic models in statistical models. Statistical models have played an important role in studying faulting mechanisms [16]. For example, Weibull [16] examined the changes in the random distribution of micro crack lengths in the sample by applying force to similar samples in a laboratory.

The start of the earthquake occurrence (movement on the cellular level, flow process, faulting occurrence) is controlled by largely random processes from one point of weakness to another, rather than by the smooth progression of a fault or fracture in a homogeneous and elastic medium. Vere-Jones [13] tried to apply the idea of a faulting occurrence and proposed a stochastic model that predicts the G-R law with a b-value around 2/3 in the critical case and the Kagan distribution in the subcritical case. The same concept of the faulting occurrence is also introduced in the ETAS model. The differences between the fractures and the intervals between the fractures are due to the limitations of the instruments' perception rather than to the physical process. Additionally, the roles of stochastic models for earthquake occurs, such as the ETAS model or the separation model for rupture, are compared with models for complex systems. In many cases, most models such as G-R law, aftershock sequences, and Omori law. Show the characteristic features of earthquake occurrence. Each model provides a different perspective depending on the conditions that can display these features. The importance of models such as the faulting occurrence for crack propagation is that they contribute to explain complex phenomena from simple structures [13].

5.1 Gutenberg-Richter (G-R) relationship

Gutenberg and Richter [17] studied the relationship between the magnitude and number of earthquakes occurring around the world. In this relationship, the frequency-magnitude distribution of earthquakes of a certain magnitude occurring in a defined region is given by the following simple relationship [18]:

$$log_{10}N = a - b * M \tag{9}$$

Here N is the number of earthquakes equal to or greater than Mc (completeness magnitude) value, and a and b are coefficients. The N value used is the cumulative number of earthquakes. The values of the coefficients a and b are determined by using the least squares method and earthquake parameters (catalog data) are used. a coefficient depends on the observation period, the size of the studied area and the level of seismic activity. b coefficient varies depending on the ratio of the number of earthquakes in the magnitude groups from small to large.

The Eq. (9), known as the Gutenberg-Richter relation, is generally nonlinear for all magnitude values. Therefore, it is necessary to define a magnitude range in which $Log_{10}N$ is linear. The lower magnitude limit Mc depends on the noise level in the considered region, and a high noise level prevents the use of very sensitive seismographs. The correlation for large earthquakes is unclear because data for large earthquakes are scarce. The graph regarding the Gutenberg-Richter relationship can be seen in **Figure 5**. T_o determines the coefficients a and b in the frequency-magnitude relationship given by Eq. (9), the curve fitting method, the least squares method, the generalized (weighted) least squares method, the maximum likelihood method, and the Utsu and Page formulas are used. Apart from these linear relationships, non-linear (second or third order) relationships can also be calculated. Frequency-magnitude relations are used successfully in defining seismicity because they are directly related to the physics of earthquake occurrence. Seismicity of regions with different tectonic

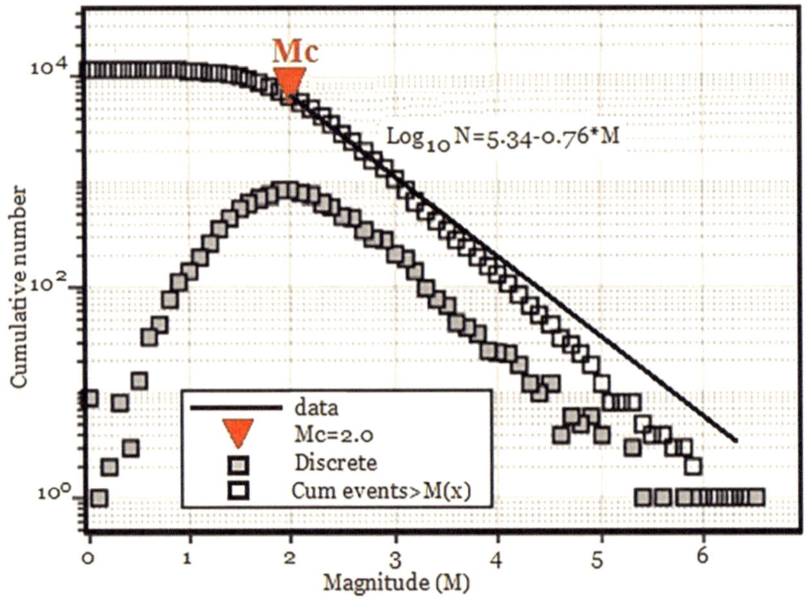

Figure 5.
The magnitude of completeness is obtained from maximum curvature and the b-value is calculated by using the weighted least squares method [19].

structures is different from each other. This difference also affects earthquake statistics. a variable depends on the observation period, the dimensions of the area studied, and the level of activity in temporal. Since the b coefficient is considered to be related to the physics of earthquake occurrences, it is an important variable in the statistical analysis of earthquakes.

Gutenberg and Richter [17] determined that b-values are 0.90 ± 002 for shallow earthquakes and 1.2 ± 0.2 for medium and deep focus earthquakes. Miyamura [20], who conducted a worldwide study, states that the b-value is between 0.4 and 1.8 and varies depending on the geological age of the seismotectonic belt. Miyamura [20] found b-values as high as 1.0–1.8 in the Pacific and Alpine orogenic belts (including island arcs) and smaller b-values in older geological units.

The low b-value is associated with a high-stress drop. Aftershocks show bigger b-values. Because of the existing tectonic stress is released by the main shock. The fact that earthquakes at ocean ridges have larger b-values than those at subduction zones can be explained by the lower stress concentration at oceanic ridges.

5.2 Båth model

The Båth model states that the difference between the magnitude of the main shock and the magnitude of the largest aftershock is approximately 1.2 units, regardless of the main shock [18, 21]. It is also referred to as Båth's law in the literature. Båth's law is generally valid for all different occurrences. It means that the average increase in the main shock magnitude from *M1* to *M1 + 1* is compatible with the increase in the largest aftershock from *Ma* to *Ma + 1*. The probability of a random earthquake with a magnitude greater than or equal to *M2* is $P(M_2) = 10^{b(M_{min}-M_2)}$ or $P(M_a+1) = (1/10^b)P(M_a)$ is possible [22]. Therefore, to verify both Båth's law and the assumption that the magnitude of each aftershock is chosen randomly, the probability of the decay of all aftershocks with a magnitude of 1.2 units below the main shock should increase in proportion to the increasing number of aftershocks as the main shock magnitude increases.

5.3 Omori model

In order to develop a suitable model of aftershock occurrences and to make an appropriate statistical interpretation of earthquakes in general, Omori [23, 24] defined the number of aftershocks per unit time interval at time *t*, which expresses the rate of decrease of aftershock activity over time and defined by his name as;

$$n(t) = K(t+c)^{-1} \tag{10}$$

Here *t* is the time after the main shock, *n(t)* is the formation rate of aftershocks in *t* unit time after the occurrence of the main shock. *K* and *c* are coefficients. In addition, the main shock occurrence time is generally taken for time *t*. But Omori [23] used a different time start in his studies and gave these times in relation to the first, second, and other time intervals ($t = 0, t = 1, \ldots$), respectively. First, the exponential function was used for data fit, but the fit with the data was not good. It was stated that Eq. (10) provided a high fit to the data and was the simplified version of the equations below.

$$n(t) = K\left\{(t+c) + k'(t+c)^2 + \ldots\right\}^{-1} \tag{11}$$

This correlation has been successfully applied to many earthquakes such as 1889 Kumamato, 1854 Nankai, 1904 Formaso, 1923 Kanto, and 1933 Sanriku.

5.4 Etas, Kagan-Jackson, Eepas models

The ETAS model is an important method of aftershock interpretation. It is widely used in data fitting, in investigating model properties, and as a descriptive method for regions experiencing exceptionally large seismic activity.

The Jackson-Kagan model is designed to provide a more realistic base model than the Poisson model. The exact prediction terms resulting from EEPAS model logarithmic regression studies are interpreted together with the Jackson-Kagan model.

The full (spatial-temporal) ETAS model [25] is written as:

$$\lambda_1(t,x,M) = f(M) \tag{12}$$

The Kagan-Jackson [26] model is written as:

$$\lambda_2(t,x,M) = f(M)H(t)\left\{\delta + A\sum_{i=t_i<t} g(x-x_i)\right\} \tag{13}$$

The EEPAS model [27];

$$\lambda_3(t,x,M) = \mu\lambda_0(t,x,M) + \sum_{t_i<t} f(M-M_i)g\big(x-x_{(t)}|M_i\big)h(t-t_i|M_i) \tag{14}$$

is written. In these equations, f, g, and h are all normalized as probability densities, $f(M)$ is the G-R relationship or one of its variables. Φ in the ETAS model given by Eq. (12) is the exponential efficiency term. μ generally constitutes the spatial structures. The ETAS model is used to clearly define after-shock activity, check that after-shocks are reduced properly, and determine subsequent after-shock activity if a second after-shock has occurred. In the Kagan-Jackson model given by Eq. (13), the coefficient A adjusts the total contribution from the term in parentheses when a new earthquake is added to the total, and $H(t)$ indicates the total ratio. The model depends significantly on the initial conditions, and the effect of unexpected events is controlled by the δ term. Even though H is constant, it is not clear that this model can be related to the stationary point process model or even that the model can produce information about the future. In the EEPAS model given by Eq. (14), λ_0 is obtained primarily from a model similar to the Jackson-Kagan model. The f, g, and h terms in the total are derived from the logarithmic relationship analysis of the temporal and spatial coordinates of an initial event and the seismic moment ratios for the expected events. Moreover, the model is rearranged sequentially, and it is not clear whether this model can be related to the stationary point process model. However, these three models are successfully applied to many earthquakes.

6. Temporal and spatial distributions of earthquakes

One of the main subjects of Statistical Seismology is the examination of the distribution of earthquakes in spatial and temporal. The spatial and temporal evaluation of earthquakes is the determination of which magnitude of earthquakes may occur in the region in question and the recurrence intervals of earthquakes of certain magnitudes.

In order to determine the effectiveness and return period of earthquakes that may occur in the future, the date, time, epicenter coordinates, focal depth, magnitude, and locations of the earthquakes that occurred in the past must be known. These values, defined as catalog data and earthquake parameters, are published up-to-date by national and international organizations such as AFAD KOERI in Turkey, USGS and IRIS in the USA, and ISC in Great Britain.

In addition to these parameters, the intensity of the earthquake, acceleration records, and the decay relationships of the signals of the recorded earthquakes must also be determined. In order to the studies to be carried out to be accurate, the examination period must be kept as long as possible and the existing data must be reliable. Detecting earthquake parameters accurately is possible with instrumental studies. The fact that earthquake recording stations are frequent enough allows earthquakes with low magnitudes to be recorded more sensitively and the parameters related to them to be determined with near-realistic accuracy.

6.1 Distribution of earthquakes as temporal

Seismicity in a region is a function of time as well as space. Therefore, when discussing about the seismicity of any region, the examination time interval must be determined. Since statistical calculations are predominant in seismicity studies, all past earthquakes in the study area are included in the calculations. The start dates of the earthquakes used in these calculations are the same as the beginning of the instrumental period.

The instrumental period started from 1900s and this period only covers a period of 124 years. Considering that the geological age is expressed in millions of years, it is seen that the instrumental period covers a very small time interval. For this reason, earthquakes that occurred before the instrumental period and are recorded in various historical records are also taken into consideration in examining time-dependent changes. Earthquakes before the instrumental period are called historical earthquakes. Historical earthquakes are important in that they contribute to the determination of the occurrence times of intense earthquakes in a region. In long-term observations, the distribution of earthquakes over time is generally examined in annual intervals (5, 10, 50, 100 years) and is expressed as earthquake frequency (number of earthquakes/year). It is important to determine the energy released in the examination of the temporal and spatial distribution of earthquakes. They are usually given as graphs (**Figure 6**).

In short-time interval observations, the number of earthquake occurrences can be taken annually, monthly, and daily. This type of short-term observation is mostly used in seismicity studies on bigger structures, such as nuclear power plants and dams. Earthquake occurrence distributions obtained as a function of time are also used to determine the stability and active periods of the study area, its periodicity, if any, whether there is a certain recurrence period, especially for bigger earthquakes, and if so, its duration. At the same time, it is used in seismic risk calculations to determine the physical variables of the source, such as energy, deformation, and stress drop of the study area, as a function of time by using relationships such as level-energy and dimension-deformation.

6.2 Distribution of earthquakes as spatial

Epicenter distribution maps are obtained by drawing the latitudes and longitudes of the examined region and marking earthquakes according to coordinates on a map of appropriate scale. In the preparation of these maps, all earthquakes that have occurred

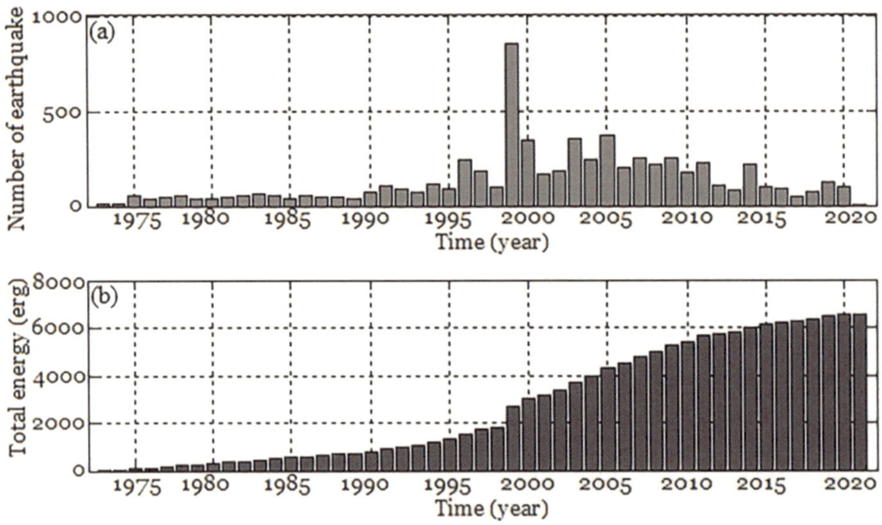

Figure 6.
(a) The distribution of earthquakes in temporal and (b) distribution of accumulated energy released by years.

within the study area can be marked, as well as mapping historical and instrumental periods separately, mapping earthquakes of a certain magnitude, or mapping where both duration and magnitude limitations are prepared. Since these maps can be prepared in different ways depending on the purpose, it should be stated which magnitude and time intervals of earthquakes are used.

Approximately 90% of all earthquakes in the world are of tectonic origin. For this reason, epicenters of earthquakes are located on or very close to active tectonic structures. Therefore, if earthquakes occurring in any region are mapped, the active tectonic structure in the region will be well revealed (**Figure 1**). For this reason, the distribution of earthquakes not only indicates the boundaries of the plates but also contributes to determine the types of these boundaries. **Figure 7** shows the earthquake

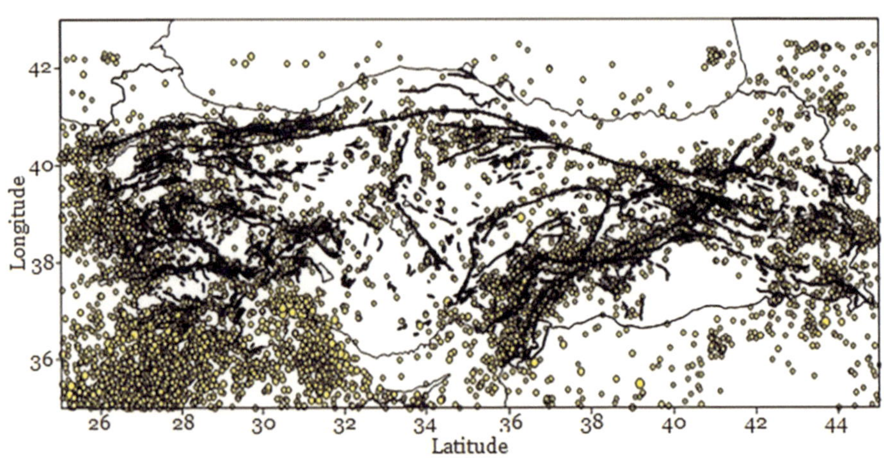

Figure 7.
Epicenter distributions of earthquakes (M_L > 3.5) that occurred in Turkey between 1990 and 2024. Yellow dots denote epicenters and black lines denote faults [28].

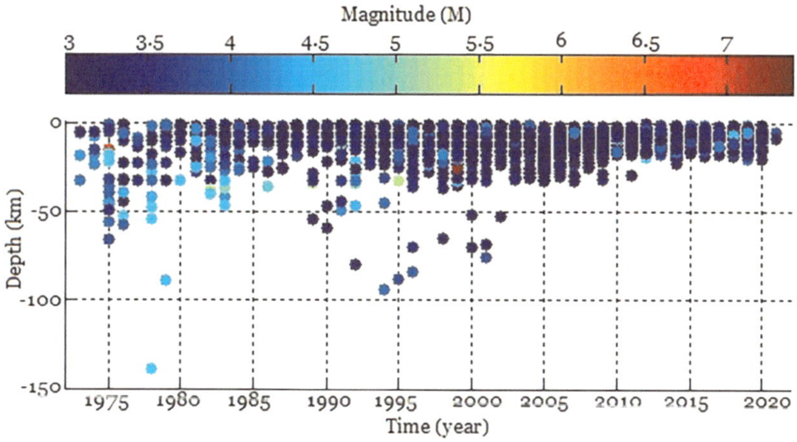

Figure 8.
Variation and magnitude of earthquake focal with depth.

distribution map of Turkey and its surroundings. Here, it is clearly seen that the belts whose epicenters are concentrated coincide with active tectonic belts.

Even though focal depths are tried to be specified on the distribution maps of earthquakes, a good representation cannot be obtained. For this reason, earthquake focal are drawn in a direction or along a narrow zone according to their depth. In graphs, earthquake magnitudes are also indicated, as in epicenter maps (**Figure 8**). The magnitudes and depths of earthquakes occurring at plate boundaries with different characteristics are also different. Particularly, the distribution of focal depths is important in recognizing the types of movements at the determined plate boundaries, receding (seafloor spreading) plate boundaries, and approaching (collision and subduction zones) plate boundaries.

7. Seismic risk estimates

The aim of seismic risk studies is to identify earthquake sources that could pose a danger in newly established residential areas by analyzing the geographical distribution of past earthquakes, including geological and tectonic structures. For this purpose, it is necessary to determine the seismotectonic active regions that can be earthquake sources, the average recurrence time interval, the faulting types, and the seismic activity history of each studying region. At the same time, the geological structures of the residential area and its surroundings should be examined by determining the value that the ground acceleration at the epicenter of an earthquake will lose until it reaches the location of the area. Since the seismogenic sources in the study area are of different types and scattered, first a general seismic risk estimate is determined for the entire region, and then different source types are examined separately (**Figures 9** and **10**).

Earthquake risk analysis is the determination of the future earthquake potential of the study area by using the earthquake history of the study area and its 50–200 km surroundings. It is also the determination of the earthquake that will affect that region in the future and the determination of the recurrence interval by using the date, epicenter, magnitude, and depth parameters of past earthquakes. Statistical methods

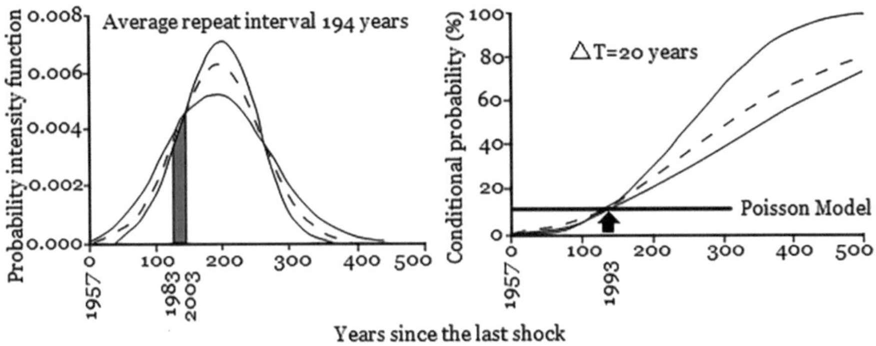

Figure 9.
The earthquake risk prediction is on the Pallet Creek segment of the San Andreas Fault zone [29].

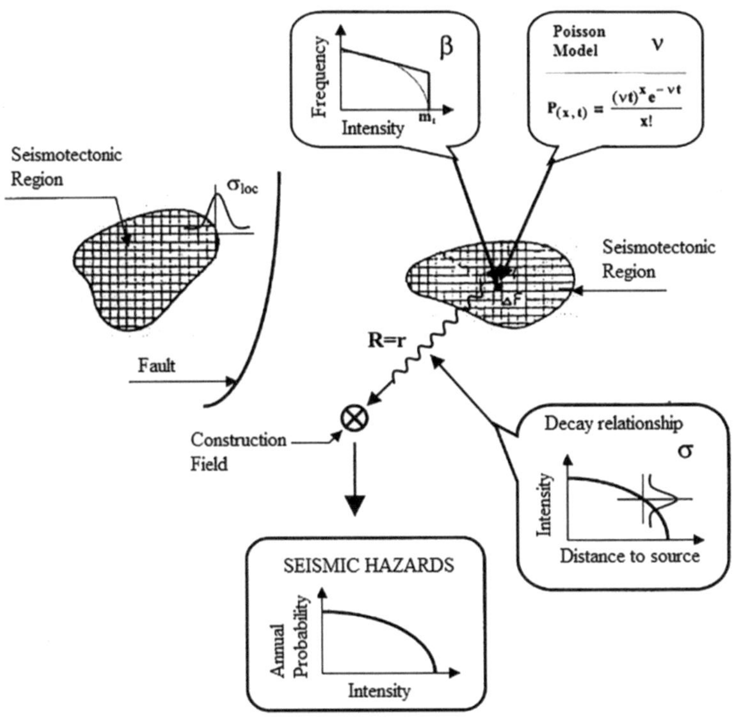

Figure 10.
Stages of earthquake risk study in a residential area [30].

are used in earthquake risk analysis. In risk analysis, methods such as least squares, Gumbel Extreme Values Distribution Model, Risk Analysis with Fisher-Tippett Type-I (Gumbel) Method, Risk Analysis with Fisher-Tippett Type-III (Weibull) Method, and Semi-Markov model are used.

Temporal distributions of earthquakes can be represented by a negative exponential expression such as $e^{-\lambda\delta t}$ according to Eq. (1). The relations calculated from the normal and cumulative occurrence numbers of earthquakes are used in the "seismic risk estimation" calculation. According to the relation (9), known as the G-R relation, it can be written as follows;

$$N(M) = 10^{a-b*M} \tag{15}$$

By dividing this by the examined time (T), Eq. (16) is obtained as;

$$N(M)/T = 10^{a-b*M/T} \tag{16}$$

Taking the logarithm of both sides of this expression;

$$log\left(\frac{N(M)}{T}\right) = a - b*M - LogT \tag{17}$$

Here the relation (18) is obtained as:

$$N(M) = 10^{a-b*M-LogT} \tag{18}$$

If $a_1 = a - LogT$ is written in this last relation, relation (19) is obtained:

$$N(M) = 10^{a_1-b*M} \tag{19}$$

The N(M) value found here is the annual average number of earthquakes that are greater than or equal to the M_C value. The difference between these negative exponential time distributions of earthquakes and 1 will express the seismic risk value. Seismic risk is obtained from the Eq. (20) as;

$$R = 1 - e^{-N(M)*T} \tag{20}$$

8. Distribution of earthquake hazards and intensity

Earthquake hazard is defined as the danger of maximum ground motion (acceleration, particle velocity, temporary or permanent displacement) occurring in a place and at a time, resulting from an earthquake large enough to cause damage and loss of life. Earthquake risk is expressed as the degree of damage, loss of property, and life that may occur due to these movements. The intensity of the earthquake, its distance, the soil properties of the residential area, the quality of construction, damage and loss of life, and the hazard conditions are included in the risk. In order to determine the risk, the possibility of an earthquake is defined as deterministically or probabilistically predicted. In order to define where the earthquake may occur, it is necessary to establish a seismotectonic model that includes the earthquake sources that will affect the residential area site and its surroundings. The errors that may occur when building a model arise from a lack of data and observations. If we know a process (as temporal and spatial) very well (100%), there will be no need to make hazard predictions, in which case descriptive models will be sufficient. It is important the size of the contribution of the margin of error to the difference between the admission and the result. Since the location, time of occurrence, magnitude, and other characteristics of possible earthquakes involve uncertainty, the earthquake hazards are given with a probability percentage based on statistical and probabilistic methods.

In earthquake risk analysis, the future earthquake potential of the study area is determined by utilizing the earthquake history of the study area and its 50–200 km surroundings. For this purpose, the characteristics of earthquakes that will affect that

Magnitude (M)	Intensity (I_o)	Damage Status	Ground acceleration (m/sec^2)
1.0–3.0	I	Only sensitive devices detect	~0.01
3.0–3.9	II	It is especially felt by people resting on the upper floors. Sensitively suspended objects may sway.	0.02–0.03
	III	It is felt inside the building, but it is not always clear whether it is an earthquake or not. Parked cars shake as if a truck has passed next to them.	0.03–0.07
4.0–4.9	IV	Felt by most people inside the building and by a few people outside. Some people wake up at night and hear pots and pans, doors and windows shaking.	0.07–0.15
	V	Everyone feels it. Plaster, windows break, long objects move	0.15–0.30
5.0–5.9	VI	Almost everyone feels it. Chimneys and plasters fall. Minor damage occurs.	0.30–0.70
	VII	Varying damages occur depending on the strength of the structure. Car drivers feel it too.	0.70–1.50
	VIII	Walls burst out of their frames, monuments, chimneys, and walls collapse. Sand and mud gush out.	1.50–3.00
6.0–6.9	IX	Buildings separate from their foundations, crack, and bend. Soil and underground pipes crack.	3.00–7.00
	X	Most of the masonry and frame structures are destroyed. The ground cracks, the rails buckle. Landslides occur.	7.00–15.00
7.0 and upper	XI	New types of buildings may remain standing, bridges may be destroyed, and underground pipes may be broken. Landslides occur. The rails buckle.	15.00–30.00
	XII	Almost everything is destroyed. Ripples are seen on the soil surface. Objects are thrown into the air.	30.00–70.00

Table 1.
Relationship between earthquake magnitude, intensity, acceleration, and damage status [18].

region in the future, such as location, magnitude, depth, intensity, number, type of fault rupture, recurrence in a certain time interval (1–50 years), and the displacement that occurred on the fault in the past are estimated using statistical methods.

The earthquake intensity is the destruction that an earthquake causes in the studying area of buildings, people, and living things. This effect depends on the characteristics of the earthquake, soil conditions, and their interactions. The intensity depends on the earthquake epicenter distance, depth, earthquake wave arrival direction, magnitude, duration, frequencies, acceleration it will create on the soil, local geological conditions' ability to magnify the earthquake effect (the dominant ground vibration period), groundwater saturation and depth, soil structure type (rocky or soil) and features, such as bedrock depth (**Table 1**).

9. Socio-economic effects of earthquakes

The degree of damage in countries where earthquakes occur varies depending on the level of development. In underdeveloped countries where a large part of the

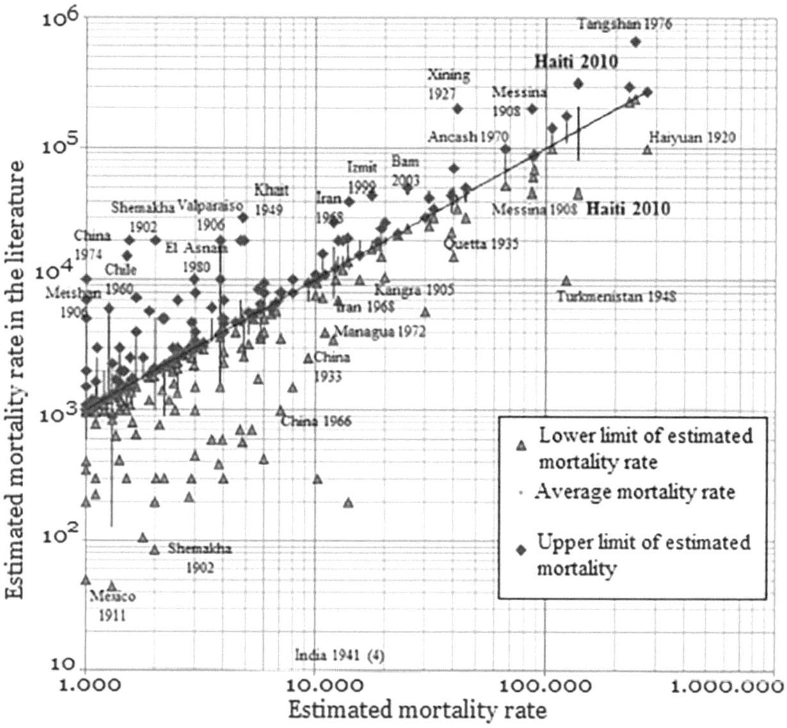

Figure 11.
Upper and lower bounds on the death toll from all earthquakes with a death toll greater than 1000 worldwide [31].

population and land is at risk of earthquakes, the damage caused by earthquakes is also high. Therefore, statistical methods have an important place in the analysis of earthquake effects. After an earthquake, along with physical losses, deaths, and injuries, economic, social, and psychological consequences also occur (**Figures 11–13**).

9.1 Loss of life and injuries

More than 34 large-scale earthquakes have occurred worldwide since the early 1900s, killing approximately 2 million people who have lost their lives as a result of earthquakes including tsunami and other collateral damages (**Figure 11**). Although the size of a disaster is calculated by taking into account the area it physically affects and various parameters, the most important point is the loss of life. If a low-magnitude earthquake causes a major loss of life, this earthquake cannot be considered minor. If there is little or no loss of life in a very large-scale earthquake, the size of the disaster will be small for humanity.

9.2 Economic impacts

According to the terminology developed to explain the economic losses of the earthquake, the economic effects can be examined under two titles. These are primary economic effects and secondary economic effects [33]. The effects of earthquakes on systems, such as transportation, energy, communication, infrastructure, and industry, are primary (direct and indirect) effects (**Figures 12–14**). The economic effects of

Figure 12.
Effect of earthquake on buildings [32].

Figure 13.
Earthquake impact on transportation structures [32].

earthquakes on public assets, product stocks, capital stocks, infrastructure, and population are direct effects. Indirect effects are derivatives of direct effects and can start immediately after earthquakes and last up to several years. Decreases in production and problems in services, such as water, transportation, and communication, lead to indirect effects [34]. The secondary effects are macroeconomic. Their effects on the country's economy. These events take place over time and have an effect on growth and employment levels, inflation rates, budget deficits, public expenditures, and the balance of payments. The most important point at this stage is the effects on national income [33].

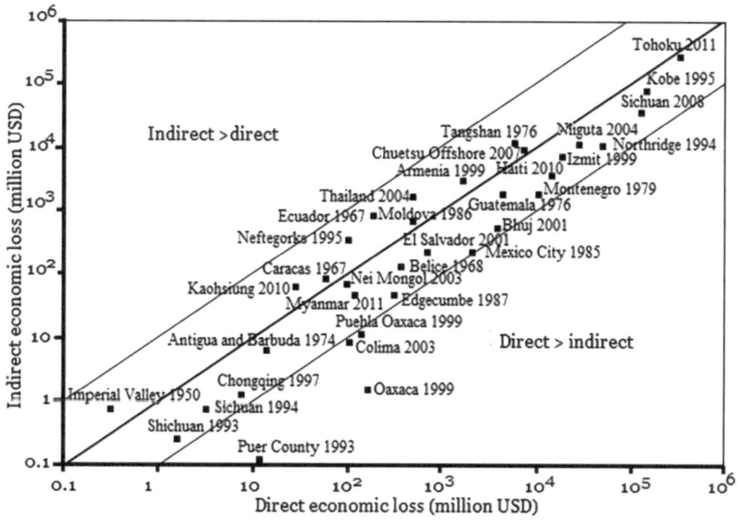

Figure 14.
Direct and indirect economic losses of earthquakes in the production sector [29].

On the other hand, earthquakes have the potential different effects on power plants. After an earthquake, significant damage occurs in industrial, service, and commercial facilities. Considering the losses in buildings, machinery, and equipment, finished and semi-finished goods stocks, a large economic resource is needed to repair the damages that occur. Similarly, huge resources are needed to repair damage to infrastructure such as bridges, highways, power and communication lines (**Figure 13**).

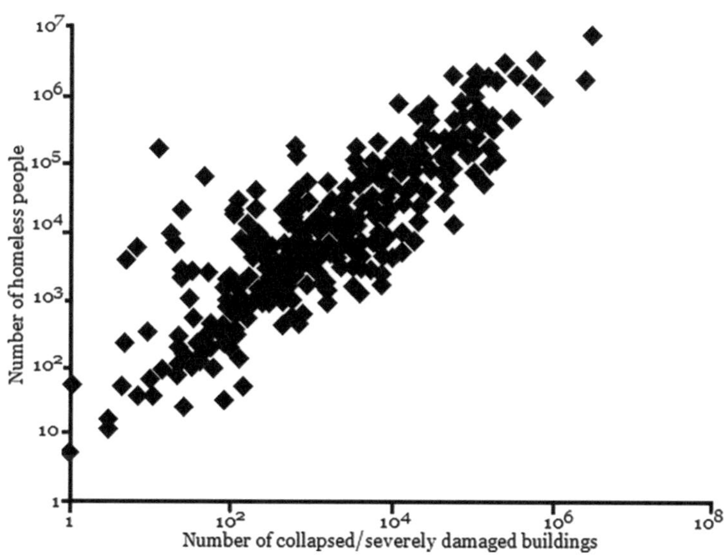

Figure 15.
Number of severely damaged buildings and homeless people in major historical earthquakes worldwide [31].

The most important damages that occur after earthquakes are the public or private sector that will protect the basic needs of the society such as health, safety, and welfare. Examples include police, fire, medical services, food, cleaning services, communication companies, and financial institutions [35].

9.3 Social and psychological effects

The effects resulting from earthquakes are evaluated as loss of life and property (**Figure 14**). As a result, negative social effects may occur. This represents the impact on production, income, savings, investment, and productivity. Those affected by earthquakes are divided into different groups; people residing in the disaster area and directly experiencing the disaster are called "primary victims." After the earthquake, shelter and food become the most important problems (**Figure 15**). The families and close circles of the primary victims can be defined as "secondary victims." Officials carrying out relief efforts in the disaster area, members of non-governmental organizations and volunteers are "tertiary victims." Another group is those who are sociologically affected and follow the disaster from the media [36].

Author details

Şakir Şahin
Geophysical Division, Faculty of Engineering and Natural Sciences, Suleyman Demirel University, Isparta, Turkey

*Address all correspondence to: sakirsahin@sdu.edu.tr

References

[1] Rhoades D, Savage M, Smith E, Gerstenberger M, Verejones D. Introduction. Special issue: Seismogenesis and earthquake forecasting: The frank Evison symposium. Pure and Applied Geophysics. 2010;**167**:619-621

[2] Tsalids GM, Papadimitriou EE, Limnios N. Statistical tools for earthquake and mining seismology: Preface to the topical issue. Acta Geophysica. 2011;**59**:657-658. DOI: 10.2478/s11600-0022-4

[3] Console R, Yamaoka K, Zhuang J. Implementation of short- and medium-term earthquake forecasts. Geophysical Journal International. 2012;**2012**:1-2. DOI: 10.1155/2012/217923

[4] Papadopoulos G, editor. 7th international workshop in statistical seismology. Research in Geophysics. 2012;**2**(1):1-3

[5] Huang Q, Gerstenberger M, Zhuang J. Current challenges in statistical seismology. Pure and Applied Geophysics. 2016;**173**:1-3

[6] NASA, DTAM project team, National Aeronautics and Space Administration, Washington, DC, USA. 2004. Available from: https://commons.wikimedia.org/wiki/File:Quake_epicenters_1963-98.png?uselang=en#Licensing

[7] Michell J. Conjectures Concerning the Cause and Observations upon the Phaenomena of Earthquakes. London: Darwin Estate and Cambridge University Library. 1760

[8] Albini P. The Premodern descriptive earthquake Catalogs of von Hoff (1840-1841), Perrey (1845-1850), and mallet (1853-1855) under a magnifying glass. Seismological Research Letters. 2023;**94**(5):2456-2468

[9] Mallet R. Great Neapolitan earthquake of 1857. The first principles of observational seismology. In: Map D. Seismic Bands of the Mediterranean, Report. Vol. 2. Royal Society of London; Chapman & Hall. 1862

[10] Oldham RD. On the propagation of earthquake motion to great distance. Philosophical Transactions of the Royal Society A. 1900;**194**(252–261):135-174. DOI: 10.1098/rsta.1900.0015

[11] Montessus de Ballore F. Sur la Repartition de l'Inestabilité Sismique en Bolivie. Comptes Rendus Hebdomadaires des Seances de l'Academie des Sciences (in Spanish). 1911;**153**:988-990

[12] Jeffreys H. Theory of Probability. 1st ed. Oxford: The Clarendon Press; 1939

[13] Vere-Jones D, Ben-Zion Y, Zuniga R. Statistical seismology. Pure and Applied Geophysics. 2005;**162**(6–7):1023-1026

[14] Knopoff L. The statistics of earthquakes in Southern California. Bulletin of the Seismological Society of America. 1969;**54**:1871-1873

[15] Parzen E. On estimation of a probability density function and mode. Annals of Mathematical Statistics. 1962; **33**(3):1065-1076. DOI: 10.1214/aoms/1177704472

[16] Weibull W. A statistical distribution of wide applicability. Journal of Applied Mechanic. 1951;**18**:293-297

[17] Gutenberg B, Richter CF. Seismicity of the Earth. 2nd ed. Princeton, New Jersey: Princeton University Press; 1954. p. 273

[18] Richter CF. Elementary Seismology. San Fransisco: W. H. Freeman and Co.; 1958

[19] Lordi AL, Neves MC, Custódio S, Dumont S. Seasonal modulation of oceanic seismicity in the Azores. Frontiers in Earth Science. 2022;**10**: 995401. DOI: 10.3389/feart.2022.995401

[20] Miyamura S. Magnitude-frequency relations and its bearing to Geotectonics. Proceedings of the Japan Academy. 1962; **38**:27-30

[21] Båth M. Lateral in homogeneities in the upper mantle. Tectonophysics. 1965; **52**:483-514

[22] Felzer KR, Becker TW, Abercrombie RE, Ekström G. Triggering of the 1999 MW 7.1 Hector mine earthquake by aftershocks of the 1992 MW 7.3 landers earthquake. Journal of Geophysical Research. 2002;**107**:B9. DOI: 10.1029/2001JB000911

[23] Omori F. On aftershocks. Reports Important Earthquake Investments Commercial. 1894a;**2**:103-138

[24] Omori F. On aftershocks of earthquakes. The Journal of the College of Science, Imperial University of Tokyo. 1894b;**7**:111-200

[25] Ogata Y. Statistical model for standard seismicity and detection of anomalies by residual analysis. Tectonophysics. 1989;**169**:159-174

[26] Kagan YY, Jackson DD. Long-term probabilistic forecasting of earthquakes. Journal of Geophysical Research. 1994; **99**(B7):13685e13700

[27] Rhoades DA, Evison FF. Long-range earthquake forecasting with every earthquake a precursor according to scale. Pure and Applied Geophysics. 2004;**161**:47-71

[28] AFAD. Earthquake catalog. Republic of Turkey, Ministry of Interior Disaster and Emergency Management Authority. Ankara, Turkey; 2024. Available from: https://deprem.afad.gov.tr/event-catalog

[29] Daniell JE. The socioeconomic impact of earthquake disaster, hazards and disaster. Series Earthquake Hazard, Risk and Disaster. Elsevier. 2014;**9**(32): 203-236

[30] Öztürk N. Probabilistic Seismic Hazard Analysis: A Sensitivity Study with Respect to Different Models [Ph. D. Thesis]. Ankara, Turkey: Middle East Technical University; 2008

[31] Wyss M, Shroder JF. Hazards and Disasters Series, Earthquake Hazard, Risk, and Disasters. 32 Jamestown Road, London NW1 7BY, UK: Academic Press is an imprint of Elsevier; 2014

[32] AA. Anadolu Agency. Ankara, Turkey. 2023. Available from: https://www.aa.com.tr/

[33] Güvel EA. Depremin Türkiye Ekonomisine Etkileri Üzerine Ekonometrik Bir Uygulama. In: 2. Ulusal İktisat Kongresi. İzmir: Dokuz Eylül University Publication; 2008

[34] Vermeiren JC. Natural disasters: Linking economics and the environment with a vengeance. In: Conference on Economics and the Environment. 1989. Available from: http://cidbimena.desastres.hn/docum/crid/Abril-Mayo2005/CD2/pdf/eng/doc1610/doc161

[35] Carrido M. An international disaster recovery business Alliance. Natural Hazards Rewiev. 2000;**1**(1):50-55

[36] Akyılmaz F, Karka O. Afetlerde Psikolojik İlkyardım. İstanbul: AFAD; 2011